D1554092

Theory of
Elastic Waves in Crystals

Theory of Elastic Waves in Crystals

Academician Fedor I. Fedorov

Director, Laboratory of Theoretical Physics
Physics Institute of the Academy of Sciences
of the Belorussian SSR
Chairman, Department of Theoretical Physics
Belorussian State University, Minsk

With a Foreword by

Hillard B. Huntington

Chairman, Physics Department
Rensselaer Polytechnic Institute
Troy, New York

Translated from Russian by

J. E. S. Bradley

PLENUM PRESS • NEW YORK • 1968

PHYSICS

In addition to this monograph, Academician Fedor I. Fedorov has written *Optics of Anisotropic Media* and is the author of 140 scientific papers dealing mainly with quantum theory, elementary particles, and the theory of the propagation of electromagnetic and elastic waves in anisotropic media. Fedorov was born in 1911 and was graduated from the Belorussian State University in 1931. He defended his doctoral dissertation on invariant methods in the optics of anisotropic media in 1954. In 1966, he was elected to membership in the Academy of Sciences of the Belorussian SSR. At present, Fedorov is Director of the Laboratory of Theoretical Physics of the Physics Institute of the Academy of Sciences of the Belorussian SSR, a post he has held since 1955, and since 1936 has also been Chairman of the Department of Theoretical Physics of the Belorussian State University in Minsk.

The Russian text, originally published by Nauka Press in Moscow in 1965, for the Department of Theoretical Physics, has been extensively revised and updated by the author for this edition.

Федоров Федор Иванович

Теория упругих волн в кристаллах

TEORIYA UPRUGIKH VOLN V KRISTALLAKH

THEORY OF ELASTIC WAVES IN CRYSTALS

Library of Congress Catalog Card Number 65-27349

A Division of Plenum Publishing Corporation
227 West 17 Street, New York, N. Y. 10011

Printed in the United States of America

Foreword

The translation into English of Academician Fedorov's excellent treatise on elastic wave propagation in solids has come at an opportune time. His systematic exposition of all aspects of this field is most lucid and straightforward. The author has gone to considerable pains to develop in his mathematical background a consistent tensor framework which acts as a unifying motif throughout the various aspects of the subject.

In many respects his approach will appear quite novel as his treatment introduces several concepts and parameters previously unfamiliar to the literature of the West. Extensive tables in the final chapters illustrate the application of these ideas to the existing body of experimental data.

The book is both extensive and comprehensive in all phases of the subject. Workers in the fields of ultrasonic propagation and elastic properties will find this treatise of great interest and direct concern.

H. B. Huntington

Rensselaer Polytechnic Institute
Troy, New York
November 1967

Preface to the American Edition

In preparing this edition I have corrected various misprints and errors appearing in the Russian edition, but I have also incorporated some substantial changes and additions, the latter representing some results I and my colleagues have recently obtained and published in Russian journals. For example, in section 32 I have added a general derivation of the equation for the section of the wave surface by a symmetry plane for cubic, hexagonal, tetragonal, and orthorhombic crystals. I have also greatly extended, supplemented, and revised the tables, as well as added several new ones. The changes in Chapter 9 (Debye temperatures of crystals) have been particularly extensive: two sections have been completely rewritten, and section 51 has been added, which contains the derivation of the general explicit formula for the Debye temperature of a hexagonal crystal.

I am indebted to Plenum Publishing Corporation for providing me with the opportunity of bringing my results to the notice of the English-speaking world, and I hope that the methods presented in this book will find acceptance there.

F. I. Fedorov

Minsk, USSR

Preface to the Russian Edition

This book is based on a course of lectures I presented in the fall of 1962 to graduate students and others specializing in crystal physics at Moscow University. The material has been considerably extended and supplemented during preparation for publication.

The theory of elastic waves in crystals is basic to ultrasonic technology, piezoelectronic devices, and many areas in solid-state physics of considerable practical and theoretical importance; there have recently been many papers on the subject. Surveys have been published in other countries (see [24], for example), but these are either review articles or single chapters in books on the theory of elasticity or the mechanics of solids. I know of no book solely concerned with a detailed and up-to-date exposition of the general theory of elastic waves in homogeneous crystalline solids.

My object here is to fill this deficiency. The method I have adopted is very different from that in all other works on the subject, since it is based on the general and systematic use of direct (coordinate-free) methods from vector and tensor calculus. These methods have proved their power in crystal optics [17] and are convenient in the study of the more complex laws governing the propagation of elastic waves in crystals. The methods are still somewhat unfamiliar to most physicists, so the second chapter (one of the longest) is entirely devoted to a detailed exposition of them.

Much of the material represents my own original work, in which especial attention is given to general methods of solving various classes of problems. Less space is given to the elucidation of specific features for particular crystals in the various symmetry classes, and the treatment then is partly illustrative.

A much larger book would have resulted from a more extensive application to problems that can be solved in this way; but in considering particular problems I have tended to expound the method in some detail in order to allow the reader to carry through similar calculations for himself.

A particular object has been clarity of exposition. Those attending the lectures at Moscow University included workers from some leading research laboratories, and their inquiries convinced me of the need for the present book. I have endeavored to make the book intelligible to scientific and technical workers specializing in practical areas of elastic waves in crystals; this has left its mark on the general style, which is far from academic. For this reason I have on occasion gone into details in the discussion and have derived some relationships in more than one way. I also hope that the book will be found of value as a textbook for graduate students and others specializing in crystal physics.

I am indebted to V. A. Koptsik, who read the book in typescript and made some valuable suggestions. I am also indebted to T. G. Bystrova for performing the calculations.

Contents

Chapter 1

General Equations of the Theory of Elasticity

1. Deformation Tensor

A body is deformed when forces are applied to it; the distances between points in the body subject to the forces are somewhat different from those between the same points in the absence of the forces. The change in distance is due to displacement of one relative to another during the deformation. Let $\mathbf{r}_0(x_i^0)$ be the radius vector of some point in the body (relative to some fixed point in space) before the deformation, and let $\mathbf{r}(x_i)$ be the same after deformation. The difference between these vectors is the displacement vector of the point and is

$$\boldsymbol{u} = \boldsymbol{r} - \boldsymbol{r}_0, \tag{1.1}$$

or in coordinate form

$$u_i = x_i - x_i^0, \quad i = 1, 2, 3. \tag{1.2}$$

The various points in the body are displaced differentially during deformation; equality in the displacement vectors of all points would simply mean displacement of the body as a whole, so $\mathbf{u}$ is a function of the coordinates of the point in the body. The deformation is specified completely by giving $\mathbf{u}$ as a function of $\mathbf{r}$.

Consider the deformation-induced change in the separation of two points infinitely close together; let the radius vector between the two before deformation be $d\mathbf{r}_0$, and after deformation $d\mathbf{r}$. From (1.1) we have that these are related by

$$d\boldsymbol{r} = d\boldsymbol{r}_0 + d\boldsymbol{u}. \tag{1.3}$$

Squaring this vector equation, we have

$$(d\boldsymbol{r})^2 = (d\boldsymbol{r}_0)^2 + 2\,d\boldsymbol{r}_0\,d\boldsymbol{u} + (d\boldsymbol{u})^2. \tag{1.4}$$

Here $\mathbf{dr}_0\mathbf{du}$ is the scalar product of vectors $\mathbf{dr}_0$ and $\mathbf{du}$. Expressed in terms of components, (1.4) becomes

$$(dx_i)^2 = (dx_i^0)^2 + 2\,dx_i^0\,du_i + (du_i)^2. \tag{1.5}$$

Here and henceforth we assume summation with respect to the repeated subscripts 1 to 3 (unless this is specifically excepted). Then, for example,

$$(dx_i)^2 = dx_i\,dx_i = \sum_{i=1}^{3} dx_i^2 = dx_1^2 + dx_2^2 + dx_3^2,$$

$$dx_i^0\,du_i = \sum_{i=1}^{3} dx_i^0\,du_i = dx_1^0\,du_1 + dx_2^0\,du_2 + dx_3^0\,du_3.$$

The components u_i of the displacement vector for a point are functions of the components x_i^0 of the radius vector for the point, so we may put

$$du_i = \frac{\partial u_i}{\partial x_k^0}\,dx_k^0 = \frac{\partial u_i}{\partial x_1^0}\,dx_1^0 + \frac{\partial u_i}{\partial x_2^0}\,dx_2^0 + \frac{\partial u_i}{\partial x_3^0}\,dx_3^0. \tag{1.6}$$

Thus (1.5) becomes

$$(dx_i)^2 = (dx_i^0)^2 + 2\,\frac{\partial u_i}{\partial x_k^0}\,dx_k^0\,dx_i^0 + \frac{\partial u_l}{\partial x_k^0}\,\frac{\partial u_l}{\partial x_i^0}\,dx_k^0\,dx_i^0. \tag{1.7}$$

It is clear that the doubly repeated (dummy) subscripts may be designated in any convenient way without affecting the expressions, so we may, for instance, permute i and k in the second term on the right in (1.7). The result is

$$\frac{\partial u_i}{\partial x_k^0}\,dx_k^0\,dx_i^0 = \frac{\partial u_k}{\partial x_i^0}\,dx_i^0\,dx_k^0.$$

Similarly, permuting i and l in the third term, we give (1.7) the form

$$(dx_i)^2 = (dx_i^0)^2 + 2\gamma_{ik}\, dx_i^0\, dx_k^0. \tag{1.8}$$

Here γ_{ik} denotes

$$\gamma_{ik} = \frac{1}{2}\left(\frac{\partial u_i}{\partial x_k^0} + \frac{\partial u_k}{\partial x_i^0} + \frac{\partial u_l}{\partial x_i^0}\frac{\partial u_l}{\partial x_k^0}\right). \tag{1.9}$$

This is a tensor (see Chapter 2) and is called the deformation tensor. The right-hand side of (1.9) is unaffected by permuting i and k, i.e.,

$$\gamma_{ik} = \gamma_{ki}. \tag{1.10}$$

The deformation tensor is therefore symmetric. It resembles any other symmetric three-dimensional tensor of second rank in having six independent components: γ_{11}, γ_{22}, γ_{33}, $\gamma_{12} = \gamma_{21}$, $\gamma_{23} = \gamma_{32}$, $\gamma_{31} = \gamma_{13}$.

The deformation of a solid is usually small, and this is the only case that will be considered here. Then we can neglect the third term in (1.9) as being of the second order of small quantities, and the expression for the deformation tensor becomes

$$\gamma_{ik} = \frac{1}{2}\left(\frac{\partial u_i}{\partial x_k} + \frac{\partial u_k}{\partial x_i}\right). \tag{1.11}$$

2. Stress Tensor

Experiment shows that an elastic body returns to its initial state when the forces that deformed it are removed, i.e., it regains its former shape (if the deformation has not been too large). This return to the normal state from the deformed state on removal of the external forces occurs in response to internal forces arising in the deformed body. These are called internal stresses, and they are absent from an undeformed body.

Let $\boldsymbol{f}$ be the volume density of the forces acting within the deformed body; then a volume dV experiences a force $\boldsymbol{f}$dV, while the total force **F** acting on an arbitrary volume V, within the body is given by

$$F = \int_V f\,dV. \tag{2.1}$$

We assume that there are no mass (bulk) forces, such as gravitation or inertial forces; then the force of (2.1) arises solely from the forces exerted on the volume by adjacent parts of the body. These forces are applied to the surface of that volume, so the resultant force must be expressed as the integral over the surface. Then, on the one hand, each component F_i equals the integral over the volume $\int_V f_i\,dV$, while on the other hand that same component must be expressed in the form of an integral over the surface bounding that volume:

$$F_i = \oint \sigma_i\,dS, \tag{2.2}$$

in which σ_i obviously denotes the component of the force acting on unit surface of that volume. But here we must remember that an infinitesimal element of the surface is itself a vector quantity:

$$d\mathbf{S} = \mathbf{n}\,dS, \qquad dS_k = n_k\,dS, \tag{2.3}$$

in which **n** is unit vector for the exterior normal to the surface. The integrand in (2.2) must therefore contain the vector n_k, and so σ_i must be put as

$$\sigma_i = \sigma_{ik} n_k, \tag{2.4}$$

so

$$F_i = \oint \sigma_{ik} n_k\,dS = \oint \sigma_{ik}\,dS_k. \tag{2.5}$$

The set of σ_{ik} forms an orthogonal second-rank tensor (see Chapter 2), because this is a necessary condition for (2.5) to be a correct covariant equation, i.e., one whose two sides represent quantities of the same type (vectors).

Hence in the absence of mass forces we have

$$\int f_i\, dV = \oint \sigma_{ik}\, dS_k. \tag{2.6}$$

Relations of this type represent Gauss' theorem, according to which

$$\oint \varphi\, dS_k = \int \frac{\partial \varphi}{\partial x_k}\, dV. \tag{2.7}$$

The right-hand side of equation (2.6) may thus be put in the form $\int (\partial \sigma_{ik}/\partial x_k)\, dV$; equality of the volume integrals must occur for any volume, so

$$f_i = \frac{\partial \sigma_{ik}}{\partial x_k}. \tag{2.8}$$

The second-rank tensor σ_{ik} is called the stress tensor; (2.5) shows that $\sigma_{ik}n_k$ is the i-th component of the force acting on unit area whose normal has the direction **n**, which may lie along one of the coordinate axes; e.g., x_3, in which case $n_1 = n_2 = 0$, $n_3 = 1$, and $\sigma_{ik}n_k = \sigma_{i1}n_1 + \sigma_{i2}n_2 + \sigma_{i3}n_3 = \sigma_{i3}$. Similarly, for $\mathbf{n} \parallel x_1$, $\sigma_{ik}n_k = \sigma_{i1}$. Hence σ_{ik} may be reckoned as the i-th component of a force acting on unit area oriented perpendicular to the x_k axis; (2.8) implies that the volume density of the forces at all points is specified directly via the stress tensor.

3. Equilibrium Conditions and the Equation of an Elastic Medium

An elastic body in static equilibrium in the absence of mass forces must have its internal stresses balanced in each element of volume, i.e., we have $f = 0$ or

$$f_i = \frac{\partial \sigma_{ik}}{\partial x_k} = 0. \tag{3.1}$$

In the general case (mass forces present) we may derive the equations of motion of a deformed body from d'Alembert's principle as a condition of equilibrium, i.e., by equating to zero the sums of all forces, including inertial ones. The external bulk force (the weight) per unit volume is $\rho \mathbf{g}$, in which ρ is the density of the medium and **g** is the vector for the acceleration due to gravity. The inertial forces per unit volume are $\rho(d^2\mathbf{r}/dt^2) = -\rho\ddot{\mathbf{r}} = -\rho(d^2u/dt^2)$. The condition for equilibrium, which is equivalent to the equations of motion, now becomes

$$\rho \boldsymbol{g} - \rho \frac{d^2 \boldsymbol{r}}{dt^2} + \boldsymbol{f} = 0 \tag{3.2}$$

or in terms of components

$$\rho \frac{d^2 x_i}{dt^2} - \rho g_i = f_i = \frac{\partial \sigma_{ik}}{\partial x_k}. \tag{3.3}$$

To this equation we must add the requirement that the sums of the moments of all forces (including inertial ones) be zero. Let **K** be the density of the bulk force; then the moment of the force acting on the volume element dV is $[\mathbf{Kr}]dV$, in which $[\mathbf{Kr}]$ is the vector product of the force by the radius vector to its point of application; in terms of components this becomes $(K_i x_l - K_l x_i)dV$, and the total moment of the bulk forces is $\int(K_i x_l - K_l x_i)dV$. Similarly, the moment of the surface force acting on a surface element dS is [see (2.2)-(2.4)] $(\sigma_i x_l - \sigma_l x_i)dS = (\sigma_{ik} x_l - \sigma_{lk} x_i)dS_k$, while the total moment of all forces is $\oint(\sigma_{ik} x_l - \sigma_{lk} x_i)dS_k$. Then we have the condition that the total moment of all forces is zero as

$$\int (K_i x_l - K_l x_i)\, dV + \oint (\sigma_{ik} x_l - \sigma_{lk} x_i)\, dS_k = 0. \tag{3.4}$$

In the present case the bulk mass force is $K_i = \rho(g_i - \ddot{x}_i)$; using (2.7) to convert the surface integral to a volume one, we get

$$\int \left\{\rho[(g_i - \ddot{x}_i)x_l - (g_l - \ddot{x}_l)x_i] + \frac{\partial}{\partial x_k}(\sigma_{ik} x_l - \sigma_{lk} x_i)\right\} dV = 0.$$

This must be so for any volume within the body, so the integrand must be zero; substituting $-\partial\sigma_{ik}/\partial x_k$ for $\rho(g_i - \ddot{x}_i)$ from (3.3), we get

$$\sigma_{ik} \frac{\partial x_l}{\partial x_k} - \sigma_{lk} \frac{\partial x_i}{\partial x_k} = 0. \tag{3.5}$$

The x_k for different k are independent variables, so $\partial x_l/\partial x_k = 0$ for $l \neq k$, but for $l = k$ we have $\partial x_1/\partial x_1 = \partial x_2/\partial x_2 = \partial x_3/\partial x_3 = 1$. Hence

$$\frac{\partial x_l}{\partial x_k} = \delta_{kl}. \tag{3.6}$$

in which δ_{kl} is Kronecker's symbol, which has the properties

$$\delta_{kl} = \delta_{lk} = 0 \quad (k \neq l); \quad \delta_{11} = \delta_{22} = \delta_{33} = 1. \tag{3.7}$$

The set of δ_{kl} forms the unit tensor

$$(\delta_{kl}) = \begin{pmatrix} 1 & 0 & 0 \\ 0 & 1 & 0 \\ 0 & 0 & 1 \end{pmatrix}. \tag{3.8}$$

Properties (3.7) mean that δ_{kl} acts as the operator for subscript permutation; (3.7) gives $\sigma_{ik}\delta_{kl} = \sigma_{il}$, i.e., summation with respect to one subscript (k) in δ_{kl} causes this subscript to be replaced by the other (l) in the expression multiplied by δ_{kl}. Hence (3.5) becomes $\sigma_{il} - \sigma_{li} = 0$ or

$$\sigma_{il} = \sigma_{li}. \tag{3.9}$$

Hence the stress tensor is symmetric, as is the deformation tensor of (1.10).

Very often the force of gravity is small relative to the other forces in (3.3); in this book I neglect it and so use the equation of motion in the form

$$\rho \frac{d^2 x_i}{dt^2} = \frac{\partial \sigma_{ik}}{\partial x_k}. \tag{3.10}$$

All points in the body have zero acceleration for static equilibrium; (3.10) shows this to be necessary in order to comply with condition (3.1),

$$\frac{\partial \sigma_{ik}}{\partial x_k} = 0.$$

Usually the deformation is produced by external forces applied to the surface of the body. Let **P** (P_i) be the density of these surface forces, which means that a force **P**dS acts on an element of surface dS. Of course, **P** varies in magnitude and direction from one point to another, and at equilibrium each element of the surface must have the external force $P_i dS$ equal to the internal stresses $\sigma_{ik} dS_k$, i.e., $\sigma_{ik} dS_k = P_i dS$. Now $dS_k = n_k\, dS$, in which

n (n_k) is the vector of the exterior normal to the surface; then we get

$$\sigma_{ik} n_k = P_i. \tag{3.11}$$

These are the boundary conditions at the surface of the body, which must be obeyed as well as the equations of equilibrium of (3.1).

4. Hooke's Law

By an elastic body we mean one that produces internal stresses in response to external forces that cause deformation, these stresses tending to remove the deformation; the body returns to its initial state when the external forces are removed, the stress and deformation vanishing simultaneously. Small deformations produce small stresses, the two increasing or decreasing together; there is a definite relation of stress to deformation, which we can establish by assuming that the stress at any point at any instant is completely determined by specifying the deformation at that point at that moment. This means that we consider only an absolutely elastic body, which is an ideal case; but many real bodies satisfy this condition sufficiently closely for reasonably small deformations. Hence we may put

$$\sigma_{ik} = \sigma_{ik}(\gamma_{lm}). \tag{4.1}$$

Hooke established three centuries ago the relation of stress to deformation for the simplest case; Hooke's law states that the stress is proportional to the strain. For the general case we see from (4.1) that we have six quantities σ_{ik} dependent components γ_{lm}, so a natural extension of Hooke's law is to suppose that we have a linear homogeneous relation between these. In fact, if we expand each σ_{ik} as a function of γ_{lm} as a Maclaurin series, we get

$$\sigma_{ik} = \sigma_{ik}(0) + \left(\frac{\partial \sigma_{ik}}{\partial \gamma_{lm}}\right)_{\gamma_{lm}=0} \gamma_{lm} + \cdots \tag{4.2}$$

We assume henceforth that all the γ_{lm} are small, so we need retain only terms up to the first degree.* On the other hand, the absolute elasticity means that the stress becomes zero in the

* This represents restriction to the linear theory of elasticity.

absence of deformation, i.e., $\sigma_{ik}(0) = 0$. This gives us our linear homogeneous relation between the components of the stress and deformation tensors:

$$\sigma_{ik} = c_{iklm}\gamma_{lm}, \tag{4.3}$$

in which, from (4.2),

$$c_{iklm} = \left(\frac{\partial \sigma_{ik}}{\partial \gamma_{lm}}\right)_{\gamma_{lm}=0}. \tag{4.4}$$

Now $\sigma = (\sigma_{ik})$ and $\gamma = (\gamma_{lm})$ are orthogonal second-rank tensors, so the set of c_{iklm} forms an orthogonal fourth-rank tensor subject to the covariant condition of (4.3), i.e., that the form of this is independent of the choice of cartesian coordinate system (see Chapter 2). Relation (4.3) is often called the generalized Hooke's law, while the c_{iklm} tensor is called the tensor of the elastic moduli or the tensor of the elastic constants.

The left side of (4.3) is unaffected if i and k are permuted, so this must be so for the right side also; hence tensor c_{iklm} is symmetric with respect to permutation of the first two subscripts, while (4.4) leads to the same conclusion for the second pair, because permutation of l and m does not affect the right part.

5. Energy of a Deformed Elastic Body

The work done by the external forces in time dt is found by considering the change in the displacements of the points in this time, $\delta\mathbf{u} = (\delta u_i) = (\partial\mathbf{u}/\partial t)dt$; then the work of the mass (bulk) forces over the entire volume is

$$\int \rho \mathbf{g}\delta\mathbf{u}\, dV = \int \rho g_i \delta u_i\, dV \tag{5.1}$$

while the work done by $\mathbf{P}$ (surface forces) is

$$\oint \delta\mathbf{u}\mathbf{P}\, dS = \oint \delta u_i P_i\, dS. \tag{5.2}$$

Condition (3.11) allows us to put the latter as

$$\oint \delta u_i \sigma_{ik} n_k\, dS = \oint \delta u_i \sigma_{ik}\, dS_k. \tag{5.3}$$

We use (2.7) to convert the surface integral into a volume one and get

$$\oint \delta u_i \sigma_{ik}\, dS_k = \int \frac{\partial}{\partial x_k}(\delta u_i \sigma_{ik})\, dV = \int \delta u_i \frac{\partial \sigma_{ik}}{\partial x_k}\, dV + \int \frac{\partial \delta u_i}{\partial x_k} \sigma_{ik}\, dV. \tag{5.4}$$

The symmetry of the σ_{ik} gives

$$\frac{\partial \delta u_i}{\partial x_k} \sigma_{ik} = \frac{1}{2}\left(\frac{\partial \delta u_i}{\partial x_k} + \frac{\partial \delta u_k}{\partial x_i}\right)\sigma_{ik} = \frac{1}{2}\delta\left(\frac{\partial u_i}{\partial x_k} + \frac{\partial u_k}{\partial x_i}\right)\sigma_{ik} = \sigma_{ik}\delta\gamma_{ik}. \tag{5.5}$$

Then, with (2.8), we have the sum of these two components of the work done by the external forces as

$$\int \delta A dV = \int \rho g_i \delta u_i\, dV + \int f_i \delta u_i\, dV + \int \sigma_{ik}\delta\gamma_{ik}\, dV =$$

$$= \int (\rho g_i + f_i)\delta u_i\, dV + \int \sigma_{ik}\delta\gamma_{ik}\, dV. \tag{5.6}$$

Here δA is the work per unit volume.

But (3.3) is $\rho g_i + f_i = \rho \ddot{x}_i$, so

$$\int \delta A\, dV = \int \rho \ddot{x}_i \delta u_i\, dV + \int \sigma_{ik}\delta\gamma_{ik}\, dV. \tag{5.7}$$

The kinetic energy per unit volume is

$$W = \frac{1}{2}\rho \dot{\boldsymbol{r}}^2 = \frac{1}{2}\rho\left(\frac{\partial \boldsymbol{u}}{\partial t}\right)^2 = \frac{1}{2}\rho\left(\frac{\partial u_i}{\partial t}\right)^2, \tag{5.8}$$

and the change in this in time dt is

$$\delta W = \frac{\partial W}{\partial t}\, dt = \rho \frac{\partial^2 u_i}{\partial t^2}\delta u_i. \tag{5.9}$$

Now $\ddot{x}_i = \partial^2 u_i/\partial t^2$, so the integrand in the first part of (5.7) is δW:

$$\int \delta A\, dV = \int \delta W\, dV + \int \sigma_{ik}\delta\gamma_{ik}\, dV.$$

The corresponding relation for unit volume is

$$\delta A = \delta W + \sigma_{ik}\delta\gamma_{ik}. \tag{5.10}$$

The law of conservation of energy (first law of thermodynamics) implies that δA together with the incoming heat δQ must equal the sum of δW and dU, the last being the change in the internal energy:

$$\delta A + \delta Q = \delta W + dU. \tag{5.11}$$

The symbol dU differs from δA and δQ in order to emphasize that U is a function of the state of the body, so dU is a total differential. The second law of thermodynamics gives us for reversible processes that

$$\delta Q = T\,dS, \tag{5.12}$$

in which T is the absolute temperature and S is the entropy of the body, the latter also being a function of the state, so dS is a total differential. Hence

$$\delta A - \delta W = dU - T\,dS. \tag{5.13}$$

Of course, δQ, dU, and dS all refer to unit volume. We put $\delta\Phi = \sigma_{ik}\delta\gamma_{ik}$, and get from (5.10) and (5.13) that

$$\delta\Phi = dU - T\,dS. \tag{5.14}$$

I consider only the case in which the effects in the body are (thermodynamically speaking) of two basic types: adiabatic and isothermal. There is no exchange of heat with the surroundings in an adiabatic process; $\delta Q = 0$, and so (5.12) gives $dS = 0$. This case occurs in rapid vibrations and so is of the greatest interest to us. The right half of (5.14) then equals dU and this is a total differential; (5.14) then becomes $\sigma_{ik} d\gamma_{ik} = dU$, so

$$\sigma_{ik} = \left(\frac{\partial U}{\partial \gamma_{ik}}\right)_S. \tag{5.15}$$

The subscript S denotes that the partial derivative is to be taken with the entropy constant. T = constant for an isothermal process, and the right side of (5.14) becomes d(U-TS), which is a total differential of (U-TS), which is called the free energy. By analogy with (5.15) we have here

$$\sigma_{ik} = \left(\frac{\partial (U - TS)}{\partial \gamma_{ik}}\right)_T. \tag{5.16}$$

In both cases $\delta\Phi$ is a total differential, so we may put

$$d\Phi = \sigma_{ik}\, d\gamma_{ik} = \frac{\partial \Phi}{\partial \gamma_{ik}}\, d\gamma_{ik}, \tag{5.17}$$

whence it follows* that

$$\sigma_{ik} = \frac{\partial \Phi}{\partial \gamma_{ik}}. \tag{5.18}$$

Function Φ is termed the elastic potential; (5.10) shows that it is the density of the potential energy of an elastically deformed body.

6. Tensor for the Elastic Moduli

Relation (5.18) is connected with the law of conservation of energy and imposes a major additional restriction on the possible values for the components of the c_{iklm} tensor; from (4.4) and (5.18)

$$c_{iklm} = \frac{\partial \sigma_{ik}}{\partial \gamma_{lm}} = \frac{\partial^2 \Phi}{\partial \gamma_{lm}\, \partial \gamma_{ik}} = \frac{\partial^2 \Phi}{\partial \gamma_{ik}\, \partial \gamma_{lm}} = c_{lmik}. \tag{6.1}$$

Hence the components of the tensor (see section 4) must satisfy the following symmetry conditions:

$$c_{iklm} = c_{kilm} = c_{ikml} = c_{lmik}, \tag{6.2}$$

i.e., the components are unaltered if a pair of subscripts are permuted internally or with the other pair.

(5.18) and (4.3) imply that the derivatives of Φ with respect to the γ_{ik} are linear homogeneous functions of the γ_{ik}, so Φ is a quadratic function of the γ_{ik}. The differential relation of (5.17) defines Φ apart from an additive constant. We assume that the elastic energy is zero in the absence of deformation ($\gamma_{ik} = 0$), in which case this constant becomes zero. Hence Φ is a homogeneous quadratic function of the γ_{ik}; Euler's theorem for homogeneous

---

* In (5.15) and (5.16), and also in many other cases, we operate with γ_{ik} and σ_{ik} as though they were independent variables, whereas this is actually not so, by virtue of the condition $\gamma_{ik} = \gamma_{ki}$. Nevertheless, it is clear that these relationships are correct; this may be shown by assuming initially that $\gamma_{ik} \neq \gamma_{ki}$ and then passing continuously to $\gamma_{ik} = \gamma_{ki}$.

functions implies from this that

$$\gamma_{lk}\frac{\partial\Phi}{\partial\gamma_{lk}}=2\Phi,$$

or, from (5.18),

$$\Phi=\frac{1}{2}\sigma_{ik}\gamma_{ik}. \tag{6.3}$$

Then (4.3) enables us to put Φ also as

$$\Phi=\frac{1}{2}c_{iklm}\gamma_{ik}\gamma_{lm}. \tag{6.4}$$

Φ is defined in terms of U and S by (5.15) and (5.16), and so the c_{iklm} are dependent on whether the process is adiabatic or isothermal; we have corresponding adiabatic and isothermal elastic moduli, which are found as the second derivatives of the Φ of (6.1) with the entropy and temperature constant, respectively.

The A_{iklm} have $3^4=81$ independent components for an arbitrary three-dimensional fourth-rank tensor. The symmetry conditions of (6.2) reduce the number of independent c_{iklm} considerably. It is convenient in considering various aspects of the c_{iklm} to pass from the three-dimensional fourth-rank tensor to a six-dimensional matrix, whose elements (as for any matrix) are governed by two subscripts. This transition is performed by replacing pairs of subscripts with the values 1, 2, 3 by one taking the values 1, 2, 3, 4, 5, 6 as follows, with allowance for the symmetry of the pairs:

$$\left.\begin{array}{lll}(11)\leftrightarrow 1, & (22)\leftrightarrow 2, & (33)\leftrightarrow 3,\\ (23)=(32)\leftrightarrow 4, & (31)=(13)\leftrightarrow 5, & (12)=(21)\leftrightarrow 6.\end{array}\right\} \tag{6.5}$$

The three-dimensional subscripts are indicated by the latin letters i, k, l, m = 1, 2, 3, while the six-dimensional ones are indicated by the greek letters α, β, ε, τ = 1, 2, 3, 4, 5, 6; the previous notation is retained for quantities such as γ_{ik}, σ_{ik}, c_{iklm}, but with the new greek subscripts in the six-dimensional representation. Hence we have

$$\gamma_\alpha\leftrightarrow\gamma_{ik},\qquad \sigma_\alpha\leftrightarrow\sigma_{ik},\qquad c_{\alpha\beta}=c_{\beta\alpha}\leftrightarrow c_{iklm}. \tag{6.6}$$

This notation may be made concrete by requiring that the elastic energy and Hooke's law take the following form in the six-dimensional formulation:

$$\Phi = \frac{1}{2} \sigma_\alpha \gamma_\alpha, \tag{6.7}$$

$$\sigma_\alpha = c_{\alpha\beta} \gamma_\beta. \tag{6.8}$$

Here also we assume summation over the duplicated greek subscripts from 1 to 6; comparison of (6.3) with (6.7) gives $\sigma_\alpha \gamma_\alpha = \sigma_{ik} \gamma_{ik}$, or in expanded form,

$$\begin{aligned} &\sigma_1\gamma_1 + \sigma_2\gamma_2 + \sigma_3\gamma_3 + \sigma_4\gamma_4 + \sigma_5\gamma_5 + \sigma_6\gamma_6 = \\ &\sigma_{11}\gamma_{11} + \sigma_{22}\gamma_{22} + \sigma_{33}\gamma_{33} + 2\sigma_{23}\gamma_{23} + 2\sigma_{31}\gamma_{31} + 2\sigma_{12}\gamma_{12}. \end{aligned} \tag{6.9}$$

It is clear that this is satisfied if [see (6.5)] we put

$$\left. \begin{aligned} &\gamma_1 = \gamma_{11}, \quad \gamma_2 = \gamma_{22}, \quad \gamma_3 = \gamma_{33}, \\ &\gamma_4 = 2\gamma_{23}, \quad \gamma_5 = 2\gamma_{31}, \quad \gamma_6 = 2\gamma_{12}, \end{aligned} \right\} \tag{6.10}$$

$$\left. \begin{aligned} &\sigma_1 = \sigma_{11}, \quad \sigma_2 = \sigma_{22}, \quad \sigma_3 = \sigma_{33}, \\ &\sigma_4 = \sigma_{23}, \quad \sigma_5 = \sigma_{31}, \quad \sigma_6 = \sigma_{12}. \end{aligned} \right\} \tag{6.11}$$

Substituting these expressions for γ_α and σ_α into (6.8) and comparing with (4.3), we see that

$$c_{\alpha\beta} = c_{iklm}, \tag{6.12}$$

in which $\alpha \leftrightarrow (ik)$, $\beta \leftrightarrow (lm)$ in accordance with (6.5). Relations (6.10)-(6.12) are generally accepted, but we should bear in mind that they are not the only ones possible. They have the disadvantage that the two quantities γ_{ik} and σ_{ik}, which are of the same type as regards geometrical properties, are differently related to the corresponding quantities γ_α and σ_α. In principle, it would be more regular and logical to have a relation which would apply for any symmetric tensor θ_{ik}:

$$\left. \begin{aligned} &\theta_1 = \theta_{11}, \qquad \theta_2 = \theta_{22}, \qquad \theta_3 = \theta_{33}, \\ &\theta_4 = \sqrt{2}\,\theta_{23}, \quad \theta_5 = \sqrt{2}\,\theta_{31}, \quad \theta_6 = \sqrt{2}\,\theta_{12}. \end{aligned} \right\} \tag{6.13}$$

It is clear that these relations, as applied to σ and γ, give us (6.9) again; but (6.12) is then replaced by a more complicated relation. The general rule of (6.13) for converting from the σ_{ik}, γ_{ik}, c_{iklm} to the σ_α, γ_α, $c_{\alpha\beta}$ is that a quantity with greek subscripts equal [in the sense of (6.5)] the quantity with latin subscrips multiplied by $(\sqrt{2})^p$, in which p is the number of greek subscripts (4, 5, or 6). This single general rule is a consequence of the universality of (6.13) and is an undoubted advantage of the latter; but the formulas contain the inconvenient factors $\sqrt{2}$, and the simple relations of (6.12) are replaced by more complex ones, so in future I shall use (6.10)-(6.12).

It is clear from this that the number of independent c_{iklm} equals the number of different elements in the six-row symmetrical square matrix

$$C = (c_{\alpha\beta}) = \begin{pmatrix} c_{11} & c_{12} & c_{13} & c_{14} & c_{15} & c_{16} \\ c_{21} & c_{22} & c_{23} & c_{24} & c_{25} & c_{26} \\ c_{31} & c_{32} & c_{33} & c_{34} & c_{35} & c_{36} \\ c_{41} & c_{42} & c_{43} & c_{44} & c_{45} & c_{46} \\ c_{51} & c_{52} & c_{53} & c_{54} & c_{55} & c_{56} \\ c_{61} & c_{62} & c_{63} & c_{64} & c_{65} & c_{66} \end{pmatrix}. \qquad (6.14)$$

Hence this number does not exceed 21 in the most general case.*

---

* There is a long history to the question of the number of independent elastic constants (elastic moduli). In the first half of the 19th century there was a controversy between Stokes and Green on the one hand (who believed there were a maximum of 21 independent constants) and Cauchy and Poisson on the other (who considered that there were at most 15). The difference was due to differences in assumptions about the microstructure of the crystal lattice and the nature of the forces in it. Voigt's experiments decided the question in favor of the 21-constant theory. Very recently Laval, Raman, and others have expressed the view that rotation and torque should be considered in the theory of elasticity, as well as deformation and stress; in that case, the σ_{ik} and γ_{ik} tensors would not be symmetric, so each would contain 9 independent components. The $c_{\alpha\beta}$ matrix would remain symmetrical with respect to α and β (because this follows from the basic law of conservation of energy) but would have 9 rows, which would raise the maximum number of elastic constants to 45. The experimental aspect of this problem is not clear; recent measurements [1] would not support it. The problem is consided in a review [2] which also gives an extensive bibliography.

From (6.7) and (6.8)

$$\Phi = \frac{1}{2} c_{\alpha\beta} \gamma_\alpha \gamma_\beta. \tag{6.15}$$

Thus Φ is quadratic in the six variables γ_α and must be positive under all conditions, since the potential energy is minimal in a state of stable equilibrium. Now $\Phi = 0$ in the nature (undeformed state), and this is the state of stable equilibrium, so $\Phi > 0$ when there is any deformation. Hence the quadratic form of (6.15) is positively defined, as is the corresponding matrix $c_{\alpha\beta}$. This implies, in particular, that the diagonal elements $c_{11}, c_{22}, \ldots, c_{66}$ must all be positive. For example, if $c_{11} < 0$, we would get, on putting $\gamma_1 \neq 0$, $\gamma_\alpha = 0$, $\alpha \geq 2$, that $\Phi = (1/2)c_{11}\gamma_1^2 < 0$.

Positive diagonal elements in (6.14) are necessary but not sufficient to give positive definition of $(c_{\alpha\beta})$; the necessary and sufficient condition for this for a symmetrical real $(c_{\alpha\beta})$ is [3] that the sequential principal minors be positive. These minors are delineated by the dotted lines in (6.14). Hence the elastic moduli of any medium always satisfy the following inequalities:

$$\left.\begin{array}{l} c_{11} > 0, \quad \begin{vmatrix} c_{11} & c_{12} \\ c_{21} & c_{22} \end{vmatrix} = c_{11}c_{22} - c_{12}^2 > 0, \\ \begin{vmatrix} c_{11} & c_{12} & c_{13} \\ c_{21} & c_{22} & c_{23} \\ c_{31} & c_{32} & c_{33} \end{vmatrix} > 0, \ \ldots, \ |c_{\alpha\beta}| > 0. \end{array}\right\} \tag{6.16}$$

Here $|c_{\alpha\beta}|$ denotes the determinant of $(c_{\alpha\beta})$. The condition on Φ obviously should not be dependent on the order in which the γ_α are numbered, but any alteration in that numbering causes a certain identical permutation of the rows and columns of $(c_{\alpha\beta})$, which can bring any principal minor to the top left corner.* Hence the condition that successive minors be positive is equivalent to the condition that all principal minors be positive. The diagonal elements are the first-order principal minors, which thus implies (as above) that these must be positive.

---

*A principal minor (of any order) is a minor obtained by striking out equal numbers of rows and columns with identical numbers.

From (6.16) we have $|c_{\alpha\beta}| \neq 0$, so $(c_{\alpha\beta})$ has an inverse, and the system of (6.8) can be solved for the γ_β. Let the inverse of $(c_{\alpha\beta})$ be $(s_{\alpha\beta})$:

$$(s_{\alpha\beta}) = (c_{\alpha\beta})^{-1}. \tag{6.17}$$

This means that

$$s_{\alpha\beta}c_{\beta\tau} = c_{\alpha\beta}s_{\beta\tau} = \delta_{\alpha\tau}, \tag{6.18}$$

in which $\delta_{\alpha\tau}$ is a six-dimensional Kronecker symbol, whose properties are $\delta_{\alpha\tau} = 1$ for $\alpha = \tau$ and $\delta_{\alpha\tau} = 0$ for $\alpha \neq \tau$. Hence $(\delta_{\alpha\tau})$ is a unit six-row matrix:

$$(\delta_{\alpha\tau}) = \begin{pmatrix} 1 & 0 & 0 & 0 & 0 & 0 \\ 0 & 1 & 0 & 0 & 0 & 0 \\ 0 & 0 & 1 & 0 & 0 & 0 \\ 0 & 0 & 0 & 1 & 0 & 0 \\ 0 & 0 & 0 & 0 & 1 & 0 \\ 0 & 0 & 0 & 0 & 0 & 1 \end{pmatrix}. \tag{6.19}$$

We multiply (6.8) by $s_{\tau\alpha}$ and use (6.18) to get $s_{\tau\alpha}\sigma_\alpha = s_{\tau\alpha}c_{\alpha\beta}\gamma_\beta = \delta_{\tau\beta}\gamma_\beta = \gamma_\tau$, i.e., the solution is of the form

$$\gamma_\tau = s_{\tau\alpha}\sigma_\alpha. \tag{6.20}$$

We put the last relation in the following form by reverting to the three-dimensional quantities γ_{ik}, σ_{lm}, s_{iklm}, $(s_{iklm} \leftrightarrow s_{\alpha\beta})$:

$$\gamma_{ik} = s_{iklm}\sigma_{lm}. \tag{6.21}$$

As $(c_{\alpha\beta})$ is symmetric, $(s_{\alpha\beta})$ is the same: $s_{\alpha\beta} = s_{\beta\alpha}$ (see Chapter 2), and hence the s_{iklm} tensor has the same symmetry properties as the c_{iklm} one:

$$s_{iklm} = s_{kilm} = s_{ikml} = s_{lmik}. \tag{6.22}$$

The s_{iklm} are termed the compliance (or flexibility) constants, a name which is natural on the basis of (6.21), which shows that the γ_{ik} increase* with the s_{iklm} for given σ_{lm}.

The relation between the $s_{\alpha\beta}$ and the s_{iklm} is not as in (6.12); comparison of (6.20) and (6.21) with (6.10) and (6.11) shows that

$$s_{\alpha\beta} = 2^p s_{iklm}, \tag{6.23}$$

in which p is the number of the subscripts in excess of 3 that appear in α and β. The difference between (6.12) and (6.23) is due to the difference between (6.10) and (6.11); a symmetric relation as of (6.13)

$$\theta_\alpha = (\sqrt{2})^p \theta_{ik}, \tag{6.24}$$

where by θ_{ik} we understand γ_{ik} and σ_{ik}, would replace (6.12) and (6.23) by the relation

$$\psi_{\alpha\beta} = (\sqrt{2})^p \psi_{iklm}, \tag{6.25}$$

where by ψ_{iklm} we understand c_{iklm} and s_{iklm}; here p is as in (6.23).

7. Crystal Symmetry

Crystals are symmetrical bodies whose properties are unaltered by certain transformations, e.g., reflection at a plane or rotation around an axis. These are termed the symmetry transformations of that crystal, and the set of these is termed the symmetry of the crystal. This set forms a group, because any transformation that is reducible to successive use of two of the symmetry transformations will not alter the properties of the body and so belongs to that set.

Any possible symmetry transformation of any crystal can be reduced to one of two basic types of transformation or to a combination of these. These basic types are rotation around an axis through an angle $2\pi/k = \varphi_k$ (k equal to 2, 3, 4, or 6) and mirror

---

* From analogous arguments based on (4.3) we call the c_{iklm} the rigidity constants.

reflection* at a plane. If the body coincides with itself after rotation through $2\pi/k$, the axis is a k-fold symmetry axis and is denoted by L^k or simply by k. The second basic element is a symmetry plane, which is denoted by P or m; the presence of such a plane means that the body coincides with itself after mirror reflection at that plane.

A body may have symmetry such as to coincide with itself when two transformations are performed simultaneously (rotation around a k-fold axis and reflection at a plane perpendicular to that axis). Here the order of the operations is irrelevant, so these operations mutually permutate or commutate. In fact, an arbitrary point ends up in the same position whichever order is used. This is termed a k-fold reflection-rotation axis, which is a new symmetry element only if both transformations together are needed to bring the body into coincidence with itself. Hence such an axis is not a new element if k is odd; performing the operation k times, we turn the body through 2π (which is equivalent to no rotation) and reflect it k times (odd), which is equivalent to a single reflection. This means that the body coincides with itself after simple reflection, so it has a symmetry plane as an independent element; but in that case it has an independent k-fold axis. Hence a reflection-rotation axis of odd order is equivalent to an independent symmetry plane and an independent axis of the same order. The only new independent symmetry elements (although they are combinations of the basic types) are reflection-rotation axes of even order. The presence of one of these implies the presence of a simple symmetry axis of half that order, for rotation through $2\pi/2k$ twice and double reflection together amount to simple rotation through $2\pi/k$. The possible reflection-rotation axes are denoted by $\bar{4}$ and $\bar{6}$ in Shubnikov's notation, with $\bar{2}$ as a special case. We take the point where the L^2 axis meets the P plane as the origin (point 0 in Fig. 1) and draw from the latter a radius vector **r** to any point A in the body.

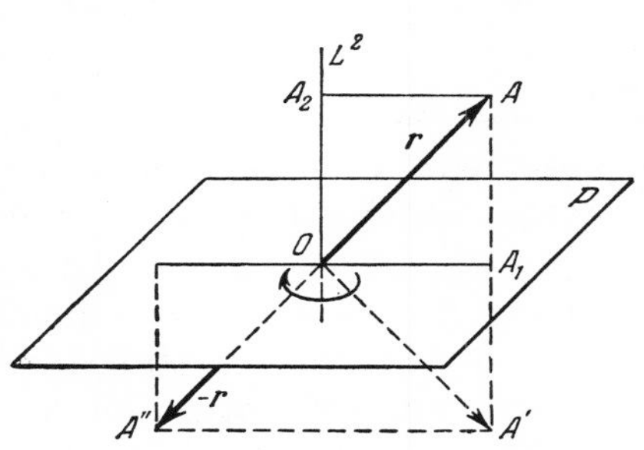

Fig. 1

*We disregard the parallel-displacement transformation, which leaves an infinite crystal lattice unaltered.

Table I. Crystal Symmetry

System	Groth class	Fedorov Institute name	Formula
Triclinic	1	Primitive	1
	2	Central	$\bar{2}$
Monoclinic	3	Axial	2
	4	Planal	m
	5	Planaxial	2 : m
Orthorhombic	6	Axial	3 : 2
	7	Planal	2 : m
	8	Planaxial	m · 2 : m
Tetragonal	9	Primitive gyroidal	$\bar{4}$
	10	Primitive	4
	11	Trigonal	$\bar{4}$ · m
	12	Axial	4 : 2
	13	Central	4 : m
	14	Planal	4 : m
	15	Planaxial	m · 4 : m
Trigonal	16	Primitive	3
	17	Central	$\bar{6}$
	18	Axial	3 : 2
	19	Planal	3 · m
	20	Planaxial	$\bar{6}$ · m
Hexagonal	21	Primitive gyroidal	3 : m
	22	Planogyroidal	m · 3 : m
	23	Primitive	6
	24	Axial	6 : 2
	25	Central	6 : m
	26	Planal	6 · m
	27	Planaxial	m · 6 : m
Cubic	28	Primitive	3/2
	29	Axial	3/4
	30	Central	$\bar{6}$/2
	31	Planal	3/$\bar{4}$
	32	Planaxial	$\bar{6}$/4

Then rotation through π around the axis changes the sign of OA_1, the projection of **r** on the symmetry plane, while reflection changes the sign of OA_2, the projection of **r** on the axis, whereas simultaneous performance of both operations causes the radius vector itself to change sign, i.e., to reverse, as OA". This transformation is known as reflection at a point or inversion, while the origin in the above case is called a center of symmetry or inversion center. Hence $\bar{2}$ is equivalent to inversion, i.e., to the transformation $\mathbf{r} \to -\mathbf{r}$ ($x_i \to -x_i$). The special symbol C is often used instead of $\bar{2}$.

The presence of one symmetry element often results in that of others, as we have seen for even-order reflection-rotation axes. If there are several symmetry elements, the different combinations of these may give rise to a series of new elements. For example, if a k-fold axis ($k > 2$) has a symmetry plane parallel to it, the latter cannot be unique, because rotation through $2\pi/k$ should not alter the properties of the body. Hence the presence of one plane implies that of $(k-1)$ others. This means that we do not have to draw up an exhaustive list of the symmetry elements in order to describe the symmetry of a body completely; it is sufficient to state only the elements sufficient to give all the others by repetition or combination. These basic elements are called the generating symmetry elements

The derivation of all the possible symmetry groups will not be considered, since this is to be found in textbooks [6]. It is known that there are 32 classes of crystal symmetry, which fall into 7 systems (Table I). The first column gives the name of the system; the second gives the number of the class in Groth's notation. The third gives the names adopted in the Fedorov Institute in 1935. The fourth gives the formula for the symmetry in Shubnikov's symbolism; the numbers 2, 3, 4, and 6 denote symmetry axes of those orders, while $\bar{4}$ is a fourfold reflection-rotation axis, m is a symmetry plane, a point (·) denotes parallelism, and a colon (:) denotes perpendicularity. The symbol 3/2 denotes twofold and threefold axes meeting at an angle α such that $\cos\alpha = 1/\sqrt{3}$.

The symmetry elements listed in the fourth column are the generating elements of their classes and hence completely define the symmetry of the class.

8. Elastic Moduli of Crystals of the Lower Systems

The presence of symmetry imposes certain conditions on the c_{iklm}, so the number of independent components become less than 21. This is most readily seen for an isotropic medium, which has the highest possible symmetry. All symmetry groups of crystals are subgroups of the three-dimensional orthogonal group, which is defined as the set of all rotations and reflections in three-dimensional space. Symmetry in a crystal means simply invariance of the properties with respect to the transforms of some subgroup of the orthogonal group, whereas the properties of an isotropic medium are, by definition, invariant with respect to all the transforms of the orthogonal group. This at once gives us the form of the c_{iklm} tensor for any isotropic medium, since it expresses the properties of the medium, and hence it is invariant with respect to all the transforms of the orthogonal group. However, there is a unique tensor that is unaffected by all orthogonal transforms: this is unit tensor, apart from a scalar factor, and so the c_{iklm} should be expressed as combinations of the components δ_{ik} of that tensor with certain coefficients. There are only three different such combinations, which contain the four subscripts i, k, l, and m, namely $\delta_{ik}\delta_{lm}$, $\delta_{il}\delta_{km}$, $\delta_{im}\delta_{kl}$, so the c_{iklm} tensor must be expressed as follows:

$$c_{iklm} = c\delta_{ik}\delta_{lm} + b\delta_{il}\delta_{km} + a\delta_{im}\delta_{kl},$$

in which a, b, and c are scalar coefficients. The symmetry of the c_{iklm} means that the right side should not alter when i and k are permuted, so $a = b$; it is readily seen that all the other symmetry properties of (6.2) then satisfy the condition, so the most general form for the elastic-modulus tensor for an isotropic medium is

$$c_{iklm} = c\delta_{ik}\delta_{lm} + a(\delta_{il}\delta_{km} + \delta_{im}\delta_{kl}). \tag{8.1}$$

This contains only two independent constants a and c.

To find the $(c_{\alpha\beta})$ of (6.14) that corresponds to (8.1) we must use (6.5) and (6.12), whereupon we find that

$$c_{11} = c_{22} = c_{33} = c + 2a, \quad c_{12} = c_{23} = c_{31} = c,$$

$$c_{44} = c_{55} = c_{66} = a = \frac{c_{11} - c_{12}}{2},$$

while all the other $c_{\alpha\beta}$ are zero. Hence for an isotropic medium

$$(c_{\alpha\beta}) = \begin{vmatrix} c_{11} & c_{12} & c_{12} & 0 & 0 & 0 \\ c_{12} & c_{11} & c_{12} & 0 & 0 & 0 \\ c_{12} & c_{12} & c_{11} & 0 & 0 & 0 \\ 0 & 0 & 0 & \frac{1}{2}(c_{11} - c_{12}) & 0 & 0 \\ 0 & 0 & 0 & 0 & \frac{1}{2}(c_{11} - c_{12}) & 0 \\ 0 & 0 & 0 & 0 & 0 & \frac{1}{2}(c_{11} - c_{12}) \end{vmatrix}. \tag{8.2}$$

Consider the changes in the c_{iklm} in response to orthogonal transforms. By definition, the c_{iklm} tensor is orthogonal, as are the γ_{ik} and σ_{ik} ones (see [4]), which means that the c_{iklm} vary in accordance with the formulas for the product of the corresponding components of four vectors, e.g., $p_i q_k u_l v_m$. Let (S_{ik}) be the matrix of some orthogonal transform. The change in the vector this produces is given by

$$p'_i = S_{ik} p_k. \tag{8.3}$$

Correspondingly the product $p_i q_k u_l v_m$ changes in accordance with the formula $(p_i q_k u_l v_m)' = S_{ii'} p_{i'} S_{kk'} q_{k'} S_{ll'} u_{l'} S_{mm'} v_{m'}$ whence we have for the c_{iklm}

$$c'_{iklm} = S_{ii'} S_{kk'} S_{ll'} S_{mm'} c_{i'k'l'm'}. \tag{8.4}$$

If $S = (S_{ik})$ belongs to the symmetry group of the crystal, c'_{iklm} must coincide with c_{iklm}, so we get that

$$c_{iklm} = S_{ii'} S_{kk'} S_{ll'} S_{mm'} c_{i'k'l'm'}, \tag{8.5}$$

which should be obeyed for all i, k, l, and m as well as for all S belonging to the symmetry group of the crystal.

The set of equations (8.5) constitutes the conditions imposed by the symmetry on the elastic moduli. Consider now the S belonging to the basic symmetry elements of all crystals (section 7).

Center of symmetry C. The corresponding transformation (inversion) alters the signs of all components of any vector, which may be put symbolically as

$$1 \to -1, \ 2 \to -2, \ 3 \to -3. \tag{8.6}$$

Here the number represents the corresponding component of the vector, so $1 \to -1$ denotes $p_1 \to -p_1$.

Symmetry plane P. Setting (say) the x_3 axis perpendicular to this plane, we find that the third component of any vectors changes sign, while the other components should not be affected. This is expressed symbolically as

$$1 \to 1, \ 2 \to 2, \ 3 \to -3. \tag{8.7}$$

Twofold axis L^2. This corresponds to rotation through π around, say, the x_3 axis. The corresponding transformation is expressed as

$$1 \to -1, \quad 2 \to -2, \quad 3 \to 3. \tag{8.8}$$

Transformations corresponding to axes of higher order are considered below.

We must note that invariance under inversion imposes no restriction on the components of any even-rank tensor, including the c_{iklm} tensor; the transform of (8.6) produces a four-fold sign change in each c_{iklm}, which thus remains unaltered. This means that we need consider only P and L^k transformations, and combinations of these, in order to determine the number of independent elastic moduli.

Triclinic crystals (system of lowest symmetry, Table I) may possess only an inversion center, so the number of elastic moduli is not reduced by the symmetry. However, we still have various choices for the coordinate system, since the orientation of the latter relative to the crystal is not specified, and this is defined by three parameters (e.g., the Euler angles). Hence a definite choice of orientation imposes three conditions on the c_{iklm}, with the result that the number of independent significant moduli cannot exceed 18. Similar arguments apply to other crystals, as we shall see; they arise because the c_{iklm} resemble the components of any other tensor that describes a property of the crystal in being dependent on the orientation of the crystal as a whole as well as on the internal structure inherent in that crystal.

We shall not discuss here the rational choice of coordinate system for triclinic or other crystals, since this will be dealt with later in relation to the properties of elastic waves (section 18).

We see from (8.6)-(8.8) that the transformation corresponding to a symmetry axis $L^2 \parallel x_3$ differs from reflection in a plane $P \perp x_3$ by the addition of inversion (section 7); but we have seen that inversion is of no importance to the c_{iklm} tensor, so a twofold axis is precisely equivalent to a symmetry plane perpendicular to it as regards conditions imposed on the c_{iklm}, and such a plane is very easy to consider. We have from (8.7) that

$$c_{iklm} = (-1)^p c_{iklm}, \tag{8.9}$$

in which p is the number of subscripts equal to 3 in the i, k, l, m. This implies that those c_{iklm} are zero which have a subscript corresponding to an L^2 axis (or to a symmetry plane normal to it) appearing an odd number of times.

A monoclinic crystal of any class has either L^2 or P, or both (Table I). Placing the x_3 axis appropriately, we have

$$(c_{\alpha\beta}) = \begin{pmatrix} c_{11} & c_{12} & c_{13} & 0 & 0 & c_{16} \\ c_{12} & c_{22} & c_{23} & 0 & 0 & c_{26} \\ c_{13} & c_{23} & c_{33} & 0 & 0 & c_{36} \\ 0 & 0 & 0 & c_{44} & c_{45} & 0 \\ 0 & 0 & 0 & c_{45} & c_{55} & 0 \\ c_{16} & c_{26} & c_{36} & 0 & 0 & c_{66} \end{pmatrix}. \tag{8.10}$$

This contains 13 parameters, but not all are significant, for we have specified only the x_3 axis of the coordinate system, which leaves an arbitrary element associated with the choice of the x_1 axis, which allows us to eliminate one modulus (see section 19). The number of significant elastic moduli for a monoclinic crystal is thus 12.

An orthorhombic crystal of any class has three mutually perpendicular directions parallel to normals to symmetry planes or to the equivalent twofold axes (Table I); these directions form a natural coordinate system, so the c_{iklm} must have the property of (8.9) in respect to each subscript, which implies that the only c_{iklm} different from zero are those in which each subscript appears an even number of times. This gives us the matrix

$$(c_{\alpha\beta}) = \begin{pmatrix} c_{11} & c_{12} & c_{13} & 0 & 0 & 0 \\ c_{12} & c_{22} & c_{23} & 0 & 0 & 0 \\ c_{13} & c_{23} & c_{33} & 0 & 0 & 0 \\ 0 & 0 & 0 & c_{44} & 0 & 0 \\ 0 & 0 & 0 & 0 & c_{55} & 0 \\ 0 & 0 & 0 & 0 & 0 & c_{66} \end{pmatrix}. \tag{8.11}$$

All 9 elastic moduli are significant, since the coordinate system is completely specified relative to the crystal.

9. Elastic Moduli of Crystals of the Higher Systems

A tetragonal crystal of any class has an L^4 axis; rotation through 90° around $x_3 \parallel L^4$ changes the components of a vector in accordance with the scheme

$$1 \to 2, \quad 2 \to -1, \quad 3 \to 3. \tag{9.1}$$

An L^4 axis is simultaneously an L^2 axis, so we use (8.10). The transformation of (9.1) gives

$$c_{26} = c_{2212} \to -c_{1121} = -c_{16}, \quad c_{11} \to c_{22}, \quad c_{13} \to c_{23}, \quad c_{44} \to c_{55},$$

$$c_{36} = c_{3312} \to -c_{3321} = -c_{36}, \quad c_{45} = c_{2331} \to -c_{1332} = -c_{54}.$$

Hence $c_{36} = c_{45} = 0$, $c_{26} = -c_{16}$, $c_{11} = c_{22}$, $c_{13} = c_{23}$, $c_{44} = c_{55}$, and $(c_{\alpha\beta})$ matrix becomes

$$(c_{\alpha\beta}) = \begin{pmatrix} c_{11} & c_{12} & c_{13} & 0 & 0 & c_{16} \\ c_{12} & c_{11} & c_{13} & 0 & 0 & -c_{16} \\ c_{13} & c_{13} & c_{33} & 0 & 0 & 0 \\ 0 & 0 & 0 & c_{44} & 0 & 0 \\ 0 & 0 & 0 & 0 & c_{44} & 0 \\ c_{16} & -c_{16} & 0 & 0 & 0 & c_{66} \end{pmatrix}, \tag{9.2}$$

which contains 7 independent parameters. However, we have considered only the symmetry associated with the fourfold axis, but the tetragonal system has seven classes (Table I), of which four ($\bar{4} \cdot m$, $4:2$, $4 \cdot m$, and $m \cdot 4 : m$) have a symmetry plane parallel to L^4 or a twofold axis perpendicular to the latter. Taking the L^2 axis, or the normal to the plane, as the x_1 axis, a condition analogous to (8.9) gives us $c_{16} = 0$, so for these four classes the matrix becomes

$$(c_{\alpha\beta}) = \begin{pmatrix} c_{11} & c_{12} & c_{13} & 0 & 0 & 0 \\ c_{12} & c_{11} & c_{13} & 0 & 0 & 0 \\ c_{13} & c_{13} & c_{33} & 0 & 0 & 0 \\ 0 & 0 & 0 & c_{44} & 0 & 0 \\ 0 & 0 & 0 & 0 & c_{44} & 0 \\ 0 & 0 & 0 & 0 & 0 & c_{66} \end{pmatrix}. \tag{9.3}$$

This contains 6 independent moduli, which are all significant, since the coordinate has been specified. The other three tetragonal classes ($\bar{4}$, 4, and 4 : m) have no elements that distinguish a direction perpendicular to L^4; but hence we can choose x_1 to simplify $(c_{\alpha\beta})$. Consider the transformation of c_{16} on rotation around $x_3 \| L^4$ through an arbitrary angle φ. The corresponding orthogonal matrix is

$$S = \begin{pmatrix} \cos\varphi & \sin\varphi & 0 \\ -\sin\varphi & \cos\varphi & 0 \\ 0 & 0 & 1 \end{pmatrix}. \tag{9.4}$$

The transformed element $c_{16} = c_{1112}$ is shown by (8.14) to be expressed by

$$c'_{16} = c'_{1112} = S_{1i} S_{1k} S_{1l} S_{2m} c_{iklm}. \tag{9.5}$$

Here (9.4) shows that i, k, *l*, and m should be given only the values 1 and 2. Summing with respect to *l*, m = 1, 2, we get

$$c'_{16} = S_{1i} S_{1k} \left[\frac{1}{2} \sin 2\varphi \left(c_{ik22} - c_{ik11} \right) + \cos 2\varphi c_{ik12} \right].$$

Now we sum with respect to i, k = 1, 2 and use the results of (9.2) that $c_{22} = c_{11}$ and $c_{26} = -c_{16}$ to get that

$$c'_{16} = c_{16} \cos 4\varphi - \frac{1}{4} (c_{11} - c_{12} - 2c_{66}) \sin 4\varphi. \tag{9.6}$$

Hence, choosing φ to be such as to satisfy

$$\tan 4\varphi = \frac{4c_{16}}{c_{11} - c_{12} - 2c_{66}}, \tag{9.7}$$

we get $c'_{16} = 0$, so (9.2) coincides with (9.3), and all tetragonal classes have only six significant elastic moduli, $(c_{\alpha\beta})$ always reducing to the form of (9.3) on appropriate choice of the coordinate system. The difference between classes 11, 12, 14, and 15 on the one hand and 9, 10, and 3 on the other (Table I) is only that the first have the direction of x_1 (or x_2) directly determined by the elements of crystallographic symmetry as being parallel to L^2 or perpendicular to P, whereas the second have the direction of x_1 defined by the condition of (9.7).

A trigonal crystal has a threefold axis; as usual, we assume $x_3 \| L^3$. Matrix (9.4) gives the transformation of coordinates corresponding to a rotation through φ around L^3. Here the components γ_{ik} (tensor of second rank) transform via a formula analogous to (8.4):

$$\gamma'_{ik} = S_{ii'} S_{kk'} \gamma_{i'k'}. \tag{9.8}$$

This relation may be expanded via (9.4) to get

$$\left.\begin{aligned}
\gamma'_{11} &= S_{1i} S_{1k} \gamma_{ik} = \gamma_{11} \cos^2\varphi + \gamma_{22} \sin^2\varphi + \gamma_{12} \sin 2\varphi, \\
\gamma'_{22} &= \gamma_{11} \sin^2\varphi + \gamma_{22} \cos^2\varphi - \gamma_{12} \sin 2\varphi, \quad \gamma'_{33} = \gamma_{33}, \\
\gamma'_{23} &= \gamma_{23} \cos\varphi - \gamma_{13} \sin\varphi, \\
\gamma'_{31} &= \gamma_{23} \sin\varphi + \gamma_{13} \cos\varphi, \\
\gamma'_{12} &= \frac{1}{2} (\gamma_{22} - \gamma_{11}) \sin 2\varphi + \gamma_{12} \cos 2\varphi.
\end{aligned}\right\} \tag{9.9}$$

Consider the six-dimensional vectors

$$\gamma = (\gamma_\alpha) = \begin{pmatrix} \gamma_1 \\ \gamma_2 \\ \gamma_3 \\ \gamma_4 \\ \gamma_5 \\ \gamma_6 \end{pmatrix}, \quad \gamma' = (\gamma'_\alpha) = \begin{pmatrix} \gamma'_1 \\ \gamma'_2 \\ \gamma'_3 \\ \gamma'_4 \\ \gamma'_5 \\ \gamma'_6 \end{pmatrix}, \tag{9.10}$$

whose components γ_α are related to the γ_{ik} by (6.10); then (9.9) may be put as

$$\gamma' = U\gamma, \tag{9.11}$$

in which U is a six-row matrix of the form

$$U = U(\varphi) = \begin{pmatrix}
\cos^2\varphi & \sin^2\varphi & 0 & 0 & 0 & \frac{1}{2}\sin 2\varphi \\
\sin^2\varphi & \cos^2\varphi & 0 & 0 & 0 & -\frac{1}{2}\sin 2\varphi \\
0 & 0 & 1 & 0 & 0 & 0 \\
0 & 0 & 0 & \cos\varphi & -\sin\varphi & 0 \\
0 & 0 & 0 & \sin\varphi & \cos\varphi & 0 \\
-\sin 2\varphi & \sin 2\varphi & 0 & 0 & 0 & \cos 2\varphi
\end{pmatrix}. \tag{9.12}$$

Hooke's law of (6.8) now becomes

$$\sigma = C\gamma, \tag{9.13}$$

in which σ is the stress 6-vector, which is analogous to (9.10), while C is the matrix of (6.14); (9.13) in the transformed coordinate system takes the form

$$\sigma' = C'\gamma', \tag{9.14}$$

in which $\sigma' = U'\sigma$, where U' differs from U in that the σ_α are related to the σ_{ik} by (6.11), which differs from (6.10). Hence

$$U' = \begin{pmatrix} \cos^2\varphi & \sin^2\varphi & 0 & 0 & 0 & \sin 2\varphi \\ \sin^2\varphi & \cos^2\varphi & 0 & 0 & 0 & -\sin 2\varphi \\ 0 & 0 & 1 & 0 & 0 & 0 \\ 0 & 0 & 0 & \cos\varphi & -\sin\varphi & 0 \\ 0 & 0 & 0 & \sin\varphi & \cos\varphi & 0 \\ -\frac{1}{2}\sin 2\varphi & \frac{1}{2}\sin 2\varphi & 0 & 0 & 0 & \cos 2\varphi \end{pmatrix}. \tag{9.15}$$

Symmetrical normalization of γ and σ in accordance with (6.24) would give U' = U. Equation (9.14) may be put as

$$U'\sigma = C'U\gamma$$

or, from (9.13), $U'C\gamma = C'U\gamma$, whence we conclude that

$$U'C = C'U. \tag{9.16}$$

We multiply the latter from the right by U^{-1} to get

$$C' = U'CU^{-1}. \tag{9.17}$$

We get U^{-1} from U by replacing φ by $-\varphi$. The last formula defines the transformation matrix C undergoes when the coordinates undergo that of (9.4) while the σ and γ tensors, respectively, undergo those of (9.12) and (9.15). We must have C' = C if the transformation U belongs to the symmetry of the crystal, i.e., from (9.16)

$$CU = U'C. \tag{9.18}$$

This is the condition imposed on matrix C by the crystal symmetry.

Condition (9.18) should be obeyed when there is a threefold axis if in the U of (9.12) and the U' of (9.15) we put $\varphi = \varphi_0 = 2\pi/3 = 120°$; substituting in (9.18) for $U(\varphi_0)$, $U'(\varphi_0)$ and the C of (6.14), and comparing the matrix elements in the two halves, we get the conditions imposed on the $c_{\alpha\beta}$. For instance, $(CU)_{11} = c_{1\alpha}U_{\alpha 1} = (U'C)_{11} = U'_{1\alpha}c_{\alpha 1}$ gives

$$c_{11}\cos^2\varphi_0 + c_{12}\sin^2\varphi_0 - c_{16}\sin 2\varphi_0 = c_{11}\cos^2\varphi_0 + c_{21}\sin^2\varphi_0 + c_{61}\sin 2\varphi_0,$$

which implies that $c_{16} = 0$. Similar calculations on the other elements give

$$C = (c_{\alpha\beta}) = \begin{pmatrix} c_{11} & c_{12} & c_{13} & c_{14} & -c_{25} & 0 \\ c_{12} & c_{11} & c_{13} & -c_{14} & c_{25} & 0 \\ c_{13} & c_{13} & c_{33} & 0 & 0 & 0 \\ c_{14} & -c_{14} & 0 & c_{44} & 0 & c_{25} \\ -c_{25} & c_{25} & 0 & 0 & c_{44} & c_{14} \\ 0 & 0 & 0 & c_{25} & c_{14} & \frac{c_{11}-c_{12}}{2} \end{pmatrix}. \tag{9.19}$$

Here there are 7 independent parameters, but the direction of x_1 has been left undefined. Classes 18, 19, and 20 (Table I) have an additional element $L^2 \perp L^3$ or P parallel to L^3; putting x_1 along L^2 or along the normal to P, we get from (8.9) that $c_{25} = c_{2231} = 0$, whereas if we treat x_2 in this way we get $c_{14} = c_{1123} = 0$. This leaves only 6 independent parameters. But the same result is obtained for classes 16 and 17, which lack additional symmetry elements, as we may see by considering the transformation of c_{14} and c_{25} on rotation around $x_3 \parallel L^3$ through some angle φ [see (9.4)]. As in (9.5), we find, using (9.19), that *

$$\left.\begin{aligned} c'_{14} &= c_{14}\cos 3\varphi + c_{25}\sin 3\varphi, \\ c'_{25} &= -c_{14}\sin 3\varphi + c_{25}\cos 3\varphi. \end{aligned}\right\} \tag{9.20}$$

It is clear that the choice $\tan 3\varphi = c_{25}/c_{14}$ makes c'_{25} zero, while $\tan 3\varphi = -c_{14}/c_{25}$ gives $c'_{14} = 0$, so suitable choice of the coordinate system gives the same expression for $(c_{\alpha\beta})$ for all classes of the trigonal system. We shall use this matrix in the form

---

* The same relationships may be derived from (9.17).

$$C=(c_{\alpha\beta})=\begin{pmatrix} c_{11} & c_{12} & c_{13} & c_{14} & 0 & 0 \\ c_{12} & c_{11} & c_{13} & -c_{14} & 0 & 0 \\ c_{13} & c_{13} & c_{33} & 0 & 0 & 0 \\ c_{14} & -c_{14} & 0 & c_{44} & 0 & 0 \\ 0 & 0 & 0 & 0 & c_{44} & c_{14} \\ 0 & 0 & 0 & 0 & c_{14} & \frac{1}{2}(c_{11}-c_{12}) \end{pmatrix}. \quad (9.21)$$

As for the tetragonal system, C contains 6 independent parameters, but it differs from (9.3).

A baseless division into two groups of the classes in the tetragonal and trigonal systems is used in many works, including [5]; one group has six independent moduli, while the other has seven. It has been shown [12]* (in agreement with the above argument) that the difference between these groups vanishes when the coordinate system is properly chosen, so the general form of the tensor for the elastic moduli is completely determined by the system. All crystals as regards elastic properties are divided into 7 systems, not 9.

A hexagonal crystal has either a reflection-rotation sixfold axis or a simple one; as regards elastic properties, this is equivalent to simultaneous presence of identically direct twofold and threefold axes. The first axis gives the $(c_{\alpha\beta})$ of (8.10), which is characteristic of a monoclinic crystal, while the second axis gives (9.21), which is characteristic of a trigonal crystal. We can write down at once the matrix having the properties of both of these:

$$C=(c_{\alpha\beta})=\begin{pmatrix} c_{11} & c_{12} & c_{13} & 0 & 0 & 0 \\ c_{12} & c_{11} & c_{13} & 0 & 0 & 0 \\ c_{13} & c_{13} & c_{33} & 0 & 0 & 0 \\ 0 & 0 & 0 & c_{44} & 0 & 0 \\ 0 & 0 & 0 & 0 & c_{44} & 0 \\ 0 & 0 & 0 & 0 & 0 & \frac{c_{11}-c_{12}}{2} \end{pmatrix}, \quad (9.22)$$

which contains 5 independent moduli.

---

* This has been shown [33] in another way, via the group properties of tensors, as the coincidence of the number of independent elastic moduli in all classes of the tetragonal and trigonal systems.

The other possible symmetry elements in the hexagonal systems (classes 23-27) add nothing to these restrictions.

Direct verification shows that the C of (9.22) satisfies (9.18) for any φ: $CU = U'C$, which means that the elastic properties of a hexagonal crystal do not alter on rotation through any angle around L^6, so L^6 is an axis of infinite order as regards the elastic properties; but in that case all directions in a plane perpendicular to $x_3 \parallel L^6$ are equivalent. Such a medium is naturally termed transversely isotropic; a hexagonal crystal is an example. A medium of this type may also be produced by subjecting an isotropic medium to subjecting an isotropic medium to some oriented field, e.g., an electric field.

There remains the cubic system. Class 28 of this has three mutually perpendicular twofold axes and also a threefold axis lying along a body diagonal of the cube. We set the x_1, x_2, and x_3 axes along the twofold axes, which gives us the C of (8.11), which is characteristic of an orthorhombic crystal. The threefold axis makes these axes all equivalent, the coordinates transforming one into another on rotation around L^3, so the elements of the C of (8.11) should not alter in response to any permutation of the subscripts 1, 2, 3. This implies that

$$c_{1111} = c_{2222} = c_{3333}, \quad c_{1122} = c_{2233} = c_{3311}, \quad c_{2323} = c_{3131} = c_{1212}$$

or

$$c_{11} = c_{22} = c_{33}, \quad c_{12} = c_{13} = c_{23}, \quad c_{44} = c_{55} = c_{66}.$$

Matrix (8.11) then becomes

$$C = (c_{\alpha\beta}) = \begin{pmatrix} c_{11} & c_{12} & c_{12} & 0 & 0 & 0 \\ c_{12} & c_{11} & c_{12} & 0 & 0 & 0 \\ c_{12} & c_{12} & c_{11} & 0 & 0 & 0 \\ 0 & 0 & 0 & c_{44} & 0 & 0 \\ 0 & 0 & 0 & 0 & c_{44} & 0 \\ 0 & 0 & 0 & 0 & 0 & c_{44} \end{pmatrix}. \tag{9.23}$$

This contains 3 independent moduli c_{11}, c_{12}, c_{44}. All other cubic classes have the same C, since they also contain the elements of class 28 and impose no further restrictions on the $c_{\alpha\beta}$.

Comparison of (9.23) with (8.2) shows that a cubic crystal differs (as regards elastic properties) from an isotropic medium, because $(c_{\alpha\beta})$ contains three independent moduli instead of two, whereas there is no difference as regards optical properties (e.g., optical activity). The reason is that the elastic properties are governed by a tensor of high (fourth) rank for the c_{iklm}, whereas the tensor ε_{ik} for the dielectric constant (which governs the optical properties) is only of second rank.

We may convert from a cubic crystal to an isotropic medium by requiring that C does not change on rotation through any angle around an axis in any direction. In fact, it is sufficient to require that the C of (9.23) should satisfy (9.18) for any φ other than a right angle. For instance, we put $\varphi=\pi/4$ and calculate both sides of $(CU)_{16}=(U'C)_{16}$ to get that $c_{44}=(c_{11}-c_{12})/2$, i.e., (9.23) coincides with matrix (8.2) for an isotropic medium.

Finally I consider briefly the restrictions imposed by the requirement that the elastic energy be positive; the condition was formulated in general form in section 6 as (6.16). I considered only the more symmetrical crystals, because (6.16) scarcely simplifies for the less symmetrical ones.

The complete necessary and sufficient conditions for a trigonal crystal are given by (9.21) and (6.16) as

$$c_{11}>|c_{12}|,\quad (c_{11}+c_{12})c_{33}>2c_{13}^2,\quad (c_{11}-c_{12})c_{44}>2c_{14}^2. \tag{9.24}$$

Those for a tetragonal crystal are given by (9.3) as

$$c_{11}>|c_{12}|,\quad (c_{11}+c_{12})c_{33}>2c_{13}^2,\quad c_{44}>0,\quad c_{66}>0. \tag{9.25}$$

Those for a hexagonal crystal give, from (9.22), the same result as for a tetragonal crystal, apart from the last inequality. For a cubic crystal we have from (9.23) that

$$c_{11}>|c_{12}|,\quad c_{11}+2c_{12}>0,\quad c_{44}>0. \tag{9.26}$$

The conditions for an isotropic medium are as for a cubic crystal, apart from the last inequality.

Chapter 2

Elements of Linear Algebra and Direct Tensor Calculus

10. Vectors and Matrices in n-Dimensional Space

Since linear algebra and direct tensor calculus are widely used in what follows, some details of the appropriate techniques are given here for the convenience of the reader. While no attempt is made at completeness or mathematical rigor, the material present is nonetheless of value in that some of the relationships given here are not to be found in the accessible literature, at least not in the form used here.

In a linear n-dimensional space* we specify the components of a vector u_β ($\beta = 1, 2, 3, \ldots, n$). Performing on this vector a linear homogeneous transformation, we obtain a new vector whose components are v_α:

$$\begin{aligned} v_1 &= A_{11}u_1 + A_{12}u_2 + \ldots + A_{1n}u_n, \\ v_2 &= A_{21}u_1 + A_{22}u_2 + \ldots + A_{2n}u_n, \\ &\cdots\cdots\cdots\cdots\cdots\cdots \\ v_n &= A_{n1}u_1 + A_{n2}u_2 + \ldots + A_{nn}u_n, \end{aligned} \tag{10.1}$$

or

$$v_\alpha = A_{\alpha\beta} u_\beta \quad (\alpha,\ \beta = 1, 2, \ldots, n). \tag{10.2}$$

---

* A linear space is a set of elements termed vectors such that a sum of these, or the result of multiplying one by a number, also belongs to that set, these operations satisfying the axioms of commutation, association, and distribution (see [8] for details). Here we envisage only real linear spaces.

Here and subsequently we envisage summation with respect to the repeated subscripts within the range of values open to them. The set of transformation coefficients forms a square table called a matrix:

$$A=(A_{\alpha\beta})=\begin{pmatrix} A_{11}A_{12} & \ldots & A_{1n} \\ A_{21}A_{22} & \ldots & A_{2n} \\ \cdot\ \cdot\ \cdot & \cdot\ \cdot & \cdot \\ A_{n1}A_{n2} & \ldots & A_{nn} \end{pmatrix}. \tag{10.3}$$

The $A_{\alpha\beta}$ are the elements of the matrix A, which as a whole may be considered as the symbol for a linear operation or transformation, which from any vector **u** (u_β) produces another vector **v** (v_α). Then (10.1) may be put in the symbolic form

$$v=Au. \tag{10.4}$$

This form has advantages over (10.1) and (10.2) in various respects. Firstly, **u**, **v**, and A appear as single objects, which usually corresponds to their physical (mechanical or geometrical) meaning. For example, **v** and **u** might be the vectors for the moment of a force and the angular acceleration in real physical three-dimensional space, or the electrical induction and the field strength, while A would correspond respectively to the tensor for the moments of inertia or for the dielectric constant. The components u_α and v_β are simply the coefficients in the expansion of vectors **u** and **v** with respect to some specified system n of linearly independent vectors that form the basis of the space. In real three-dimensional space these would be the projections of vectors **u** and **v** on the axes of a rectangular cartesian coordinate system. Hence the components of fixed **u** and **v** may vary in accordance with the choice of basis (coordinate system), and so a given physical property expressed by a vector may be represented by different sets of numbers (components) in view of the arbitrary element in the choice of basis. It will appear from what follows that the same applies to the elements $A_{\alpha\beta}$, because these also are dependent on the choice of basis. Clearly, such a position is not entirely satisfactory; but (10.4) frees us from this disadvantage, for it does not contain u_α, v_β, or $A_{\alpha\beta}$.

Moreover, (10.4) is simpler and more obvious than (10.1). It is also found that some of the properties of vectors and matrices can be used in performing various operations with resort to the components, which thus obviates the choice of any particular basis. Such calculations are frequently simpler and less laborious than ones in terms of components. A distinctive feature of this book is that the main exposition is given by direct (coordinate-free) methods, which do not involve the use of components. The object of this chapter is to survey this method, which is utilized elewhere in the book.

Although (10.2) and (10.4) express essentially the same relation, there are essential differences between them. The u_β, v_α, and $A_{\alpha\beta}$ of (10.2) are ordinary numbers, so all the laws applicable to numbers apply to them also (commutation, association, distribution, etc.); in particular, we can interchange the factors on the right in (10.2) without rendering it wrong:

$$v_\alpha = u_\beta A_{\alpha\beta}. \tag{10.5}$$

On the other hand, such an operation is impermissible in (10.4), because **u, v,** and A are not ordinary numbers but more complex objects, namely ordered sets of numbers.* All the same, suitable definitions allow us in (10.5) to retain the meaning while permuting the quantities on the right.

We introduce the following general conditions: if vectors or matrices stand together without any signs between them, this denotes convolution, i.e., summation of both quantities over similar subscripts. For example, if two vectors stand together, the expression has the meaning:

$$uv = u_\gamma v_\gamma \tag{10.6}$$

and represents the scalar product of the two vectors. Another example is (10.4), whose meaning is revealed by (10.2). The repeated β subscripts involved in the summation of (10.2) are termed dummy, while the α are free; the former may be denoted in any convenient way, since they are used up in the summation, e.g., $A_{\alpha\beta}u_\beta = A_{\alpha\gamma}u_\gamma$. If an expression contains several such pairs of subscripts, each pair must be denoted by a different letter to avoid misunderstanding. As regards the free subscripts, the following rule will be used everywhere in what follows: the set of free subscripts on the left must be the same as the set on the right.

* These sets in general are hypercomplex numbers.

Rule for dummy subscripts: These must always be adjacent in conversion from the subscript form to the direct form or vice versa.

Rule for free subscripts: These must always be identical in the right and left halves.

Consider some examples of the use of these rules. The relation

$$v = uA, \tag{10.7}$$

in which **v** and **u** are vectors and A is a matrix* is to be put as follows in terms of subscripts:

$$v_\alpha = u_\beta A_{\beta\alpha}. \tag{10.8}$$

We have already seen that we can place the factors in (10.8) in either order. Comparison of (10.8) with (10.5) shows that $A_{\alpha\beta}$ has been replaced by $A_{\beta\alpha}$; in general, $A_{\alpha\beta} \neq A_{\beta\alpha}$, so (10.5) does not agree with (10.8). From (10.3) we see that the first subscript in $A_{\alpha\beta}$ defines the line in which $A_{\alpha\beta}$ lies, while the second defines the column; hence interchange of the subscripts is equivalent to interchange of the rows and columns. This operation is called transposition, the product being a transposed matrix. Transposition will be denoted by ~ over the symbol for the matrix. Hence by definition

$$\tilde{A}_{\alpha\beta} = A_{\beta\alpha}. \tag{10.9}$$

Then (10.8) may be put as

$$v_\alpha = A_{\beta\alpha} u_\beta = \tilde{A}_{\alpha\beta} u_\beta. \tag{10.10}$$

Here both subscripts are in accordance with the above two rules, so we may convert directly to the coordinate-free form as

$$v = \tilde{A}u. \tag{10.11}$$

---

* A vector is taken as any quantity dependent on a single subscript, while a matrix is a quantity governed by two subscripts, all subscripts taking the same set of values.

But (10.10) is the same as (10.8), so (10.11) coincides with (10.7). Hence

$$uA = \tilde{A}u. \tag{10.12}$$

Obviously, double transposition leaves any matrix unaltered, which is expressed by

$$\tilde{\tilde{A}} = A. \tag{10.13}$$

Thus if we put $B = \tilde{A}$, then $\tilde{B} = A$, so (10.12) becomes

$$u\tilde{B} = Bu. \tag{10.14}$$

Relations (10.12) and (10.14) are true for arbitrary A and B; both have the general property that any matrix multiplied from the left (right) by a vector gives the same result as the transposed matrix multiplied from the right (left) by the same vector. In other words, we can transfer the vector from left to right in a product of a vector by a matrix if the latter is replaced by the transposed matrix.

A very important particular type of matrix is a symmetric matrix, which coincides with its transposed form. For such a matrix A

$$\tilde{A} = A, \qquad A_{\alpha\beta} = A_{\beta\alpha}. \tag{10.15}$$

The opposite case is an antisymmetric matrix, which changes sign on transposition. For such a matrix B

$$\tilde{B} = -B, \qquad B_{\alpha\beta} = -B_{\beta\alpha}. \tag{10.16}$$

The last equation implies that the diagonal elements, which have $\alpha = \beta$, are zero in such a matrix. From (10.12) we have for a symmetric matrix (and only for such a matrix) that

$$uA = Au, \qquad (A = \tilde{A}) \tag{10.17}$$

for any vector **u**. Similarly, for an antisymmetric matrix with any*

---

* There are certain **u** that allow (10.17) and (10.18) to be obeyed although (10.15) and (10.16), respectively, are not.

vector **u** we have

$$uB = -Bu \qquad (B = -\tilde{B}). \tag{10.18}$$

The sum of matrices is defined as the matrix whose elements equal the sums of the corresponding elements of the constituent matrices.

Multiplication of a matrix by a number is equivalent to multiplication of all elements by that number.

Any matrix C may be represented as the sum of a symmetric matrix and an antisymmetric matrix:

$$C = A + B, \quad A = \frac{1}{2}(C + \tilde{C}) = \tilde{A}, \quad B = \frac{1}{2}(C - \tilde{C}) = -\tilde{B}. \tag{10.19}$$

We multiply (10.2) by w_α (and sum with respect to α):

$$v_\alpha w_\alpha = A_{\alpha\beta} u_\beta w_\alpha. \tag{10.20}$$

To pass to the direct form we should place the factor on the right in accordance with the rule for dummy subscripts, i.e., as $w_\alpha A_{\alpha\beta} u_\beta$; on the left the rule is obeyed for any order of the factors. The result is

$$vw = wv = wAu = w_\alpha A_{\alpha\beta} u_\beta. \tag{10.21}$$

From (10.9) we may put $A_{\alpha\beta} u_\beta w_\alpha = \tilde{A}_{\beta\alpha} u_\beta w_\alpha = u_\beta \tilde{A}_{\beta\alpha} w_\alpha$, so as well as (10.21) we have

$$vw = u\tilde{A}w. \tag{10.22}$$

For a symmetric A and any **u** and **w**

$$uAw = wAu \qquad (A = \tilde{A}). \tag{10.23}$$

The expression $A_{\alpha\beta} u_\beta w_\alpha$ is the bilinear form of u_β and w_α, the $A_{\alpha\beta}$ being the coefficients of this form. We see from (10.21) and (10.22) that this form may be written in two ways, which differ in the order of the vectors and in transposition of the form matrix A.

We transform the vector **v** in (10.4) via a matrix B to get

$$w = Bv = BAu. \tag{10.24}$$

Transferring to subscripts via the above rules, we have

$$w_\gamma = B_{\gamma\alpha} v_\alpha = B_{\gamma\alpha} A_{\alpha\beta} u_\beta. \tag{10.25}$$

We put

$$C = BA, \qquad C_{\gamma\beta} = B_{\gamma\alpha} A_{\alpha\beta}. \tag{10.26}$$

Clearly, vector **w** is derived from vector **u** by a linear transformation via matrix C, which is termed the product of matrices B and A; (10.26) gives a definition of the product of two matrices. In general, $BA \neq AB$, so the product of matrices is dependent on the order of the factors; but in particular cases we may have

$$BA = AB. \tag{10.27}$$

Then it is said that matrices A and B commute one with the other. We also see from (10.26) that the product of three or more matrices is associative:

$$(AB)C = A(BC). \tag{10.28}$$

Transposition of both parts of (10.26) gives

$$C_{\beta\gamma} = B_{\beta\alpha} A_{\alpha\gamma} = \tilde{A}_{\gamma\alpha} \tilde{B}_{\alpha\beta}.$$

In direct form this states that $\tilde{C} = \tilde{A}\tilde{B}$, so

$$\widetilde{BA} = \tilde{A}\tilde{B}. \tag{10.29}$$

From (10.28) we get for the product of three matrices that

$$\widetilde{CBA} = \widetilde{C(BA)} = \widetilde{BA}\tilde{C} = \tilde{A}\tilde{B}\tilde{C}. \tag{10.30}$$

Hence the result of transposing the product of any number of matrices is equal to the product of the transposed matrices in reverse order; (10.29) shows, in particular, that if $\tilde{A} = A$, $\tilde{B} = B$, then this still does not imply that BA is a symmetric matrix, since

$\tilde{B}\tilde{A} = AB \neq BA$. The product of two symmetric matrices will be a symmetric matrix only if these commute; any matrix commutes with itself, so any integral power of a symmetric matrix is also a symmetric matrix. For antisymmetric matrices the even powers are symmetric, while the odd ones are antisymmetric.

The matrix of the form

$$E = \begin{Vmatrix} 1 & 0 & 0 & \dots & 0 \\ 0 & 1 & 0 & \dots & 0 \\ 0 & 0 & 1 & \dots & 0 \\ \cdot & \cdot & \cdot & \cdot & \cdot \\ 0 & 0 & 0 & \dots & 1 \end{Vmatrix} \tag{10.31}$$

is termed unit matrix; its elements coincide with the values of Kronecker's symbol: $E_{\alpha\beta} = \delta_{\alpha\beta}$ ($\delta_{\alpha\beta} = 1$ for $\alpha = \beta$, $\delta_{\alpha\beta} = 0$ for $\alpha \neq \beta$). Any vector u is unaltered by multiplication from right or left by unit matrix:*

$$Eu = uE = u. \tag{10.32}$$

The same is true for multiplication of any matrix A by unit matrix:

$$EA = AE = A. \tag{10.33}$$

Hence multiplication by E in no way differs from that by one. Similarly, multiplication of any vector or matrix by a number k gives the same result as multiplication by the matrix

$$\begin{Vmatrix} k & 0 & 0 & \dots & 0 \\ 0 & k & 0 & \dots & 0 \\ 0 & 0 & k & \dots & 0 \\ \cdot & \cdot & \cdot & \cdot & \cdot \\ 0 & 0 & 0 & \dots & k \end{Vmatrix} = k \begin{Vmatrix} 1 & 0 & 0 & \dots & 0 \\ 0 & 1 & 0 & \dots & 0 \\ 0 & 0 & 1 & \dots & 0 \\ \cdot & \cdot & \cdot & \cdot & \cdot \\ 0 & 0 & 0 & \dots & 1 \end{Vmatrix} = kE. \tag{10.34}$$

A matrix such as (10.34) is called scalar. No distinction will be drawn between a number k and the corresponding scalar matrix. For example, the sum of a matrix A and a scalar matrix kE may be written as A + k instead of A + kE.

Denoting the determinant of matrix A by $|A| = |A_{\alpha\beta}|$, we see that the matrix is not unique if $|A| \neq 0$. If $|A| = 0$, the matrix is

---

* Conversely, a matrix is unit matrix if it leaves a vector unaltered.

unique. The general theory of linear algebraic equations [16] indicates that system (10.1) and (10.2) may be solved for the u_β (considered as unknowns) if A is not unique; the u_β are expressed linearly and homogeneously in terms of the v_α, which may be put as

$$u_\beta = B_{\beta\alpha} v_\alpha, \quad u = Bv, \tag{10.35}$$

in which B is some matrix, which is reciprocal with respect to A, this being expressed by

$$B = A^{-1}. \tag{10.36}$$

Conversely, A is reciprocal with respect to B, i.e., $A = B^{-1}$, because (10.4) serves to solve (10.35) for **v**. Hence

$$(A^{-1})^{-1} = A. \tag{10.37}$$

We should always remember that the matrix reciprocal to A exists when, and only when $|A| \neq 0$. Multiplying both parts of (10.35) by matrix A, we get, with (10.4), that

$$ABv = Au = v. \tag{10.38}$$

This must be so for any vector v, as is clear from the derivation, so matrix AB does not alter any vector, and hence it is, by definition, unit matrix: $AB = E$ or $AA^{-1} = 1$. Here 1 has been written instead of unit matrix, as noted above. We may, from (10.36) and (10.37), also put $B^{-1}B = 1$; these relationships are true for any matrices A and B that are not unique, and hence for any matrix S of this class we have

$$SS^{-1} = S^{-1}S = 1. \tag{10.39}$$

In terms of subscripts this is

$$S_{\alpha\beta} S^{-1}_{\beta\gamma} = S^{-1}_{\alpha\beta} S_{\beta\gamma} = \delta_{\alpha\gamma}. \tag{10.40}$$

These relations may be considered as the definitions of S^{-1}, whence we readily see that

$$(AB)^{-1} = B^{-1}A^{-1}. \tag{10.41}$$

In fact, $(AB)(B^{-1}A^{-1}) = A(BB^{-1})A^{-1} \pm AA^{-1} = 1$. Similarly, it is easily shown that

$$(ABC)^{-1} = C^{-1}B^{-1}A^{-1}, \tag{10.42}$$

etc. Hence the matrix reciprocal to the product of several matrices equals the product of the matrices reciprocal to the cofactors taken in the reverse order. Transposing both parts of (10.39), we get from (10.29) that

$$(\tilde{S})^{-1} = \widetilde{S^{-1}}, \tag{10.43}$$

i.e., a matrix the reciprocal of the transposed one is obtained by transposing the reciprocal matrix.

Suppose we perform a linear transformation of the basic in an n-dimensional space in which are defined the vectors **u, v, w,** ... and the matrices A, B, ... This means that in place of one system of n linearly independent vectors we take another system of n linearly independent vectors, each vector in the second system being a linear combination of the vectors in the first. To the linear transformation of the basis there corresponds a linear transformation of the components of each vector specified in that space. A very important point is that the components of any vector undergo the same linear transformation on this change of basis, whose matrix we denote by S. Let the components of **u, v, w** ... in the new basis be denoted by primes; then we have the following relations between the new and old components:

$$u'_\alpha = S_{\alpha\beta}u_\beta, \quad v'_\alpha = S_{\alpha\beta}v_\beta, \quad w'_\alpha = S_{\alpha\beta}w_\beta, \ \dots \text{ etc.},$$

or in coordinate-free form*

$$u' = Su, \quad v' = Sv, \quad w' = Sw. \tag{10.44}$$

Matrix S is bound not to be unique; if $|S| = 0$, the various rows of S would not all be linearly independent, and this would mean that we had converted in n-dimensional space from n linearly independent vectors to a smaller number of other linearly independent vectors, which cannot be the basis of a complete space.

* Note that **u'** denotes the same vector as **u**, but with other components referred to a new frame of reference. The prime to **u** indicates precisely this feature.

A change of basis must produce certain changes in any matrix A appearing in a relationship such as (10.1) or (10.2) (i.e., performing a linear transformation of one vector into another); the matrix elements must alter. Denoting the new elements by primes, we may put $v'_\alpha = A'_{\alpha\beta} u'_\beta$ in the new basis, or

$$v' = A'u'. \tag{10.45}$$

Substituting from (10.44), we have $Sv = A'Su$, or multiplying both sides by S^{-1} (which is real, since $|S| \neq 0$),

$$v = S^{-1}A'Su. \tag{10.46}$$

A comparison with (10.4) shows that $S^{-1}A'Su = Au$. The last is true for any u, so $S^{-1}A'S = A$. Multiplying both parts of this from the left by S and from the right by S^{-1}, and using (10.39), we get

$$A' = SAS^{-1}. \tag{10.47}$$

This shows how the matrix is altered when all the vectors are affected by the transformation of (10.44). The conversion from A to A' in (10.47) is a congruent transformation; matrices A and A' are inversely congruent or equivalent.

In a particular case, matrix A may commute with matrix S ($SA = AS$); then (10.47) [see (10.39)] becomes

$$A' = ASS^{-1} = A. \tag{10.48}$$

Hence a matrix A that commutes with S is not altered by this transformation; conversely, if $A' = A$, we have $AS = SA$ from (10.47) after multiplying from the right by S, so A commutes with S. Hence commutation of A with S is the necessary and sufficient condition for invariance of matrix A with respect to the corresponding transformation of the basis.

These properties relating to transformation of the basis have in essence been used already in section 9 of Chapter 1 in discussing the elastic-modulus matrix for various crystals.

The sum of all the diagonal elements (the trace) plays a major part in many operations with matrices; the trace of A is denoted by A_t:

$$A_t = A_{\alpha\alpha}. \tag{10.49}$$

Consider the trace of the product of two matrices:

$$(AB)_t = (AB)_{\alpha\alpha} = A_{\alpha\beta}B_{\beta\alpha} = B_{\beta\alpha}A_{\alpha\beta} = (BA)_{\beta\beta} = (BA)_t. \tag{10.50}$$

This trace is independent of the order of the matrices, and from (10.50) we may put the trace of the product of three matrices as

$$(ABC)_t = ((AB)C)_t = (CAB)_t = (BCA)_t. \tag{10.51}$$

The following is true in general: the trace of the product of any number of matrices is unaltered by permutation of the factors. Further, the trace of a transposed matrix equals that of the initial one, because diagonal elements are unmoved by transposition:

$$\tilde{A}_t = A_t. \tag{10.52}$$

An important property of the trace is that it is unaffected by a congruent transformation, such as (10.47); from (10.47) and (10.50),

$$A'_t = (SAS^{-1})_t = (S^{-1}SA)_t = A_t. \tag{10.53}$$

Thus the trace is an *invariant* of the matrix in respect of any congruent transformation. This is the only invariant of a matrix that is expressed as a linear combination of its elements.

The trace of a matrix sum is equal to the sum of the trace of the components, which is implied by the linearity of the trace:

$$(A + B)_t = A_t + B_t. \tag{10.54}$$

Multiplication of matrix A by a number k multiplies all the elements by k, including the sum of the diagonal ones:

$$(kA)_t = kA_t. \tag{10.55}$$

The trace of an antisymmetric matrix is zero, because all diagonal elements are zero. The trace of the product of symmetric and antisymmetric matrices is also zero; for, let $\tilde{A} = A$, $\tilde{B} = -B$; then (10.29), (10.50), (10.52), and (10.55) give

$$(AB)_t = \widetilde{(AB)}_t = (\tilde{B}\tilde{A})_t = (-BA)_t = -(BA)_t = -(AB)_t.$$

But $(AB)_t$ equals itself with the sign reversed, so

$$(AB)_t = A_{\alpha\beta}B_{\beta\alpha} = 0 \qquad (A = \tilde{A},\ B = -\tilde{B}). \tag{10.56}$$

11. Three-Dimensional Tensors and Dyads

The previous section dealt with vectors and matrices in a space with an arbitrary number of dimensions. Here we restrict the discussion to three-dimensional space, in which a rectangular cartesian system is used as basis. A vector in this space is represented by a directed part of a straight line, while the components of it are the orthogonal projections on the coordinate axes. A three-dimensional vector is denoted here by bold face, e.g., $\mathbf{u}$, while the components are denoted by the same letters with latin subscripts, e.g., $\mathbf{u} = (u_k) = (u_1, u_2, u_3)$. The components of a given vector are, in general, affected by reflection and rotation, the new components being expressed in terms of the old ones by linear homogeneous transformations. All transformations of components involving rotation or reflection have the property that the sum of the squares of the components (the square of the length) is unaltered. Linear transformations of this type are called orthogonal transformations. Let $S = (S_{ik})$ be the matrix of an orthogonal transformation; then the transformed components u'_l are expressed as follows*

$$u'_l = S_{lk}u_k, \qquad \mathbf{u}' = S\mathbf{u}. \tag{11.1}$$

Orthogonality in S means that for any $\mathbf{u}$ the scalar square $\mathbf{u}^2 = u_k^2 = u_k u_k$ is equal to the square $\mathbf{u}'^2 = u'^2_l$. To find the condition imposed by this property on S, we substitute in $\mathbf{u}'^2 = \mathbf{u}^2$ expression (11.1) for $\mathbf{u}'$. This simple calculation may be done directly (without resort to subscripts); as we have seen in the previous section, we should insure a suitable order for the quantities in a product subject to

---

* Here and subsequently the prime in $\mathbf{u}'$ means that the components are primed: $\mathbf{u}' = (u'_k)$.

the condition that quantities standing together are considered as convoluted with respect to adjacent subscripts. Hence it would be wrong to write $\mathbf{u}'^2$ as SuSu, since such an expression has no definite meaning. In fact, the first of the vectors **u** in this has S adjacent to right and left, so from the basic condition of (10.6) it must be convoluted with both. But **u**, being a vector, has only one subscript, when means that this is impossible. To write down correctly the direct form for $\mathbf{u}'^2$ we must first alter somewhat the form given to the first factor in $\mathbf{u}'$. From (10.14),

$$u' = Su = u\tilde{S}, \tag{11.2}$$

in which $\tilde{S}$ is the transposed form of S. Then the orthogonality condition may be put as

$$u'^2 = u'u' = u\tilde{S}Su = u^2 = u1u. \tag{11.3}$$

Here the **1** denotes unit matrix. The equality applies for all **u**, so we must have

$$\tilde{S}S = 1. \tag{11.4}$$

This is the orthogonality condition for S; comparison with (10.39), the definition of reciprocal matrix, shows that an orthogonal matrix may be defined via the condition

$$\tilde{S} = S^{-1}. \tag{11.5}$$

Hence a transposed orthogonal matrix equals the reciprocal matrix. Then, multiplying by S from the left, we have that (11.4) is accompanied by the condition

$$S\tilde{S} = 1, \tag{11.6}$$

which, however, is not independent, because it is implied by (11.4) and (10.39); conversely, (11.4) follows from (11.6) and (10.39).

We form the determinants of the matrices in both parts of (11.6). The determinant of a matrix product equals the product of the determinants of the matrices: $|S\tilde{S}| = |S| \cdot |\tilde{S}|$. Further, $|S| = |\tilde{S}|$, and finally, the determinant of the unit matrix on the right is equal to one. Hence

$$|S|^2 = 1, \qquad |S| = \pm 1. \tag{11.7}$$

The positive sign in (11.7) corresponds to orthogonal transformations termed pure rotations, while the negative sign corresponds to rotation with reflection.

An orthogonal second-rank tensor is a square three-line matrix whose elements transform as do the corresponding products of the components of two vectors in response to an orthogonal coordinate transformation. Let $\alpha = (\alpha_{ik})$ be a tensor; then, by definition, each of its components* α_{ik} should alter as does the product $u_i v_k$. But (11.1) shows that $u_i' v_k' = S_{il} u_l S_{kn} v_n = S_{il} S_{kn} u_l v_n$, so

$$\alpha'_{ik} = S_{il} S_{kn} \alpha_{ln}. \tag{11.8}$$

From (10.9), this may be put as

$$\alpha'_{ik} = S_{il} \alpha_{ln} \tilde{S}_{nk} = (S \alpha \tilde{S})_{ik},$$

or

$$\alpha' = S \alpha \tilde{S}. \tag{11.9}$$

But S is an orthogonal matrix, so from (11.5)

$$\alpha' = S \alpha S^{-1}. \tag{11.10}$$

The definition of a second-rank orthogonal tensor accords with the general rule (10.47) for the transformation of matrices in response to change of basis.

By analogy with the above, an orthogonal tensor of rank n may be defined as a set of 3^n components $\alpha_{k_1 k_2 \ldots k_n}$ that for orthogonal coordinate transformations behave as does the product of the components of n vectors, $u_{k_1} v_{k_2} \ldots w_{k_n}$. An example is the fourth-rank tensor for the elastic moduli c_{iklm} (Chapter 1), whose components transform via the formula

---

* A second-rank tensor is a matrix with a certain law of alteration for the components in response to coordinate transformations. The α_{ik} are distinguished from the elements of a matrix by (usually) being called the components of the tensor.

$$c'_{iklm} = S_{ii'}S_{kk'}S_{ll'}S_{mm'}c_{i'k'l'm'}, \tag{11.11}$$

which we have already used in (8.5).

By definition, the product $u_i v_k$ transforms as do the components of a second-rank tensor, so the set of quantities

$$\alpha_{ik} = u_i v_k \tag{11.12}$$

may be considered as a form of tensor, which is called a dyad; far from all tensors may be represented as dyads, for if we put (11.12) in expanded form

$$\alpha = (u_i v_k) = \begin{pmatrix} u_1 v_1 & u_1 v_2 & u_1 v_3 \\ u_2 v_1 & u_2 v_2 & u_2 v_3 \\ u_3 v_1 & u_3 v_2 & u_3 v_3 \end{pmatrix}, \tag{11.13}$$

we see that all rows (and columns) are inversely proportional, so this is a tensor of a very simple type; but it plays a basic part in the theory of tensors, because we shall see that any tensor can be represented as the sum of several dyads.

We put (11.12) in the coordinate-free form

$$\alpha = \boldsymbol{u} \cdot \boldsymbol{v}, \; \alpha_{ik} = (\boldsymbol{u} \cdot \boldsymbol{v})_{ik} = u_i v_k. \tag{11.14}$$

If the point between **u** and **v** had been omitted, the general condition of section 10 would have made this the scalar product $\mathbf{uv} = u_k v_k$, i.e., a pure number; hence the point in (11.14) plays a vital part, as showing that the components of **u** and **v** standing together are not summed with respect to identical subscripts but appear independently. We shall use this point in all cases where we need to stress the quantities (vectors, tensors, matrices) standing together are not convoluted with respect to adjacent subscripts.*

---

* The point is used in the opposite sense in many works of reference; here our point is disjunctive, in the sense that the quantities separated by the point enter independently, whereas many workers use the point as the symbol of convolution, i.e., an operation of combining together here denoted simply by putting the symbols side by side. For example, in [4] the scalar product of two vectors is put as $\mathbf{u} \cdot \mathbf{v}$, while a dyad is put as **uv**. Convolution of two vectors with a matrix, as in the **wAu** of (10.21), would then be put

The **u** of (11.14) is the first (left-hand) vector of the dyad, while **v** is the second (right-hand) one; (11.13) shows that the subscripts to the components of the first vector coincide with the numbers of the rows in the tensor, while those of the second coincide with the column numbers. Transposition of a dyad is equivalent to permutation of the vectors in the dyad:

$$\tilde{\alpha} = \widetilde{\boldsymbol{u} \cdot \boldsymbol{v}} = \boldsymbol{v} \cdot \boldsymbol{u}. \tag{11.15}$$

The trace of the dyad is

$$\alpha_t = (\boldsymbol{u} \cdot \boldsymbol{v})_t = u_k v_k = \boldsymbol{uv} \tag{11.16}$$

and equals the scalar product of the constituent vectors. We have $\tilde{\alpha} = \alpha$ if the dyad is a symmetric tensor, i.e.,

$$\boldsymbol{u} \cdot \boldsymbol{v} = \boldsymbol{v} \cdot \boldsymbol{u}. \tag{11.17}$$

We multiply this equation scalarly from the left by any vector **w** such that $\mathbf{wu} \neq 0$ and $\mathbf{wv} \neq 0$. In the present notation, the result of this multiplication is obtained simply by writing in **w** on the left in both parts of (11.17):

$$\boldsymbol{wu} \cdot \boldsymbol{v} = \boldsymbol{wv} \cdot \boldsymbol{u}. \tag{11.18}$$

Then **wu** and **wv** are the scalar products, i.e., numbers. Let $\mathbf{wv}/\mathbf{wu} = k$; then (11.18) becomes

$$\boldsymbol{v} = k\boldsymbol{u}. \tag{11.19}$$

Here k and **u** are put side by side (no point) because k (a pure number) cannot be convoluted with the vector **u**. (11.19) shows that a dyad is symmetric only when the vectors in it are inversely proportional (collinear).

---

as $\mathbf{w} \cdot A \cdot \mathbf{u}$, etc. This notation is widely used, but its inconvenience is obvious. If we were to be logical, all products, e.g., the $A' = SAS^{-1}$ of (10.47) would have to be written as $A' = S \cdot A \cdot S^{-1}$, because here we have convolution with respect to adjacent subscripts. Products with convolution are incomparably more common than dyads in vector-tensor relationships generally, so it would be essential (without any real need) to infect all formulas with spot plague. The relationships would then lose a great deal in compactness and legibility. My notation is free from these disadvantages.

Multiplication of (11.14) from the right by any vector gives a vector parallel to **u**:

$$\alpha w = u \cdot v w = k u, \quad k = v w. \tag{11.20}$$

Hence multiplication by a dyad projects any vector on one particular direction. A dyad is therefore called a linear tensor.

The sum of two dyads gives a tensor of more general form:

$$\alpha = u^{(1)} \cdot v^{(1)} + u^{(2)} \cdot v^{(2)}. \tag{11.21}$$

Multiplication of this tensor by any vector **w** gives

$$\alpha w = u^{(1)} \cdot v^{(1)} w + u^{(2)} \cdot v^{(2)} w = k_1 u^{(1)} + k_2 u^{(2)}, \tag{11.22}$$

i.e., **w** is converted to a linear combination of the vectors $\mathbf{u}^{(1)}$ and $\mathbf{u}^{(2)}$. But the possible linear combinations of two independent vectors define a plane parallel to these, so (11.21) is termed a planal tensor. Here it is supposed that the first vectors $\mathbf{u}^{(1)}$ and $\mathbf{u}^{(2)}$ of the dyads are not inversely parallel, and that the second ones $\mathbf{v}^{(1)}$ and $\mathbf{v}^{(2)}$ are also not parallel, i.e., are linearly independent.

Consider finally the tensor equal to the sum of three dyads:

$$\alpha = u^{(1)} \cdot v^{(1)} + u^{(2)} \cdot v^{(2)} + u^{(3)} \cdot v^{(3)} = u^{(k)} \cdot v^{(k)}. \tag{11.23}$$

This reduces to (11.21) if there is a linear relation between the $\mathbf{u}^{(k)}$, or $\mathbf{v}^{(k)}$; for instance, let $\mathbf{u}^{(3)} = k_1\mathbf{u}^{(1)} + k_2\mathbf{u}^{(2)}$. Then

$$\alpha = u^{(1)} \cdot v^{(1)} + u^{(2)} \cdot v^{(2)} + (k_1 u^{(1)} + k_2 u^{(2)}) \cdot v^{(3)}. \tag{11.24}$$

The scalar factors k_1 and k_2 in the dyad may be assigned to either of the constituent vectors: $k(\mathbf{u} \cdot \mathbf{v}) = k\mathbf{u} \cdot \mathbf{v} = \mathbf{u} \cdot k\mathbf{v}$, so (11.24) may be put as

$$\alpha = u^{(1)} \cdot (v^{(1)} + k_1 v^{(3)}) + u^{(2)} \cdot (v^{(2)} + k_2 v^{(3)}) = u^{(1)} \cdot v^{(1)\prime} + u^{(2)} \cdot v^{(2)\prime}, \tag{11.25}$$

which is a planal tensor. Thus (11.23) gives a new type of tensor only if all the $\mathbf{u}^{(k)}$ and all the $\mathbf{v}^{(k)}$ are linearly independent, in which case (11.23) is called a complete tensor.

The sum of four or more dyads in three-dimensional space does not give a tensor more general than (11.23), for in such a space there cannot exist more than three linearly independent vectors. All the first (or all the second) vectors of the dyads may thus be expressed as linear combinations of not more than three vectors; a calculation analogous to (11.24) and (11.25) shows that the sum of any number of dyads may be reduced to a sum of not more than three. This shows that (11.23) is the general expression for any second-rank three-dimensional tensor, though the representation of any tensor in the form of (11.23) is known in advance not to be unique, for any triplet of linearly independent vectors may be taken as the $\mathbf{u}^{(k)}$ or as the $\mathbf{v}^{(k)}$. The corresponding conversion merely requires us to express the $\mathbf{u}^k$ (or the $\mathbf{v}^{(k)}$), as linear combinations of the new vectors and then to group as one dyad all the dyads having the same first (or second) vector. Hence there are infinitely many expressions for any tensor as a sum of three dyads.

12. The Levi-Civita Tensor and Its Applications

Consider a third-rank tensor ε_{ikl} having the properties

$$\varepsilon_{123} = 1, \quad \varepsilon_{ikl} = -\varepsilon_{kil} = -\varepsilon_{ilk}. \tag{12.1}$$

The latter two equations imply also that $\varepsilon_{ikl} = -\varepsilon_{ikl}$, so the tensor is antisymmetric with respect to permutation of any pair of subscripts. This is unit completely antisymmetric tensor of the third rank, or the Levi-Civita tensor. It and unit tensor (the Kronecker symbol δ_{ik}) play a special part among three-dimensional tensors; the conditions of (12.1) imply that any component of this tensor is zero if it has two subscripts the same, so the $3^3 = 27$ components are all zero except those with all three subscripts different, which are equal to +1 or -1, depending on whether 123 is converted to ikl via an even or odd number of transpositions,* so we have

$$\varepsilon_{123} = \varepsilon_{231} = \varepsilon_{312} = 1, \quad \varepsilon_{213} = \varepsilon_{132} = \varepsilon_{321} = -1. \tag{12.2}$$

The product of two Levi-Civita tensors may be expressed as follows in terms of the components of δ_{ik} (unit tensor):

$$\varepsilon_{ijk}\varepsilon_{lmn} = \begin{vmatrix} \delta_{il} & \delta_{im} & \delta_{in} \\ \delta_{jl} & \delta_{jm} & \delta_{jn} \\ \delta_{kl} & \delta_{km} & \delta_{kn} \end{vmatrix} =$$

$$= \delta_{il}\delta_{jm}\delta_{kn} + \delta_{im}\delta_{jn}\delta_{kl} + \delta_{jl}\delta_{km}\delta_{in} - \delta_{in}\delta_{jm}\delta_{kl} - \delta_{il}\delta_{jn}\delta_{km} - \delta_{im}\delta_{jl}\delta_{kn}. \tag{12.3}$$

---

* Transposition = permutation of two subscripts.

This may be demonstrated by considering the case when any two subscripts from ijk or from $l$mn coincide; the properties of ε_{ijk} show that this gives zero on the left. We obtain the same result on the right, since, for example, the first two rows of the determinant coincide for i = j, while the second and third columns coincide for m = n, and so on. Interchange of any two of ijk causes interchange of the corresponding columns of the determinant, while interchange in $l$mn causes column interchange. The sign of the determinant is thereby reversed. Finally, from (12.1)

$$\varepsilon_{123}\varepsilon_{123} = \begin{vmatrix} \delta_{11} & \delta_{12} & \delta_{13} \\ \delta_{21} & \delta_{22} & \delta_{23} \\ \delta_{31} & \delta_{32} & \delta_{33} \end{vmatrix} = \begin{vmatrix} 1 & 0 & 0 \\ 0 & 1 & 0 \\ 0 & 0 & 1 \end{vmatrix} = 1.$$

Hence the two parts of (12.3) coincide for this combination of the subscripts, while both parts alter identically in response to all permutations of the subscripts, so equality always occurs.

Let us put n = k in (12.3); δ_{ik} has the properties $\delta_{kk} = 3$, $\delta_{jk}\delta_{kl} = \delta_{jl}$, etc., which gives, after simple computations:

$$\varepsilon_{ijk}\varepsilon_{lmk} = \delta_{il}\delta_{jm} - \delta_{im}\delta_{jl} = \begin{vmatrix} \delta_{il} & \delta_{im} \\ \delta_{jl} & \delta_{jm} \end{vmatrix}. \tag{12.4}$$

Convoluting with respect to the pair of subscripts j and m, we get

$$\varepsilon_{ijk}\varepsilon_{ljk} = 2\delta_{il}. \tag{12.5}$$

Convoluting with respect to i and l, we get

$$\varepsilon_{ijk}\varepsilon_{ijk} = \varepsilon_{ijk}^2 = 3! = 6. \tag{12.6}$$

The last result is obvious, for on the left we have the sum of the squares of all the ε_{ijk}, of which only six are different from zero and equal ±1, as (12.2) shows.

Consider $\varepsilon_{lmn}\alpha_{1l}\alpha_{2m}\alpha_{3n}$, in which α_{ik} are the components of an arbitrary tensor. From (12.3) we have

$$\varepsilon_{lmn}\alpha_{1l}\alpha_{2m}\alpha_{3n} = \varepsilon_{lmn}\varepsilon_{123}\alpha_{1l}\alpha_{2m}\alpha_{3n} = \begin{vmatrix} \delta_{l1} & \delta_{l2} & \delta_{l3} \\ \delta_{m1} & \delta_{m2} & \delta_{m3} \\ \delta_{n1} & \delta_{n2} & \delta_{n3} \end{vmatrix} \alpha_{1l}\alpha_{2m}\alpha_{3n}.$$

We multiply the first line of the determinant by α_{1l}, the second by α_{2m}, and the third by α_{3n}; $\delta_{l1}\alpha_{1l} = \alpha_{11}$, $\delta_{l2}\alpha_{1l} = \alpha_{12}$, etc., so

$$\varepsilon_{lmn}\alpha_{1l}\alpha_{2m}\alpha_{3n} = \begin{vmatrix} \alpha_{11} & \alpha_{12} & \alpha_{13} \\ \alpha_{21} & \alpha_{22} & \alpha_{23} \\ \alpha_{31} & \alpha_{32} & \alpha_{33} \end{vmatrix} = |\alpha|. \tag{12.7}$$

Hence the Levi-Civita tensor allows us to put the expression for the determinant of any tensor in the form of (12.7). If in (12.7) we take a different sequence for the first subscripts of the tensor components (e.g., $\varepsilon_{lmn}\alpha_{2l}\alpha_{1m}\alpha_{3n}$), this leads to permutation of the rows of the determinant (here the first two rows), which gives $+|\alpha|$ or $-|\alpha|$ depending on whether the sequence of the first subscripts in $\alpha_{il}\alpha_{jm}\alpha_{kn}$, i.e., (i, j, l) is derived from (1, 2, 3) by an even or odd permutation. But ε_{ijk} gives us the required sign directly, so we can replace (12.7) by the somewhat more general expression

$$\varepsilon_{lmn}\alpha_{il}\alpha_{jm}\alpha_{kn} = |\alpha|\,\varepsilon_{ijk}. \tag{12.8}$$

Now we can elucidate how the Levi-Civita tensor changes in response to orthogonal transformations of the cartesian coordinate system. By definition [see (11.8)] the ε_{ijk}, being the components of an orthogonal tensor, must then alter as follows:

$$\varepsilon'_{ijk} = S_{il}S_{jm}S_{kn}\varepsilon_{lmn}, \tag{12.9}$$

in which S is the matrix of the orthogonal transformation. Comparison of (12.9) and (12.8) shows that

$$\varepsilon'_{ijk} = |S|\,\varepsilon_{ijk}. \tag{12.10}$$

In section 11 we saw that $|S| = 1$ for pure rotations, so the Levi-Civita tensor is unaltered by such transformations. A process including reflection gives $|S| = -1$, and (12.10) shows that all components of the Levi-Civita tensor reverse sign.

We multiply (12.8) by ε_{ijk} to get a further relation defining the determinant of a three-dimensional matrix via the Levi-Civita tensor:

$$|\alpha| = \frac{1}{6}\,\varepsilon_{ijk}\varepsilon_{lmn}\alpha_{il}\alpha_{jm}\alpha_{kn}. \tag{12.11}$$

Multiplication of (12.8) by ε_{ijs} gives

$$\frac{1}{2}\varepsilon_{ijs}\varepsilon_{lmn}\alpha_{il}\alpha_{jm}\alpha_{kn} = |\alpha|\,\delta_{ks}. \tag{12.12}$$

Consider matrix $\bar{\alpha}$, whose components are of the form

$$\bar{\alpha}_{ns} = \frac{1}{2}\varepsilon_{ijs}\varepsilon_{lmn}\alpha_{il}\alpha_{jm}. \tag{12.13}$$

Then (12.12) may be put as

$$\alpha_{kn}\bar{\alpha}_{ns} = |\alpha|\,\delta_{ks}. \tag{12.14}$$

In coordinate-free form this becomes

$$\alpha\bar{\alpha} = |\alpha|. \tag{12.15}$$

If α has a reciprocal (i.e., $|\alpha| \neq 0$), we multiply from the left by α^{-1} to get

$$\bar{\alpha} = |\alpha|\,\alpha^{-1}. \tag{12.16}$$

The $\bar{\alpha}$ defined by (12.13) or (12.16) is inverse (under multiplication) with respect to α; (12.16) is meaningful only for $|\alpha| \neq 0$, but all the properties of $\bar{\alpha}$ implied by this relation persist also for the general case, so we will use (12.16) widely in deriving various properties of the inverse matrix.

The determinant of a matrix β multiplied by a number k is

$$|k\beta| = k^3\,|\beta|. \tag{12.17}$$

But $|\alpha|$ is a number, so we have from (12.16)

$$|\bar{\alpha}| = |\alpha|^3\,|\alpha^{-1}| = |\alpha|^2, \tag{12.18}$$

because $|\alpha^{-1}| = |\alpha|^{-1}$, which follows from $\alpha\alpha^{-1}=1$. The matrix inverse to the reciprocal one is given by (12.16) as

$$\overline{\alpha^{-1}} = |\alpha^{-1}|\,(\alpha^{-1})^{-1} = \frac{\alpha}{|\alpha|} = (\bar{\alpha})^{-1}, \tag{12.19}$$

i.e., the operations of passing to the reciprocal and inverse matrices are commutative. For $\bar{\bar{\alpha}}$, the doubly inverse matrix, we have

$$\bar{\bar{\alpha}} = |\bar{\alpha}|\,(\bar{\alpha})^{-1} = |\alpha|^2\,|\alpha|^{-1}\,\alpha = |\alpha|\,\alpha. \tag{12.20}$$

The inverse matrix of $k\alpha$ is

$$\overline{k\alpha} = |k\alpha|\,(k\alpha)^{-1} = k^3\,|\alpha|\,\frac{1}{k}\,\alpha^{-1} = k^2\bar{\alpha}. \tag{12.21}$$

From (10.43) we see that

$$\tilde{\bar{\alpha}} = \bar{\tilde{\alpha}}. \tag{12.22}$$

The inverse matrix of the product of two matrices is given by (12.16) and (10.41) as

$$\overline{\alpha\beta} = |\alpha\beta|\,(\alpha\beta)^{-1} = |\beta|\,\beta^{-1}\,|\alpha|\,\alpha^{-1} = \bar{\beta}\bar{\alpha}. \tag{12.23}$$

Hence (10.41) applies, the rule for the matrix reciprocal to the product of two matrices.

Also, (12.11) gives us the expression for the determinant of the sum of two matrices, which is of importance in what follows:

$$|\alpha + \beta| = \frac{1}{6}\,\varepsilon_{ljk}\varepsilon_{lmn}\,(\alpha_{il} + \beta_{il})(\alpha_{jm} + \beta_{jm})(\alpha_{kn} + \beta_{kn}).$$

Expanding the brackets and using (12.11), we have

$$|\alpha + \beta| = |\alpha| + |\beta| + K_1 + K_2,$$

in which

$$K_1 = \frac{1}{6}\,\varepsilon_{ljk}\varepsilon_{lmn}\,(\alpha_{il}\alpha_{jm}\beta_{kn} + \alpha_{il}\alpha_{kn}\beta_{jm} + \alpha_{jm}\alpha_{kn}\beta_{il}), \tag{12.24}$$

and K_2 differs from K_1 by permutation of α and β. All three terms in (12.24) are equal, as may be seen by altering the notation for the dummy subscripts and using the properties of the Levi-Civita tensor; so from (12.13)

$$K_1 = \frac{1}{2}\,\varepsilon_{ljk}\varepsilon_{lmn}\alpha_{il}\alpha_{jm}\beta_{kn} = \bar{\alpha}_{nk}\beta_{kn} = (\bar{\alpha}\beta)_t.$$

Interchanging α and β we get $K_2 = (\bar{\beta}\alpha)_t = (\alpha\bar{\beta})_t$. Finally

$$|\alpha + \beta| = |\alpha| + (\bar{\alpha}\beta)_t + (\alpha\bar{\beta})_t + |\beta|. \tag{12.25}$$

The Levi-Civita tensor allows us to put the vector product of two vectors as

$$[\mathbf{uv}]_i = \varepsilon_{ikl} u_k v_l. \tag{12.26}$$

This expression may be verified directly: $[\mathbf{uv}]_1 = \varepsilon_{1kl} u_k v_l = \varepsilon_{123} u_2 v_3 + \varepsilon_{132} u_3 v_2 = u_2 v_3 - u_3 v_2$. The right-hand side of (12.26) may be considered as the result of multiplying a vector $\mathbf{v}$ by a second-rank tensor $\mathbf{u}^\times$, whose components are defined as follows:

$$u_{il}^\times = \varepsilon_{ikl} u_k. \tag{12.27}$$

The properties of the ε_{ikl} make $\mathbf{u}^\times$ antisymmetric:

$$\tilde{u}^\times = - u^\times. \tag{12.28}$$

The definition of (12.27) shows that the components of this tensor are expressed linearly and homogeneously in terms of those of the vector $\mathbf{u}$. Tensor $\mathbf{u}^\times$ is termed the dual of vector $\mathbf{u}$. In expanded form we have (12.27) as

$$u^\times = \begin{pmatrix} 0 & -u_3 & u_2 \\ u_3 & 0 & -u_1 \\ -u_2 & u_1 & 0 \end{pmatrix}. \tag{12.29}$$

Hence the Levi-Civita tensor allows us to relate any vector to an antisymmetric second-rank tensor, a relationship that is reversible, since any such tensor $(\gamma = -\tilde{\gamma})$ may be expressed in terms of a vector $\mathbf{u}$ via

$$u_i = \frac{1}{2} \varepsilon_{ikl} \gamma_{lk}. \tag{12.30}$$

Expanding this, we get

$$u_1 = \gamma_{32}, \quad u_2 = \gamma_{13}, \quad u_3 = \gamma_{21}. \tag{12.31}$$

Comparison of (12.29) with (12.31) shows that $\gamma = \mathbf{u}^\times$. The same result is obtained if we multiply both parts of (12.30) by ε_{min}. From (12.3) follows $\varepsilon_{min}\varepsilon_{ikl} = -\varepsilon_{mni}\varepsilon_{kli} = \delta_{ml}\delta_{nk} - \delta_{mk}\delta_{nl}$, so that $\varepsilon_{min} u_i = (1/2)(\delta_{ml}\delta_{nk} - \delta_{mk}\delta_{nl})\gamma_{lk} = (1/2)(\gamma_{mn} - \gamma_{nm}) = \gamma_{mn}$, since $\gamma_{nm} = -\gamma_{mn}$. Comparison with (12.27) shows that $\gamma_{mn} u_{mn}^\times$, so (12.27) implies that

$$u_k = \frac{1}{2} \varepsilon_{kli} u_{il}^\times. \tag{12.32}$$

The $\mathbf{u} = (u_i)$ defined by (12.30) is termed the dual of γ; the operation is linear, as the definition implies, which means that

$$(a\boldsymbol{u} + b\boldsymbol{v})^{\times} = a\boldsymbol{u}^{\times} + b\boldsymbol{v}^{\times}, \tag{12.33}$$

in which a and b are scalar factors.

The determinant of $\mathbf{u}^{\times}$ is zero, because $\mathbf{u}^{\times}$ is antisymmetric. The determinant of any matrix is unaffected by transposition, i.e., $|\mathbf{u}^{\times}| = |\tilde{\mathbf{u}}^{\times}| = |-\mathbf{u}^{\times}|$. But $|k\alpha| = k^3|\alpha|$, so $|-\mathbf{u}^{\times}| = (-1)^3|\mathbf{u}^{\times}| = -|\mathbf{u}^{\times}|$, i.e., $|\mathbf{u}^{\times}| = -|\mathbf{u}^{\times}| = 0$. These arguments are correct for any antisymmetric matrix with an odd number of rows and columns, whose determinant is therefore zero.

Tensor $\mathbf{u}^{\times}$ allows us to put (12.26) as $[\mathbf{uv}]_i = \overset{\times}{u}_{il} v_l$, or

$$[\boldsymbol{uv}] = \boldsymbol{u}^{\times}\boldsymbol{v}. \tag{12.34}$$

This equality may also be considered as a definition of tensor $\mathbf{u}^{\times}$ dual to $\mathbf{u}$ if it is true for any $\mathbf{v}$. Instead of (12.34) we may also put

$$[\boldsymbol{uv}] = \boldsymbol{u}\boldsymbol{v}^{\times}. \tag{12.35}$$

Consider the product of the two matrices $\mathbf{u}^{\times}$ and $\mathbf{v}^{\times}$. Using (12.27) and (12.4), we get

$$(\boldsymbol{u}^{\times}\boldsymbol{v}^{\times})_{in} = \boldsymbol{u}^{\times}_{il}\boldsymbol{v}^{\times}_{ln} = \varepsilon_{ikl}u_k\varepsilon_{lmn}v_m =$$

$$(\delta_{im}\delta_{kn} - \delta_{in}\delta_{km})\,u_k v_m = v_i u_n - u_k v_k \delta_{in}, \tag{12.36}$$

or in direct form

$$\boldsymbol{u}^{\times}\boldsymbol{v}^{\times} = \boldsymbol{v} \cdot \boldsymbol{u} - \boldsymbol{vu}. \tag{12.37}$$

Multiplying both parts of this from the right by an arbitrary vector $\mathbf{w}$ and using (12.34), we get

$$\boldsymbol{u}^{\times}\boldsymbol{v}^{\times}\boldsymbol{w} = \boldsymbol{u}^{\times}[\boldsymbol{vw}] = [\boldsymbol{u}[\boldsymbol{vw}]] = \boldsymbol{v} \cdot \boldsymbol{uw} - \boldsymbol{vu} \cdot \boldsymbol{w}. \tag{12.38}$$

This is a proof for a double vector product known from elementary vector algebra. For the particular case $\mathbf{v} = \mathbf{u}$ we have from (12.37) that

$$\boldsymbol{u}^{\times^2} = \boldsymbol{u} \cdot \boldsymbol{u} - \boldsymbol{u}^2. \tag{12.39}$$

The expression for the tensor dual to the vector product $[\mathbf{uv}]^{\times}$ is derived from (12.34) and (12.38). For any vector $\mathbf{A}$

$$[uv]^{\times} A = [\,[uv]\,A] = [A\,[vu]\,] = (v \cdot u - u \cdot v)\,A,$$

so

$$[uv]^{\times} = v \cdot u - u \cdot v. \tag{12.40}$$

The product $\mathbf{u}^{\times}\mathbf{v}^{\times}\mathbf{w}^{\times}$, of three tensors may be put in two ways: $\mathbf{u}^{\times}(\mathbf{v}^{\times}\mathbf{w}^{\times}) = (\mathbf{u}^{\times}\mathbf{v}^{\times})\mathbf{w}^{\times}$ on the basis of the associative principle. From (12.34) and (12.37),

$$u^{\times}v^{\times}w^{\times} = [uw] \cdot v - u^{\times} \cdot wv = v \cdot [uw] - vu \cdot w^{\times}. \tag{12.41}$$

A useful formula is obtained by expanding the product of four matrices $\mathbf{a}^{\times}\mathbf{c}^{\times}\mathbf{b}^{\times}\mathbf{d}^{\times}$, which also can be put in two ways:

$$a^{\times}(c^{\times}b^{\times})\,d^{\times} = (a^{\times}c^{\times})(b^{\times}d^{\times}).$$

From (12.37) we obtain after simple steps that

$$[ab] \cdot [cd] = [ab]\,[cd] + ad \cdot c \cdot b + bc \cdot d \cdot a - ac \cdot d \cdot b - bd \cdot c \cdot a. \tag{12.42}$$

In particular,

$$[ab] \cdot [ab] = [ab]^2 + ab \cdot (a \cdot b + b \cdot a) - a^2 \cdot b \cdot b - b^2 \cdot a \cdot a. \tag{12.43}$$

Formulas involving vector products known from elementary vector algebra are

$$[ab]\,[cd] = ac \cdot bd - ad \cdot bc, \tag{12.44}$$

$$[ab]^2 = a^2b^2 - (ab)^2. \tag{12.45}$$

Formula (12.43), with (12.45), gives us for the particular case $\mathbf{a} = \mathbf{n}_1$, $\mathbf{b} = \mathbf{n}_2$, $\mathbf{n}_1^2 = \mathbf{n}_2^2 = 1$, $\mathbf{n}_1\mathbf{n}_2 = 0$, that

$$[n_1n_2] \cdot [n_1n_2] = 1 - n_1 \cdot n_1 - n_2 \cdot n_2, \tag{12.46}$$

or

$$n_1 \cdot n_1 + n_2 \cdot n_2 + n_3 \cdot n_3 = 1, \tag{12.47}$$

in which

$$n_3 = [n_1 n_2], \quad n_i n_k = \delta_{ik} \qquad (i, k = 1, 2, 3). \tag{12.48}$$

Formula (12.47) gives the expression for unit tensor in terms of a triple of inversely orthogonal unit vectors $\mathbf{n}_i$.*

Consider the expression for the tensor inverse to $\mathbf{u}^{\times}$. From (12.13), (12.27), and (12.4) we get

$$(\overline{u^{\times}})_{ns} = \frac{1}{2}\varepsilon_{ijs}\varepsilon_{lmn}u^{\times}_{il}u^{\times}_{jm} = \frac{1}{2}\varepsilon_{ijs}\varepsilon_{ikl}u_k\varepsilon_{lmn}\varepsilon_{jpm}u_p$$

$$= \frac{1}{2}(\delta_{jk}\delta_{sl} - \delta_{jl}\delta_{ks})\,u_k\,(\delta_{lp}\delta_{nj} - \delta_{lj}\delta_{np})\,u_p$$

$$= \frac{1}{2}(u_j\delta_{sl} - u_s\delta_{jl})(u_l\delta_{nj} - u_n\delta_{lj}) = u_n u_s$$

or

$$\overline{u^{\times}} = u \cdot u. \tag{12.49}$$

Tensor α in $\bar{\alpha}[\mathbf{uv}]$ may be extracted from the vector product; from (12.13), (12.26), and (12.4) we can put

$$(\bar{\alpha}[uv])_n = \frac{1}{2}\varepsilon_{ijs}\varepsilon_{lmn}\alpha_{il}\alpha_{jm}\varepsilon_{spq}u_p v_q =$$

$$= \frac{1}{2}\varepsilon_{lmn}(\delta_{ip}\delta_{jq} - \delta_{iq}\delta_{jp})\,\alpha_{il}\alpha_{jm}u_p v_q =$$

$$= \frac{1}{2}\varepsilon_{lmn}(\alpha_{pl}u_p\alpha_{qm}v_q - \alpha_{ql}v_q\alpha_{pm}u_p) = \varepsilon_{lmn}(u\alpha)_l(v\alpha)_m,$$

or

$$\bar{\alpha}[uv] = [u\alpha, v\alpha] = [\tilde{\alpha}u, \tilde{\alpha}v]. \tag{12.50}$$

For α symmetric or antisymmetric

$$\bar{\alpha}[uv] = [\alpha u, \alpha v] \qquad (\tilde{\alpha} = \pm\alpha). \tag{12.51}$$

If α is such that $|\alpha| \neq 0$, we can, from (12.16), put (12.50) as

$$\alpha^{-1}[uv] = \frac{1}{|\alpha|}[\tilde{\alpha}u, \tilde{\alpha}v]. \tag{12.52}$$

---

* The subscript to the quantity in bold type, $\mathbf{n}_i$, serves to distinguish the vectors; it should not be confused with the subscript to the same letter in ordinary type, which serves to distinguish components of a single vector $\mathbf{n}$.

Here we put $\alpha^{-1}=\tilde{\beta}$, $\tilde{\alpha}=\beta^{-1}$, $1/|\alpha| = |\beta|$, to get from (10.43)

$$\tilde{\beta}[\boldsymbol{uv}] = |\beta|\,[\beta^{-1}\boldsymbol{u},\ \beta^{-1}\boldsymbol{v}]. \tag{12.53}$$

In section 11 we noted that a general tensor may be expressed as the sum of not more than three dyads; now we show that any tensor α can be represented as the sum of three dyads, in which the first or second vectors may be three arbitrary linearly independent vectors $\mathbf{a}_k$ (k=1, 2, 3). We introduce the three vectors

$$\bar{\boldsymbol{a}}_k = \frac{\varepsilon_{klm}\,[\boldsymbol{a}_l \boldsymbol{a}_m]}{2\boldsymbol{a}_1\,[\boldsymbol{a}_2\boldsymbol{a}_3]}^{*} \qquad (12.54)^{*}$$

or

$$\bar{\boldsymbol{a}}_1 = \frac{[\boldsymbol{a}_2\boldsymbol{a}_3]}{V}, \quad \bar{\boldsymbol{a}}_2 = \frac{[\boldsymbol{a}_3\boldsymbol{a}_1]}{V}, \quad \bar{\boldsymbol{a}}_3 = \frac{[\boldsymbol{a}_1\boldsymbol{a}_2]}{V}, \quad V = \boldsymbol{a}_1\,[\boldsymbol{a}_2\boldsymbol{a}_3]. \tag{12.55}$$

The independence of the $\mathbf{a}_k$ implies that $V = \alpha_1[\alpha_2\alpha_3] \neq 0$. It is easily seen that

$$\bar{\boldsymbol{a}}_k \boldsymbol{a}_l = \delta_{kl}. \tag{12.56}$$

We call the triplets $\bar{\mathbf{a}}_k$ and $\mathbf{a}_k$ inversely reciprocal one to another. Let

$$\alpha \boldsymbol{a}_k = \boldsymbol{b}_k\,; \tag{12.57}$$

then tensor α may be put as

$$\alpha = \boldsymbol{b}_k \cdot \boldsymbol{a}_k = \boldsymbol{b}_1 \cdot \boldsymbol{a}_1 + \boldsymbol{b}_2 \cdot \boldsymbol{a}_2 + \boldsymbol{b}_3 \cdot \boldsymbol{a}_3. \tag{12.58}$$

In fact, (12.56) implies that $\alpha\bar{\mathbf{a}}_l = \mathbf{b}_k \cdot \mathbf{a}_k\mathbf{a}_l = \mathbf{b}_k\delta_{kl} = \mathbf{b}_l$, as (12.57) shows should be the case. Only one tensor can satisfy (12.57); if there were another tensor α' such that $\alpha'\bar{\mathbf{a}}_l = \mathbf{b}_l$, we would have $(\alpha-\alpha')\bar{\mathbf{a}}_l = 0$ for $l=1, 2, 3$; but in that case the tensor $\alpha-\alpha'$ would give zero on multiplication by any vector, because the latter can always be put as a linear combination of three independent vectors $\bar{\mathbf{a}}_l$. But a tensor that gives zero on multiplication by any vector

* (12.54) remains true if $\mathbf{a}_k$ and $\bar{\mathbf{a}}_k$ are interchanged.

must itself be zero,* so there is only one tensor that satisfies (12.57).

If we use the same $\bar{\mathbf{a}}_k$ to define vectors $\mathbf{b}'_k$ via

$$\boldsymbol{b}'_k = \bar{\boldsymbol{a}}_k \alpha, \tag{12.59}$$

then we get, instead of (12.58),

$$\alpha = \boldsymbol{a}_k \cdot \boldsymbol{b}'_k. \tag{12.60}$$

If $\tilde{\alpha} = \alpha$, then $\mathbf{b}'_k = \mathbf{b}_k$.

Hence any tensor may be put as the sum of not more than three dyads, in which the left-hand (or right-hand) vectors may be three arbitrary linearly independent vectors.

(12.47) gives a representation of unit tensor as a sum of dyads constructed from three inversely orthogonal unit vectors, while (12.58) and (12.60) imply that unit tensor may be represented via any three linearly independent vectors $\mathbf{a}_k$. It is sufficient to put $\alpha = 1$ in (12.58) and (12.60) in order to find these representations; then $\mathbf{b}_k = \mathbf{b}'_k = \bar{\mathbf{a}}_k$ and we get

$$1 = \boldsymbol{a}_k \cdot \bar{\boldsymbol{a}}_k = \bar{\boldsymbol{a}}_k \cdot \boldsymbol{a}_k, \tag{12.61}$$

or, in expanded form

$$1 = \frac{\boldsymbol{a}_1 \cdot [\boldsymbol{a}_2\boldsymbol{a}_3] + \boldsymbol{a}_2 \cdot [\boldsymbol{a}_3\boldsymbol{a}_1] + \boldsymbol{a}_3 \cdot [\boldsymbol{a}_1\boldsymbol{a}_2]}{\boldsymbol{a}_1 [\boldsymbol{a}_2\boldsymbol{a}_3]}, \tag{12.62}$$

since

$$1 = \frac{[\boldsymbol{a}_2\boldsymbol{a}_3] \cdot \boldsymbol{a}_1 + [\boldsymbol{a}_3\boldsymbol{a}_1] \cdot \boldsymbol{a}_2 + [\boldsymbol{a}_1\boldsymbol{a}_2] \cdot \boldsymbol{a}_3}{\boldsymbol{a}_1 [\boldsymbol{a}_2\boldsymbol{a}_3]}. \tag{12.63}$$

As for (12.25), we can obtain an expression for the tensor inverse to the sum of two tensors α and β; from (12.13)

$$\overline{(\alpha + \beta)}_{ns} = \frac{1}{2} \varepsilon_{ijs} \varepsilon_{lmn} (\alpha_{il} + \beta_{il})(\alpha_{jm} + \beta_{jm});$$

---

*We consider here for simplicity vectors and tensors that are zero, whereas it would be more accurate to speak of zero vectors (vectors all of whose components are zero).

expanding and using (12.3), we get simply

$$\overline{\alpha+\beta} = \bar{\alpha} + \bar{\beta} + \alpha\beta + \beta\alpha - \alpha_t\beta - \beta_t\alpha + \alpha_t\beta_t - (\alpha\beta)_t. \qquad (12.64)$$

Hence we have

$$\overline{(\alpha+\beta)}_t = \bar{\alpha}_t + \bar{\beta}_t + \alpha_t\beta_t - (\alpha\beta)_t. \qquad (12.65)$$

We multiply (12.64) from the left by $\alpha+\beta$ and get (12.25) on the left, in accordance with (12.15); comparison with the result of multiplying the right-hand side of (12.64) by $\alpha+\beta$ gives us the identity

$$(\alpha^2 - \alpha_t\alpha)(\beta - \beta_t) + (\alpha\beta - (\alpha\beta)_t)\,\alpha + \beta\bar{\alpha} - (\beta\bar{\alpha})_t \equiv 0. \qquad (12.66)$$

Multiplication by $\alpha+\beta$ from the right gives us the analogous identity

$$(\beta - \beta_t)(\alpha^2 - \alpha_t\alpha) + (\alpha\beta - (\alpha\beta)_t)\,\alpha + \bar{\alpha}\beta - (\bar{\alpha}\beta)_t \equiv 0. \qquad (12.67)$$

We saw in section 11 that all tensors fall into three groups: linear (dyads), planal (sums of two dyads), and complete (sums of three independent dyads).* The following are some properties of these types. The determinant of the dyad of (11.13) is zero, because all the rows and columns are proportional; (12.13) readily shows that the tensor inverse to the dyad is also zero. This is also clear from the fact that (12.13) shows that the components of the tensor inverse to α are proportional to the second-order minors of tensor α, while all such minors of the dyad $\mathbf{a}\cdot\mathbf{b}$ are also zero, because the rows and columns are proportional one to another. Thus

$$|\boldsymbol{a}\cdot\boldsymbol{b}| = \overline{\boldsymbol{a}\cdot\boldsymbol{b}} = 0. \qquad (12.68)$$

It is also clear that, conversely, if the tensor inverse to some tensor α is zero, then α itself is a simple dyad:

$$\alpha = \boldsymbol{a}\cdot\boldsymbol{b} \quad \text{for} \quad \bar{\alpha} = 0. \qquad (12.69)$$

Any planal tensor may, as shown in section 11, be represented as $\alpha = \mathbf{a}_1\cdot\mathbf{b}_1 + \mathbf{a}_2\cdot\mathbf{b}_2$; (12.25) and (12.68) give

---

* Dyads are considered as independent if the left-hand vectors are linearly independent, as are the right-hand ones.

$$|a_1 \cdot b_1 + a_2 \cdot b_2| = 0. \tag{12.70}$$

In turn, (12.64) gives

$$\overline{a_1 \cdot b_1 + a_2 \cdot b_2} = a_1 \cdot b_1 a_2 \cdot b_2 + a_2 \cdot b_2 a_1 \cdot b_1 - a_1 b_1 \cdot a_2 \cdot b_2 -$$
$$- a_2 b_2 \cdot a_1 \cdot b_1 + a_1 b_1 \cdot a_2 b_2 - (a_1 \cdot b_1 a_2 \cdot b_2)_t. \tag{12.71}$$

The $\mathbf{b}_1\mathbf{a}_2$ in the last term is a scalar and may be extracted from the trace, which gives $\mathbf{b}_1\mathbf{a}_2(\mathbf{a}_1 \cdot \mathbf{b}_2)_t = \mathbf{b}_1\mathbf{a}_2 \cdot \mathbf{a}_1\mathbf{b}_2$, while (12.44) gives the last two terms together as equal to $[\mathbf{b}_1\mathbf{b}_2]\,[\mathbf{a}_1\mathbf{a}_2]$. Comparison to the right side of (12.71) with (12.42) gives

$$\overline{a_1 \cdot b_1 + a_2 \cdot b_2} = [b_1 b_2] \cdot [a_1 a_2]. \tag{12.72}$$

Hence the tensor inverse to a planal one is a dyad. The important conclusion may be put in a somewhat more general form:

$$\bar{\alpha} = a \cdot b \quad \text{for} \quad |\alpha| = 0. \tag{12.73}$$

The conditions $|\alpha| = 0$ and $\bar{\alpha} \neq 0$, from (12.68), define a planal tensor.

A complete tensor is

$$\alpha = a_k \cdot b_k = a_1 \cdot b_1 + a_2 \cdot b_2 + a_3 \cdot b_3,$$
$$(a_1 [a_2 a_3] \neq 0, \ b_1 [b_2 b_3] \neq 0). \tag{12.74}$$

The conditions of completeness are given in parentheses and amount to linear independence of all left-hand vectors in the dyad, as well as of all right-hand ones. The determinant of α is found from (12.25), (12.68), (12.70), and (12.72) as

$$|\alpha| = |(a_1 \cdot b_1 + a_2 \cdot b_2) + a_3 \cdot b_3| = ((\overline{a_1 \cdot b_1 + a_2 \cdot b_2})\, a_3 \cdot b_3)_t =$$
$$= ([b_1 b_2] \cdot [a_1 a_2]\, a_3 \cdot b_3)_t = a_1 [a_2 a_3] \cdot b_1 [b_2 b_3]. \tag{12.75}$$

But $|\alpha| \neq 0$, so a complete tensor is not unique (in the sense of section 10); conversely, $|\alpha| \neq 0$ is the necessary and sufficient condition for completeness in tensor α, i.e., for the possibility of representation in the form of (12.74). As for any such tensor, the complete tensor α has a reciprocal α^{-1}, which may be put as

$$\alpha^{-1} = \bar{b}_l \cdot \bar{a}_l. \tag{12.76}$$

in which $\bar{\mathbf{b}}_l$ and $\bar{\mathbf{a}}_l$ are given by (12.54) and are the vectors reciprocal to the $\mathbf{b}_k$ and $\mathbf{a}_k$, respectively. Multiplying the α of (12.74) by the α^{-1} of (12.76) from the left and right, and using (12.57) and (12.61), we get

$$\alpha\alpha^{-1} = \boldsymbol{a}_k \cdot \boldsymbol{b}_k\bar{\boldsymbol{b}}_l \cdot \bar{\boldsymbol{a}}_l = \boldsymbol{a}_k \cdot \delta_{kl} \cdot \bar{\boldsymbol{a}}_l = \boldsymbol{a}_k \cdot \bar{\boldsymbol{a}}_k = 1,$$

$$\alpha^{-1}\alpha = \bar{\boldsymbol{b}}_l \cdot \bar{\boldsymbol{a}}_l\boldsymbol{a}_k \cdot \boldsymbol{b}_k = \bar{\boldsymbol{b}}_k \cdot \boldsymbol{b}_k = 1.$$

The tensor inverse to the complete tensor α of (12.74) is most simply found from (12.16); (12.75), (12.76), and (12.55) give

$$\overline{\boldsymbol{a}_1 \cdot \boldsymbol{b}_1 + \boldsymbol{a}_2 \cdot \boldsymbol{b}_2 + \boldsymbol{a}_3 \cdot \boldsymbol{b}_3} =$$

$$= [\boldsymbol{b}_2\boldsymbol{b}_3] \cdot [\boldsymbol{a}_2\boldsymbol{a}_3] + [\boldsymbol{b}_3\boldsymbol{b}_1] \cdot [\boldsymbol{a}_3\boldsymbol{a}_1] + [\boldsymbol{b}_1\boldsymbol{b}_2] \cdot [\boldsymbol{a}_1\boldsymbol{a}_2]. \tag{12.77}$$

This general formula gives (12.72) $\mathbf{a}_3 = 0$ (or $\mathbf{b}_3 = 0$); a consequence of (12.68), (12.72), and (12.77) is that the tensor inverse to a certain tensor never can be planal; if it differs from zero, it must be a linear tensor or a complete tensor.

13. Eigenvalues and Eigenvectors of a Second-Rank Tensor

The unit tensor is the sole second-rank tensor which, when multiplied by any vector, leaves it unchanged; any other second-rank tensor α on multiplication by a vector $\mathbf{a}$ gives another vector $\mathbf{b}$ which in general differs from $\mathbf{a}$: $\alpha\mathbf{a} = \mathbf{b}$, $\mathbf{b} \neq \mathbf{a}$. But, no matter what α may be,* there are always several vectors (or at least one) that on multiplication by α will retain the same direction and change only in length. For such a vector the multiplication by the tensor is equivalent to multiplication by a number. Let $\mathbf{u}$ be a vector with this property with respect to a tensor α; then we have

$$\alpha\boldsymbol{u} = \lambda\boldsymbol{u}, \tag{13.1}$$

in which λ is some number. In components this is

$$\alpha_{ik}u_k = \lambda u_i. \tag{13.2}$$

The $\mathbf{u}$ that satisfies (13.1) is an eigenvector (principal vector) of tensor α, while λ is the corresponding eigenvalue (principal value,

*We assume that $0 \neq \alpha \neq 1$.

number) of tensor α. The directions of the eigenvectors are called the principal axes of α The main terms to be used here are eigenvector and eigenvalue.

The equation $(\alpha-\lambda)\mathbf{u}=0$ of (13.1) is a system of three linear homogeneous equations in the components of the vector $\mathbf{u}=(u_k)$ having the matrix $\alpha-\lambda=(\alpha_{ik}-\lambda\delta_{ik})$; for the system to have a solution $\mathbf{u}\neq 0$ it is necessary and sufficient that the determinant of the system be zero, i.e., that

$$|\lambda-\alpha|=P(\lambda)=0. \tag{13.3}$$

This is the characteristic equation of the matrix α, while the polynomial $P(\lambda)$ is the characteristic polynomial of matrix α. Formula (12.25) allows us to write a general expression for this polynomial for a three-dimensional tensor α. Here we must remember that the number λ (see section 10) should be considered as multiplied by unit tensor. Then $|\lambda|=\lambda^3$, $\bar{\lambda}=\lambda^2$ [see (12.21)] and

$$P(\lambda)=|\lambda-\alpha|=\lambda^3-\alpha_t\lambda^2+\bar{\alpha}_t\lambda-|\alpha|=0. \tag{13.4}$$

A very important feature is that this polynomial does not alter in response to any congruent transformation S as in (10.47). Let $\alpha'=S\alpha S^{-1}$; we can always replace λ by $S\lambda S^{-1}$, so $|\lambda-\alpha'|=|S(\lambda-\alpha)S^{-1}|=|S|\cdot|\lambda-\alpha|\cdot|S^{-1}|=|\lambda-\alpha|$, and the polynomial is invariant with respect to all congruent transformations. This means that the coefficients α_t, $\bar{\alpha}_t$, and $|\alpha|$ of that polynomial are also invariant in congruent transformations; they are sometimes called the first (α_t), second ($\bar{\alpha}_t$), and third ($|\alpha|$) invariants of the tensor α.

There are always three roots to the cubic equation of (13.4), including perhaps coincident ones. These roots are the eigenvalues of tensor α. It can be shown that each of the distinct roots of the characteristic equation $\lambda=\lambda_0$ corresponds to at least one eigenvector $\mathbf{u}$ that satisfies $(\alpha-\lambda_0)\mathbf{u}=0$; now $|\alpha-\lambda_0|=0$, so the tensor $(\alpha-\lambda_0)$ may be either planal or linear (section 12). In the first case, (12.73) shows that $\overline{\alpha-\lambda_0}=\mathbf{a}\cdot\mathbf{b}\neq 0$, while (12.15) gives $(\alpha-\lambda_0)\overline{(\alpha-\lambda_0)}=(\alpha-\lambda_0)\mathbf{a}\cdot\mathbf{b}=|\alpha-\lambda_0|=0$. Let $\mathbf{c}=(\alpha-\lambda_0)\mathbf{a}$; then $\mathbf{c}\cdot\mathbf{b}=0$. The definition of (11.13) shows that a dyad may equal zero only when the left-hand (or right-hand) vector is zero; but $\mathbf{b}\neq 0$ and $\mathbf{a}\neq 0$, otherwise we would have $\overline{\alpha-\lambda_0}=0$, which means that

$\mathbf{c} = (\alpha - \lambda_0)\mathbf{a} = 0$, i.e., $\mathbf{u} = \mathbf{a} \neq 0$. Thus we may, from (12.73) consider that if

$$|\alpha - \lambda_0| = 0, \qquad \overline{\alpha - \lambda_0} \neq 0, \tag{13.5}$$

then

$$\overline{\alpha - \lambda_0} = C\boldsymbol{u} \cdot \boldsymbol{b}, \qquad (\alpha - \lambda_0)\,\boldsymbol{u} = 0, \tag{13.6}$$

with

$$C = \frac{(\overline{\alpha - \lambda_0})_t}{\boldsymbol{ub}}. \tag{13.7}$$

We can assume that $\mathbf{b} = \mathbf{u}$ and $\mathbf{u}^2 = 1$ for a symmetrical matrix α, and subject to (13.5)

$$\boldsymbol{u} \cdot \boldsymbol{u} = \frac{\overline{\alpha - \lambda_0}}{(\overline{\alpha - \lambda_0})_t} \qquad (\boldsymbol{u}^2 = 1). \tag{13.8}$$

In the second case (tensor $\alpha - \lambda_0$ is linear), the tensor equals the dyad:* $\alpha - \lambda_0 = \mathbf{a} \cdot \mathbf{b} \neq 0$. In that case any vector $\mathbf{u}$ orthogonal to $\mathbf{b}$ ($\mathbf{ub} = 0$) would be a solution to (13.1), i.e., there are infinitely many such vectors, so from (12.69) we have that, if

$$|\alpha - \lambda_0| = 0, \ \overline{\alpha - \lambda_0} = 0, \ \alpha - \lambda_0 = \boldsymbol{a} \cdot \boldsymbol{b}, \ \boldsymbol{ab} = (\alpha - \lambda_0)_t, \tag{13.9}$$

then

$$(\alpha - \lambda_0)\,\boldsymbol{u} = 0, \qquad \boldsymbol{ub} = 0 \tag{13.10}$$

and

$$\alpha\boldsymbol{a} = (\lambda_0 + \boldsymbol{a} \cdot \boldsymbol{b}\,)\,\boldsymbol{a} = (\lambda_0 + \boldsymbol{ab})\,\boldsymbol{a}. \tag{13.11}$$

For a symmetrical matrix α and subject to the conditions of (13.9)

$$\alpha - \lambda_0 = C\boldsymbol{a} \cdot \boldsymbol{a}, \qquad \boldsymbol{a}^2 = 1, \qquad C = (\alpha - \lambda_0)_t \tag{13.12}$$

and

$$(\alpha - \lambda_0)\,\boldsymbol{u} = 0, \qquad \boldsymbol{ua} = 0; \qquad \alpha\boldsymbol{a} = (\lambda_0 + C)\,\boldsymbol{a}. \tag{13.13}$$

---

*If $\alpha - \lambda_0 = 0$, tensor α would differ only by a factor λ_0 from the unit tensor, and (13.1) would be satisfied by any vector $\mathbf{u}$.

This shows that (13.1) always has a solution $\mathbf{u} \neq 0$, so any tensor α must have at least one eigenvector. It also shows how the eigenvectors may be determined if the eigenvalues are known.

In general, the root of (13.4) (eigenvalue) and the eigenvector may both be complex; the latter may be put as $\mathbf{u} = \mathbf{a} + i\mathbf{b}$, in which **a** and **b** are ordinary real three-dimensional vectors. However, in this book we consider principally real eigenvectors and real eigenvalues, because the stress and deformation tensors are real symmetric tensors, and these can readily be shown to have only real eigenvalues. For example, let $\tilde{\alpha} = \alpha$ and $\alpha^* = \alpha$, in which the asterisk denotes the complex conjugate. We multiply (13.1) from the left by the conjugate vector $\mathbf{u}^*$; all operations with complex vectors are, without exception, the same as those for real vectors, the only difference being that the components are complex numbers. Hence we get

$$\mathbf{u}^* \alpha \mathbf{u} = \lambda \mathbf{u}^* \mathbf{u}. \tag{13.14}$$

The complex conjugate of this equation is

$$\mathbf{u} \alpha \mathbf{u}^* = \lambda^* \mathbf{u} \mathbf{u}^*. \tag{13.15}$$

Now α is a symmetric matrix, so $\mathbf{u}^* \alpha \mathbf{u} = \mathbf{u} \alpha \mathbf{u}^*$ from (10.23). The scalar product of two vectors is independent of the order of the factors, so $\mathbf{u}^* \mathbf{u} = \mathbf{u} \mathbf{u}^*$; hence comparison of (13.14) with (13.15) gives $\lambda = \lambda^*$, i.e., the eigenvalues are real.

The eigenvalues of a real symmetric matrix appear linearly and homogeneously in (13.1), so they are defined apart from an arbitrary factor, which may be taken as complex; but if $\alpha = \tilde{\alpha} = \alpha^*$, all the eigenvectors of α may always be taken as real, which is clear from the fact that the eigenvectors are determined from the dyad $\overline{\alpha - \lambda_0} = \mathbf{a} \cdot \mathbf{b}$, since $\alpha - \lambda_0 = \mathbf{a} \cdot \mathbf{b}$, and these are real, because α and λ_0 are real.

Appropriate choice of factor enables us always to make a real eigenvector **a** unit one; in fact, if $\mathbf{u}^2 \neq 1$, then the vector $\mathbf{u}' = C\mathbf{u}$, with $C = 1/\sqrt{\mathbf{u}^2}$, is a unit vector ($\mathbf{u}'^2 = 1$). Reduction of the length of an eigenvector to unity is called normalization, and the resulting eigenvector is said to be normalized.

Let the eigenvectors $\mathbf{u}_1$ and $\mathbf{u}_2$ of a real symmetric tensor α belong to two distinct eigenvalues $\lambda_1 \neq \lambda_2$:

$$\alpha u_1 = \lambda_1 u_1, \quad \alpha u_2 = \lambda_2 u_2. \tag{13.16}$$

We multiply the first equation scalarly from the left by $\mathbf{u}_2$, the second by $\mathbf{u}_1$, and subtract the second from the first. The tensor α is symmetric, so $\mathbf{u}_2\alpha\mathbf{u}_1 = \mathbf{u}_1\alpha\mathbf{u}_2$ (10.23), , and we get $(\lambda_1 - \lambda_2)u_1u_2 = 0$. But $\lambda_1 \neq \lambda_2$, so

$$u_1 u_2 = 0. \tag{13.17}$$

Thus the eigenvectors of a real symmetric tensor are orthogonal if they relate to different eigenvalues; hence that tensor has three inversely orthogonal eigenvectors if all the eigenvalues are different, and we may always take these vectors as unit ones, which may be used as a coordinate system called the system of principal axes of the tensor. We denote by $\mathbf{e}_i$ this triplet of orthonormalized (orthogonal and normalized) eigenvectors:

$$\alpha e_i = \lambda_i e_i, \quad e_i e_k = \delta_{ik}. \tag{13.18}$$

There is no summation with respect to i on the right in the first of these equations, although this subscript occurs twice. On the other hand, the need for this comment drops out if we note that this subscript occurs only once (i.e., is free) on the left, because a subscript must have the same character in both parts of an equation. The $\mathbf{e}_i$ are uniquely defined by (13.18), because multiplication of any one of them by -1 leaves the equations still true, which means that any of the $\mathbf{e}_i$ may be given either of two inversely opposite directions. Thus we may always choose the $\mathbf{e}_i$ to satisfy

$$[e_1 e_2] = e_3. \tag{13.19}$$

In that case the coordinate system will be right-handed. We can now use (12.58), which defines a tensor from the result of multiplying it by three linearly independent vectors. We have from (12.54), (12.55), (13.18), and (13.19) that $\bar{\mathbf{a}}_k = \mathbf{a}_k = \mathbf{e}_k$, $\mathbf{b}_k = \lambda_k \mathbf{e}_k$, $\mathbf{e}_1[\mathbf{e}_2\mathbf{e}_3] = 1$, so

$$\alpha = b_k \cdot a_k = \lambda_1 e_1 \cdot e_1 + \lambda_2 e_2 \cdot e_2 + \lambda_3 e_3 \cdot e_3. \tag{13.20}$$

If we take the $\mathbf{e}_i$ as the axes of a coordinate system, we have $\mathbf{e}_1 = (1, 0, 0)$, $\mathbf{e}_2 = (0, 1, 0)$, $\mathbf{e}_3 = (0, 0, 1)$, the components of the vectors being given in parentheses. We expand the dyads in accordance with (11.13) to get the following matrix expression for a tensor α:

$$\alpha=\begin{pmatrix}\lambda_1 & 0 & 0\\ 0 & \lambda_2 & 0\\ 0 & 0 & \lambda_3\end{pmatrix}. \tag{13.21}$$

This form for the tensor (matrix) is termed diagonal. The tensor takes a diagonal form (is diagonalized in the set of principal axes), and the diagonal elements equal the eigenvalues of the tensor.

Consider now the case of two coincident eigenvalues for a real symmetric tensor, e.g., $\lambda_1=\lambda_2$; then (13.20) gives

$$\alpha=\lambda_1(e_1\cdot e_1+e_2\cdot e_2)+\lambda_3 e_3\cdot e_3 \tag{13.22}$$

or in matrix form in the system of principal axes,

$$\alpha=\begin{pmatrix}\lambda_1 & 0 & 0\\ 0 & \lambda_1 & 0\\ 0 & 0 & \lambda_3\end{pmatrix}.$$

It is readily seen that any vector that is an arbitrary linear combination of the vectors $\mathbf{e}_1$ and $\mathbf{e}_2$ ($\mathbf{u}=a_1\mathbf{e}_1+a_2\mathbf{e}_2$) is an eigenvector of (13.22) for the eigenvalue λ_1; any vector lying in a plane parallel to $\mathbf{e}_1$ and $\mathbf{e}_2$ remains collinear with itself on multiplication by tensor α.

Finally, all three eigenvalues of a real symmetric tensor may coincide: $\lambda_1=\lambda_2=\lambda_3=\lambda$; then from (12.47) and (13.20),

$$\alpha=\lambda(e_1\cdot e_1+e_2\cdot e_2+e_3\cdot e_3)=\lambda 1=\lambda. \tag{13.23}$$

Here α is a multiple of unit tensor.

The tensor of (13.22) may be put in more compact and convenient form by the use of (12.47). Adding and subtracting $\lambda_1\mathbf{e}_3\cdot\mathbf{e}_3$ on the right, we get

$$\alpha=\lambda_1+(\lambda_3-\lambda_1)e_3\cdot e_3. \tag{13.24}$$

This representation clearly distinguishes the direction of $\mathbf{e}_3$, all directions perpendicular to this being equivalent. Hence if the tensor of (13.23) may be termed isotropic, that of (13.24) may be termed transversely isotropic; we shall also call it uniaxial [17], because in it there occurs a unique direction (axis), namely $\mathbf{e}_3$.

The general real symmetric tensor of (13.20) may be put in a different form, which is very convenient and useful. Let $\lambda_1 < \lambda_2 < \lambda_3$; on the right in (13.20) we add and subtract $\lambda_2(\mathbf{e}_1 \cdot \mathbf{e}_1 + \mathbf{e}_3 \cdot \mathbf{e}_3)$ to get

$$\alpha = \lambda_2 + (\lambda_3 - \lambda_2)\, \boldsymbol{e}_3 \cdot \boldsymbol{e}_3 - (\lambda_2 - \lambda_1)\, \boldsymbol{e}_1 \cdot \boldsymbol{e}_1. \tag{13.25}$$

We introduce the scalars

$$k_3 = \left| \sqrt{\frac{\lambda_3 - \lambda_2}{\lambda_3 - \lambda_1}} \right|, \qquad k_1 = \left| \sqrt{\frac{\lambda_2 - \lambda_1}{\lambda_3 - \lambda_1}} \right| \tag{13.26}$$

and the vectors

$$\boldsymbol{c}' = k_3 \boldsymbol{e}_3 + k_1 \boldsymbol{e}_1, \qquad \boldsymbol{c}'' = k_3 \boldsymbol{e}_3 - k_1 \boldsymbol{e}_1, \tag{13.27}$$

with $\mathbf{c}'^2 = \mathbf{c}''^2 = 1$. It is readily verified that these allow the tensor of (13.25) to be put in the form

$$\alpha = \lambda_2 + \frac{\lambda_3 - \lambda_1}{2} (\boldsymbol{c}' \cdot \boldsymbol{c}'' + \boldsymbol{c}'' \cdot \boldsymbol{c}'). \tag{13.28}$$

The tensor of (13.24) is now found as the particular case $\mathbf{c}' = \mathbf{c}'' = \mathbf{e}_3$. The vectors $\mathbf{c}'$ and $\mathbf{c}''$ sometimes have a definite geometrical or other significance, the decisive feature being the quantity characterized by the tensor α [9, 17]. We call $\mathbf{c}'$ and $\mathbf{c}''$ the axes of the symmetric real tensor α (not to be confused with the principal axes), and the general real symmetric tensor α of (13.20) or (13.28) is correspondingly called biaxial.

Real symmetric tensors may be divided into three groups in accordance with the relations between the eigenvalues:

1. Isotropic (multiples of unit tensor):

$$\alpha = a. \tag{13.29}$$

2. Uniaxial (transversely isotropic):

$$\alpha = a + b\boldsymbol{c} \cdot \boldsymbol{c}, \qquad c^2 = 1. \tag{13.30}$$

3. Biaxial:

$$\alpha = a + b(\boldsymbol{c}' \cdot \boldsymbol{c}'' + \boldsymbol{c}'' \cdot \boldsymbol{c}'), \qquad c'^2 = c''^2 = 1. \tag{13.31}$$

These tensors are very frequently encountered, so a list of their properties is given in Table II, including the eigenvalues and eigenvectors, invariants, and expressions for the mutual and reciprocal tensors. The last column gives the corresponding data for the tensor in the form of (13.20). One consequence clear from this is that the trace equals the sum of the eigenvalues, while the determinant is the product of the eigenvalues. The invariant $\bar{\alpha}_t$ is the sum of the products of pairs of eigenvalues.

The following is an interesting theorem for the case of a three-dimensional tensor. In (12.15) we replace α by $\lambda - \alpha$ (λ an arbitrary number), considering that expression as a definition of the inverse matrix:

$$(\lambda - \alpha)(\overline{\lambda - \alpha}) = |\lambda - \alpha|. \tag{13.32}$$

The inverse tensor $\overline{\lambda - \alpha}$ may be calculated from (12.64), from which we get by simple manipulation

$$\overline{\alpha - \lambda} = \overline{\lambda - \alpha} = \lambda^2 + \lambda(\alpha - \alpha_t) + \bar{\alpha}. \tag{13.33}$$

We substitute this expression into the left-hand side of (13.32) and multiply out, while the right-hand side we rewrite from (13.4) as

$$\lambda^3 - \lambda^2\alpha_t + \lambda[\bar{\alpha} - \alpha(\alpha - \alpha_t)] - |\alpha| \equiv \lambda^3 - \alpha_t\lambda^2 + \bar{\alpha}_t\lambda - |\alpha|.$$

This is an identity, because it is true for any number λ and tensor α; it implies that

$$\bar{\alpha} - \bar{\alpha}_t \equiv \alpha(\alpha - \alpha_t). \tag{13.34}$$

Multiplying both sides by α and transferring everything to the left-hand side, we have

$$\alpha^3 - \alpha_t\alpha^2 + \bar{\alpha}_t\alpha - |\alpha| \equiv 0. \tag{13.35}$$

Comparison with (13.4) shows that the difference lies in the replacement of λ by α, which transforms the numerical characteristic equation of (13.4) for the tensor α to the matrix identity (13.35). This is the proof of the Hamilton-Cayley theorem for the case of three-dimensional matrices; the theorem states that any matrix is a root of its characteristic equation, and it applies for a square matrix of any order [8].

(13.35) shows that any matrix α in three-dimensional space satisfies an equation of third degree, so any power of it may be

Table II. Some Properties of Real Symmetric Tensors

Tensor α	$\alpha = a$	$\alpha = a + bc \cdot c$, $c^2 = 1$		$\alpha = a + b(c' \cdot c'' + c'' \cdot c')$, $c'^2 = c''^2 = 1$		$\alpha = \lambda_1 e_1 \cdot e_1 + \lambda_2 e_2 \cdot e_2 + \lambda_3 e_3 \cdot e_3$, $e_i e_k = \delta_{ik}$		
Eigenvalues (coincident values in parentheses)	a (3)	a (2)	$a+b$ (1)	a (1)	$a \pm b(1 \pm c'c'')$ (1)	λ_1 (1)	λ_2 (1)	λ_3 (1)
Eigenvectors	Any vectors	$\perp c$	c	$[c'c'']$	$c' \pm c''$	e_1	e_2	e_3
$\bar{\alpha}$	a^2	$a(a+b-bc \cdot c)$		$\frac{\lvert\alpha\rvert}{a} + b^2(c' \cdot c' + c'' \cdot c'') - b(a + bc'c'') \times$ $\times (c' \cdot c'' + c'' \cdot c')$		$\lambda_2\lambda_3 e_1 \cdot e_1 +$ $+ \lambda_3\lambda_1 e_2 \cdot e_2 + \lambda_1\lambda_2 e_1 \cdot e_1$		
α^{-1}	$\frac{1}{a}$	$\frac{\alpha + b - bc \cdot c}{a(a+b)}$		$\frac{1}{a} + \frac{b^2(c' \cdot c' + c'' \cdot c'') - b(a + bc'c'')(c' \cdot c'' + c'' \cdot c')}{a[(a + bc'c'')^2 - b^2]}$		$\frac{1}{\lambda_1} e_1 \cdot e_1 + \frac{1}{\lambda_2} e_2 \cdot e_2 + \frac{1}{\lambda} e_3 \cdot e_3$		
α_c	$3a$	$3a + b$		$3a + 2bc'c''$		$\lambda_1 + \lambda_2 + \lambda_3$		
$\bar{\alpha}_c$	$3a^2$	$3a^2 + 2ab$		$3a^2 + 4abc'c'' - b^2[c'c'']^2$		$\lambda_2\lambda_3 + \lambda_3\lambda_1 + \lambda_1\lambda_2$		
$\lvert\alpha\rvert$	a^3	$a^2(a+b)$		$a[(a + bc'c'')^2 - b^2]$		$\lambda_1\lambda_2\lambda_3$		

represented as a linear combination of α^2, α, and a scalar matrix. However, it often happens that a matrix satisfies an algebraic equation whose degree is lower than that of the characteristic equation. The minimal equation is the algebraic equation of lowest degree satisfied by matrix α, while the polynomial on the left in that equation is the minimal polynomial. Consider that polynomial for the uniaxial tensor of (13.30) as an example. It is readily seen that

$$(\alpha - a)^2 = b^2 (\mathbf{c} \cdot \mathbf{c})^2 = b^2 \mathbf{c} \cdot \mathbf{c} = b(\alpha - a),$$

and so the equation

$$\alpha^2 - (2a + b)\alpha + a(a + b) = (\alpha - a)(\alpha - a - b) = 0 \tag{13.36}$$

is the minimal equation for a uniaxial tensor. The minimal polynomial in this case is of second degree, whereas the characteristic polynomial for the matrix is $|\lambda - \alpha| = \lambda^3 - (3a+b)\lambda^2 + a(3a+2b)\lambda - a^2(a+b)$ [see (13.4) and Table II], which is of third degree, as for any three-dimensional matrix. The minimal polynomial is $\alpha - a = 0$ for the isotropic tensor of (13.29), i.e., is of first degree. The degree of the minimal polynomial can never exceed that of the characteristic equation; in addition, all the distinct eigenvalues of the matrix are obliged to be roots of the minimal polynomial.

Taking the trace of both parts of (13.34), we get the useful relation

$$\overline{\alpha}_t = \frac{1}{2}\left((\alpha_t)^2 - (\alpha^2)_t\right). \tag{13.37}$$

A real symmetric tensor α is used to construct the quadratic form

$$A = \mathbf{u}\alpha\mathbf{u} = u_i \alpha_{ik} u_k. \tag{13.38}$$

The numerical value of A may increase without limit if the length of **u** can do the same. We restrict the length of **u** by the condition

$$\mathbf{u}^2 = 1, \tag{13.39}$$

which represents a sphere of unit radius in the space of **u**. Then Weierstrass' theorem [11] indicates that A, being a continuous function of the variables (u_1, u_2, u_3) that vary within a bounded closed region, must attain its greatest and least values within that region. To find these extremal values of A we use Lagrange's method of finding a local extremum and construct the function $F = A - \lambda\,(\mathbf{u}^2 - 1)$, in which λ is an undetermined Lagrangian multiplier. We equate to zero the derivatives of F with respect to the u_l:

$$\frac{\partial F}{\partial u_l} = \frac{\partial}{\partial u_l}\left[u_i \alpha_{ik} u_k - \lambda\,(u_i^2 - 1)\right] =$$

$$= \delta_{il}\alpha_{ik}u_k + u_i\alpha_{ik}\delta_{kl} - \lambda 2u_i\delta_{il} = 2\,(\alpha_{lk}u_k - \lambda u_l) = 0. \qquad (13.40)$$

This calculation may also be done by the direct method (without resort to subscripts) if we use the derivative $\partial/\partial\mathbf{u}$ with respect to a vector $\mathbf{u}$, which for a scalar B is a vector whose components are equal to the derivatives of B with respect to the corresponding components of $\mathbf{u}$, i.e.,

$$\frac{\partial B}{\partial u} = \left(\frac{\partial B}{\partial u_k}\right). \qquad (13.41)$$

The rules for an ordinary derivative are applied when $\partial/\partial\mathbf{u}$ is formed for an expression in which a scalar product of $\mathbf{u}$ appears; since $\frac{\partial}{\partial x}x^2 = 2x$, then $\frac{\partial}{\partial u}u^2 = 2u$. Similarly, $\frac{\partial}{\partial u}u\alpha u = \alpha u + u\alpha = 2\alpha u$, if $\alpha = \tilde{\alpha}$ [see (10.17)]. Hence

$$\frac{\partial F}{\partial u} = 2\,(\alpha u - \lambda u) = 0, \qquad (13.42)$$

which is equivalent to (13.40); but (13.42) is simply (13.1), the equation for the eigenvalues and eigenvectors of tensor α. We multiply (13.42) scalarly by $\mathbf{u}$ and use (13.39) to get that

$$u\alpha u = \lambda. \qquad (13.43)$$

The extremal values of $A = \mathbf{u}\alpha\mathbf{u}$ subject to $\mathbf{u}^2 = 1$ are thus equal to the eigenvalues of the tensor λ, so $A = \mathbf{u}\alpha\mathbf{u}$ has its absolute minimum on the sphere $\mathbf{u}^2 = 1$ for the least (λ_1) of the three eigenvalues $\lambda_1 \leq \lambda_2 \leq \lambda_3$ of tensor α, while the largest (λ_3) of these gives the absolute maximum of the form under the same conditions. The corresponding eigenvectors $\mathbf{u}_1$ and $\mathbf{u}_3$ define the points on $\mathbf{u}^2 = 1$ at which A has those values.

As regards λ_2, it can be shown that it gives a minimum in A if $\mathbf{u}^2 = 1$, $\mathbf{u}\mathbf{u}_1 = 0$ or a maximum if $\mathbf{u}^2 = 1$, $\mathbf{u}\mathbf{u}_3 = 0$ [3, 11].

The physical properties described by a tensor are often characterized in terms of tensor surfaces. The equation of such a surface for a tensor α is

$$r\alpha r = 1. \tag{13.44}$$

This is a surface of second order. We introduce the unit vector $\mathbf{n}$ having the direction of the radius vector: $\mathbf{n}=\mathbf{r}/|\mathbf{r}|$, $\mathbf{n}^2=1$, in case (13.44) may be replaced by

$$r^2 = \frac{1}{n\alpha n}. \tag{13.45}$$

This gives the length of the radius vector as a function of direction; if $\mathbf{n}\alpha\mathbf{n}$ becomes zero for certain $\mathbf{n}$, then $|\mathbf{r}| = \infty$ for these directions, and the surface has infinitely remote points. But $\mathbf{n}\alpha\mathbf{n}$ never becomes zero if α is a positive definite tensor, so the surface of (13.44) is finite. An ellipsoid is the general surface of second degree having all of its points at a finite distance from the origin, and hence this is the tensor surface for such a tensor in the general case. Taking tensor α in the form of (13.20) and putting $\mathbf{r}\mathbf{e}_i = x_i$, we get from (13.44) that

$$\lambda_1 x_1^2 + \lambda_2 x_2^2 + \lambda_3 x_3^2 = 1. \tag{13.46}$$

Comparison with the canonical equation for an ellipsoid,

$$\frac{x_1^2}{a^2} + \frac{x_2^2}{b^2} + \frac{x_3^2}{c^2} = 1, \tag{13.47}$$

shows that the semiaxes equal the reciprocals of the square roots of the eigenvalues: $a=1/\sqrt{\lambda_1}$, $b=1/\sqrt{\lambda_2}$, $c=1/\sqrt{\lambda_3}$, while the directions of the principal axes of the ellipsoid coincide with those of the eigenvectors of tensor α. Two eigenvalues coincide for the uniaxial tensor of (13.30); in that case, the tensor surface is an ellipsoid of rotation, while the $\mathbf{c}$ appearing in (13.30) is the axis of the ellipsoid. Finally, the tensor surface is a sphere for the isotropic tensor of (13.29). Differential geometry shows [28] that a surface given in the form $f(x_i) = C$ has the vector $\mathbf{N}$ of the normal to it parallel to the vector for the gradient of f:

$$N_i = k\frac{\partial f}{\partial x_i}, \tag{13.48}$$

or in direct form

$$N = k\frac{\partial f}{\partial r}. \qquad (13.49)$$

Applying this to (13.44), we find that the vector for the normal to that surface is given very simply by

$$N \| \alpha r. \qquad (13.50)$$

14. Tensor Relations in a Plane

Covariant tensor relations analogous to those above may also be derived for two-dimensional space (a plane); some of these do not differ in form from those considered above, but many become much simpler. In this section I present some of the basic formulas of direct tensor calculus for a plane, which will be used later on (section 24 and elsewhere).

Firstly we note that nearly all of the results of section 11 may be transferred to the case of two dimensions; there are some obvious changes due to the reduced number of dimensions. For instance, (11.13) will become for a two-dimensional (planar) dyad

$$u \cdot v = (u_i v_k) = \begin{pmatrix} u_1 v_1 & u_1 v_2 \\ u_2 v_1 & u_2 v_2 \end{pmatrix}. \qquad (14.1)$$

A general tensor in a plane may be represented as a sum of two (not three) dyads:

$$\alpha = u_1 \cdot v_1 + u_2 \cdot v_2. \qquad (14.2)$$

This allows us to use nearly all the relationships of section 11 for the two-dimensional case.

On the other hand, the relationships of section 12 are much altered. Unit antisymmetric tensor ε_{ik} in the plane is of second rank, in accordance with the number of dimensions, and is defined by

$$\varepsilon_{12} = 1, \quad \varepsilon_{ik} = -\varepsilon_{ki}, \quad i,\ k = 1,2. \qquad (14.3)$$

It resembles any second-rank tensor, and differs from ε_{ikl}, in that it may be represented via a square matrix:

$$\varepsilon = \begin{pmatrix} 0 & 1 \\ -1 & 0 \end{pmatrix}. \qquad (14.4)$$

The formula corresponding to (12.3) here becomes

$$\varepsilon_{ik}\varepsilon_{ln} = \begin{vmatrix} \delta_{il} & \delta_{in} \\ \delta_{kl} & \delta_{kn} \end{vmatrix} = \delta_{il}\delta_{kn} - \delta_{in}\delta_{kl}, \qquad (14.5)$$

in which δ_{ik} is the two-dimensional Kronecker symbol. In all the relations in this section we assume summation with respect to the repeated subscripts, but within the limits 1 to 2; hence $\delta_{kk} = 2$, although, as before $\delta_{ik}\delta_{kl} = \delta_{il}$. Then, convoluting the subscripts in (14.5) either partly or completely, we have

$$\varepsilon_{ik}\varepsilon_{lk} = \delta_{il}, \quad \varepsilon_{ik}\varepsilon_{ik} = \varepsilon_{ik}^2 = 2. \qquad (14.6)$$

These are the two-dimensional analogs of (12.4)-(12.6). It is readily seen that, by analogy with (12.7),

$$\varepsilon_{ik}\alpha_{1l}\alpha_{2k} = \alpha_{11}\alpha_{22} - \alpha_{12}\alpha_{21} = \begin{vmatrix} \alpha_{11} & \alpha_{12} \\ \alpha_{21} & \alpha_{22} \end{vmatrix} = |\alpha| \qquad (14.7)$$

and from (14.5) that

$$\varepsilon_{ln}\alpha_{il}\alpha_{kn} = |\alpha|\varepsilon_{ik}, \qquad (14.8)$$

$$|\alpha| = \frac{1}{2}\varepsilon_{ik}\varepsilon_{ln}\alpha_{il}\alpha_{kn}. \qquad (14.9)$$

We multiply (14.8) by ε_{is} and use (14.5) and (14.6) to get

$$\varepsilon_{is}\varepsilon_{ln}\alpha_{il}\alpha_{kn} = |\alpha|\delta_{ks}. \qquad (14.10)$$

By analogy with (12.13) we introduce the matrix

$$\bar{\alpha}_{ns} = \varepsilon_{is}\varepsilon_{ln}\alpha_{il}, \qquad (14.11)$$

which from (14.10) satisfies (12.14)-(12.16):

$$\alpha_{kn}\bar{\alpha}_{ns} = |\alpha|\delta_{ks}, \quad \alpha\bar{\alpha} = |\alpha|, \quad \bar{\alpha} = |\alpha|\alpha^{-1}. \qquad (14.12)$$

the last of these being meaningful for $|\alpha| \neq 0$. The $\bar{\alpha}$ tensor of (14.11) is termed inverse (under multiplication) with respect to the two-dimensional tensor α, by analogy with (12.13).*

If k is a number, relationships true for two-dimensional matrices are

$$|k\alpha| = k^2|\alpha|, \quad \overline{k\alpha} = k\bar{\alpha}, \quad |\bar{\alpha}| = |\alpha|, \tag{14.13}$$

$$\overline{\alpha^{-1}} = (\bar{\alpha})^{-1} = \frac{\alpha}{|\alpha|}, \quad \bar{\bar{\alpha}} = \alpha, \quad \bar{\tilde{\alpha}} = \tilde{\bar{\alpha}}, \quad \overline{\alpha\beta} = \bar{\beta}\bar{\alpha}, \tag{14.14}$$

which are derived as for (12.17)-(12.23). We use (14.5) in (14.11) to get

$$\bar{\alpha}_{ns} = (\delta_{il}\delta_{sn} - \delta_{in}\delta_{sl})\alpha_{il} = \alpha_t\delta_{ns} - \alpha_{ns}, \tag{14.15}$$

or

$$\bar{\alpha} = \alpha_t - \alpha. \tag{14.16}$$

This implies that

$$\bar{\alpha}_t = 2\alpha_t - \alpha_t = \alpha_t. \tag{14.17}$$

We multiply (14.16) by α and use (14.12) to get

$$|\alpha| = \alpha\bar{\alpha} = \bar{\alpha}\alpha = \alpha_t\alpha - \alpha^2. \tag{14.18}$$

Taking the trace, we get

$$|\alpha| = \frac{1}{2}((\alpha_t)^2 - (\alpha^2)_t). \tag{14.19}$$

Then $|\alpha|$ in two-dimensional space may be represented by formula (13.37), which expresses $\bar{\alpha}_t$ in three-dimensional space.

We apply (14.16) to the sum of the two-dimensional matrices

---

* The concept of inverse matrix is applicable for a matrix of any order: matrix $\bar{A}$ inverse to A has as its elements $\bar{A}_{ik}$ the algebraic complements of the corresponding elements of the transposed matrix $\tilde{A}$ [16].

α and β:

$$\overline{\alpha+\beta}=\bar{\alpha}+\bar{\beta}=\alpha_t+\beta_t-(\alpha+\beta). \tag{14.20}$$

This simple expression replaces the cumbrous (12.64) in the two-dimensional case. If β is a scalar matrix (number) k, we have

$$\overline{\alpha+k}=\alpha_t-\alpha+k=\bar{\alpha}+k. \tag{14.21}$$

We multiply (14.20) by $\alpha+\beta$ and use (14.16) and (14.18) to get

$$|\alpha+\beta|=(\alpha_t+\beta_t)(\alpha+\beta)-(\alpha+\beta)^2=|\alpha|+|\beta|+\bar{\alpha}\beta+\bar{\beta}\alpha. \tag{14.22}$$

Hence $\bar{\alpha}\beta+\bar{\beta}\alpha=|\alpha+\beta|-|\alpha|-|\beta|$ is always a scalar in two dimensions. Taking the trace of the latter matrix equation and using the result of (14.16),

$$(\bar{\alpha}\beta)_t=(\bar{\beta}\alpha)_t=\alpha_t\beta_t-(\alpha\beta)_t, \tag{14.23}$$

we get that

$$|\alpha+\beta|=|\alpha|+|\beta|+(\bar{\alpha}\beta)_t=|\alpha|+|\beta|+\alpha_t\beta_t-(\alpha\beta)_t. \tag{14.24}$$

This differs from (12.25) only in the absence of the term $(\alpha\bar{\beta})_t$.

The characteristic equation of a two-dimensional matrix α is given by (14.24) as having the following form [compare (13.4)]:

$$|\lambda-\alpha|=\lambda^2-\alpha_t\lambda+|\alpha|=0. \tag{14.25}$$

The Hamilton-Cayley theorem (section 13) indicates that replacement of λ by α should give a matrix identity; (14.18) shows that this is so. From (14.25) we have [see (14.19)] the eigenvalues of an arbitrary two-dimensional tensor as

$$\lambda=\frac{\alpha_t\pm\sqrt{(\alpha)_t^2-4|\alpha|}}{2}=\frac{\alpha_t\pm\sqrt{2(\alpha^2)_t-(\alpha_t)^2}}{2}. \tag{14.26}$$

From (14.16) and (14.18) we have for the dyad tensor $\mathbf{a}\cdot\mathbf{b}$ that

$$\overline{\boldsymbol{a}\cdot\boldsymbol{b}}=\boldsymbol{ab}-\boldsymbol{a}\cdot\boldsymbol{b}, \quad |\boldsymbol{a}\cdot\boldsymbol{b}|=0 \tag{14.27}$$

and from (14.24) we have for the general case of any tensor $\alpha=\mathbf{a}_1\cdot\mathbf{b}_1+\mathbf{a}_2\cdot\mathbf{b}_2$ that

$$|\boldsymbol{a}_1 \cdot \boldsymbol{b}_1 + \boldsymbol{a}_2 \cdot \boldsymbol{b}_2| = (\overline{\boldsymbol{a}_1 \cdot \boldsymbol{b}_1}\ \boldsymbol{a}_2 \cdot \boldsymbol{b}_2)_t = \boldsymbol{a}_1\boldsymbol{b}_1 \cdot \boldsymbol{a}_2\boldsymbol{b}_2 - \boldsymbol{a}_1\boldsymbol{b}_2 \cdot \boldsymbol{a}_2\boldsymbol{b}_1. \tag{14.28}$$

By analogy with (13.20), a two-dimensional real symmetric tensor may be put as

$$\alpha = \lambda_1 \boldsymbol{e}_1 \cdot \boldsymbol{e}_1 + \lambda_2 \boldsymbol{e}_2 \cdot \boldsymbol{e}_2, \quad \boldsymbol{e}_i \boldsymbol{e}_k = \delta_{ik}, \tag{14.29}$$

in which λ_1 and λ_2 are eigenvalues and $\mathbf{e}_1$ and $\mathbf{e}_2$ are the corresponding inversely orthogonal unit eigenvectors. From (12.47) we have

$$\boldsymbol{e}_1 \cdot \boldsymbol{e}_1 + \boldsymbol{e}_2 \cdot \boldsymbol{e}_2 = 1, \tag{14.30}$$

so from (14.29) we get $\alpha = \lambda_1 + (\lambda_2 - \lambda_1)\mathbf{e}_2 \cdot \mathbf{e}_2 = \lambda_2 + (\lambda_1 - \lambda_2)\mathbf{e}_1 \cdot \mathbf{e}_1$. Hence any symmetrical real tensor α in a plane may be put in the form of (13.30), which is not so for the three-dimensional case:

$$\alpha = a + b\boldsymbol{c} \cdot \boldsymbol{c}, \quad \boldsymbol{c}^2 = 1. \tag{14.31}$$

It is readily seen that

$$\alpha_t = \overline{\alpha}_t = 2a + b, \quad |\alpha| = a(a+b); \tag{14.32}$$

$$\overline{\alpha} = a + b - b\boldsymbol{c} \cdot \boldsymbol{c}, \quad \alpha^{-1} = \frac{1}{a} - \frac{b\boldsymbol{c} \cdot \boldsymbol{c}}{a(a+b)}. \tag{14.33}$$

The eigenvectors and eigenvalues are determined (Table II) as for the three-dimensional case.

The equation

$$\boldsymbol{r}\alpha\boldsymbol{r} = 1 \tag{14.34}$$

defines the tensor curve on a plane for a symmetric real tensor α. This curve is an ellipse for a positive definite tensor α, the squares of the semiaxes equalling the reciprocals of the eigenvalues of α [compare (13.46) and (13.47)]:

$$a^2 = \frac{1}{\lambda_1}, \quad b^2 = \frac{1}{\lambda_2}. \tag{14.35}$$

The geometrical normal to the curve of (14.34) is given by (13.50) as $\mathbf{N} \| \alpha\mathbf{r}$.

Consider the two-dimensional equation

$$\alpha u = 0 \tag{14.36}$$

for a symmetric tensor α. There is a nontrivial solution for **u** only [see (14.18)] subject to the condition

$$|\alpha| = \alpha\bar{\alpha} = \alpha(\alpha_t - \alpha) = 0. \tag{14.37}$$

Hence the columns of tensor $\bar{\alpha} = \alpha_t - \alpha$ must be proportional to **u** (see the analogous argument in section 13). We assume here that $\alpha_t \neq 0$, otherwise (14.37) would give $\alpha^2 = 0$, which implies $\alpha = 0$. Hence for $|\alpha| = 0$ and $\alpha_t \neq 0$ there exists only one normalized solution to (14.36) for **u**, and $\bar{\alpha}$ has the form $\bar{\alpha} = C\mathbf{u} \cdot \mathbf{u}$, by virtue of its symmetry. Assuming $\mathbf{u}^2 = 1$ and taking the trace, we get the $C = \bar{\alpha}_t = \alpha_t$ of (14.17). To sum up, we may say that, subject to the conditions

$$\alpha u = 0, \quad |\alpha| = 0, \quad \alpha_t \neq 0, \quad \tilde{\alpha} = \alpha \tag{14.38}$$

the unit vector **u** is defined by

$$u \cdot u = \frac{\bar{\alpha}}{\alpha_t} = 1 - \frac{\alpha}{\alpha_t}. \tag{14.39}$$

In this case matrix α/α_t is projective, i.e., satisfies the condition*

$$\left(\frac{\alpha}{\alpha_t}\right)^2 = \frac{\alpha}{\alpha_t}, \tag{14.40}$$

whose correctness follows directly from (14.18) for $|\alpha| = 0$. Consider a unit vector **u'** perpendicular to **u** in the plane where α is given; (14.30) shows that then we must have $\mathbf{u} \cdot \mathbf{u} + \mathbf{u}' \cdot \mathbf{u}' = 1$. Comparison with (14.39) gives

$$\frac{\alpha}{\alpha_t} = u' \cdot u', \quad u'u = 0, \quad u'^2 = 1. \tag{14.41}$$

Thus (14.39) actually gives both eigenvectors of a two-dimensional tensor α ($|\alpha| = 0$).

---

* A square matrix β of any order is termed projective if $\beta^2 = \beta$ [8].

Chapter 3

General Laws of Propagation of Elastic Waves in Crystals

15. Plane Waves and Christoffel's Equation

The laws of propagation of elastic waves in crystals follow from the general equations of motion derived in section 3 for an elastically deformed medium. Here the force of gravity may be neglected, so we may use (3.10)

$$\rho \ddot{u}_i = \frac{\partial \sigma_{ik}}{\partial x_k}. \tag{15.1}$$

Here the components of the displacement vector appear on the left, while those of the deformation tensor appear on the right. To obtain an equation containing only one unknown, namely the displacement vector, we use (1.11) and (4.3) to get

$$\sigma_{ik} = c_{iklm}\gamma_{lm}, \tag{15.2}$$

$$\gamma_{lm} = \frac{1}{2}\left(\frac{\partial u_m}{\partial x_l} + \frac{\partial u_l}{\partial x_m}\right). \tag{15.3}$$

These give

$$\rho \ddot{u}_i = c_{iklm}\frac{\partial \gamma_{lm}}{\partial x_k} = \frac{1}{2}c_{iklm}\left(\frac{\partial^2 u_m}{\partial x_k \partial x_l} + \frac{\partial^2 u_l}{\partial x_k \partial x_m}\right).$$

We put m for l and l for m in the dummy subscripts in the expression for $c_{iklm}(\partial^2 u_l/\partial x_k \partial x_m)$ to get $c_{ikml}(\partial^2 u_m/\partial x_k \partial x_l)$. The symmetry of the c_{iklm} tensor with respect to the second pair of subscripts means that the two terms on the right are equal, so the equations of motion are

$$\rho \ddot{u}_i = c_{ijlm} \frac{\partial^2 u_m}{\partial x_j \partial x_l}. \tag{15.4}$$

These are linear homogeneous second-order differential equations in the displacement vector.

We envisage only plane monochromatic elastic waves, for which the displacement vector may be put as

$$\boldsymbol{u} = \boldsymbol{u}^0 e^{i(\boldsymbol{kr} - \omega t)}. \tag{15.5}$$

Here $\mathbf{u}^0$ is a constant vector (independent of coordinates and time), termed the vector amplitude. The exponential factor is $e^{i\varphi}$, with

$$\varphi = \boldsymbol{kr} - \omega t, \tag{15.6}$$

and is termed the phase factor, φ being the phase variable. The basic parameter of the wave is the circular frequency $\omega = 2\pi\nu$, in which ν is the frequency (number of complete oscillations of the displacement vector per second); however, ω is often called simply the frequency. The period T (length of one oscillation) is related to ν by $T = 1/\nu$.

The wave of (15.5) is plane because φ has a linear dependence on the coordinates. In general, the shape of the wave is determined by that of the surfaces of equal phase. The condition for constancy in the φ of (15.6) is

$$\varphi = \boldsymbol{kr} - \omega t = C = \text{const}. \tag{15.7}$$

We put the constant vector $\mathbf{k}$ in the form

$$\boldsymbol{k} = k\boldsymbol{n}, \quad \boldsymbol{n}^2 = 1, \quad k = |\boldsymbol{k}|, \tag{15.8}$$

and get from (15.7)

$$\boldsymbol{nr} = \frac{C + \omega t}{k} = \zeta. \tag{15.9}$$

This implies that the equation for a surface of equal phase at any given instant is a plane perpendicular to the unit vector $\mathbf{n}$, ζ being the distance of this plane from the origin (Fig. 2). This phase plane moves parallel to itself, since $\mathbf{n}$ is a specified vector and ζ increases linearly with time. The speed v of this plane is given by

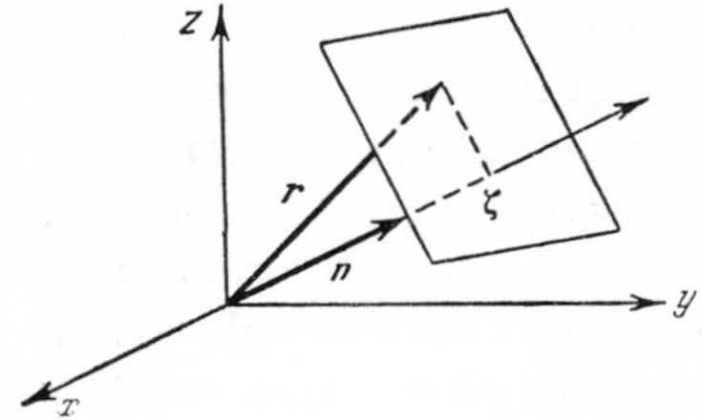

Fig. 2

$$v = \frac{d\zeta}{dt} = \frac{\omega}{k}, \qquad (15.10)$$

which is termed the phase velocity of the wave. The vector

$$\boldsymbol{v} = v\boldsymbol{n} = \frac{\omega}{k}\boldsymbol{n} \qquad (15.11)$$

is the phase-velocity vector. Vector **n** is termed the wave or phase normal, while vector **k** of (15.8) is termed the wave vector, whose length is given by (15.10) as

$$k = \frac{\omega}{v}. \qquad (15.12)$$

The distance traveled by the phase plane in one period is the wavelength λ, being the product of the phase velocity and the period:

$$\lambda = vT = \frac{v}{\nu} = \frac{2\pi v}{\omega}. \qquad (15.13)$$

Then from (15.12) we have

$$k = \frac{2\pi}{\lambda}. \qquad (15.14)$$

Expression (15.5) gives a complex displacement, although an actual displacement must be represented by a real quantity. This difficulty is resolved by noting that (15.4) is a linear homogeneous equation, while all operations on the unknown vector **u** contain no imaginary expressions. Hence the vector of (15.5) will satisfy (15.4) only if its real and imaginary parts taken separately also do. If $\mathbf{u}^0$ is real, then

$$\boldsymbol{u} = \boldsymbol{u}^0(\cos\varphi + i\sin\varphi) = \boldsymbol{u}' + i\boldsymbol{u}'', \qquad (15.15)$$

and $\mathbf{u}' = \mathbf{u}^0\cos(\mathbf{kr} - \omega t)$ and $\mathbf{u}'' = \mathbf{u}^0\sin(\mathbf{kr} - \omega t)$ are real solutions of (15.4) in the form of plane monochromatic waves. Hence we may always take either of them, e.g., u', as our actual solution. If we have $\mathbf{u}^0 = \mathbf{u}_1^0 + i\mathbf{u}_2^0$, the solution of (15.5) may also be expanded as real and imaginary parts, each of which will be a solution of (15.4), although the expansion is somewhat more complicated than (15.15).

Hence the complex solution of (15.5) always gives real solutions; but it is very much more convenient to operate with the complex exponential of (15.5) than with the trigonometric real functions $\mathbf{u}'$ and $\mathbf{u}''$ of (15.15), while the complex character of (15.15) does not prevent us from deriving from it all the properties of actual waves. We need to consider the real solution instead of (15.5) only when the quantity of interest is a nonlinear function of $\mathbf{u}$. This need arises primarily in respect of energy relationships, because the elastic energy of (6.4) is a quadratic function of the deformation tensor and hence of the displacement vector.

In most treatments in this book we may treat $\mathbf{u}^0$ as real; there are a few cases when we must deal with a complex $\mathbf{u}^0$, which are encountered in sections 22 and 23.

Of course, (15.5) does not satisfy (15.4) for any values of $\mathbf{u}^0$, $\mathbf{k}$, and ω; substituting (15.5) into (15.4) and remembering that $\frac{\partial}{\partial x_j} e^{i\mathbf{kr}} = \frac{\partial}{\partial x_j} e^{ik_l x_l} = ik_j e^{i\mathbf{kr}}$, we get

$$\rho\omega^2 u_i = c_{ijlm} k_j k_l u_m. \tag{15.16}$$

In place of c_{ijlm} we introduce the tensor

$$\lambda_{ijlm} = \frac{1}{\rho} c_{ijlm}, \tag{15.17}$$

which is termed the reduced elastic-modulus tensor. Further, we use (15.8) to replace k_j by kn_j and employ (15.10) to rewrite (15.16) in the form

$$\lambda_{ijlm} n_j n_l u_m - v^2 u_i = 0 \tag{15.18}$$

or

$$(\lambda_{ijlm} n_j n_l - v^2 \delta_{im}) u_m = 0. \tag{15.19}$$

Consider the second-rank tensor

$$\Lambda = \Lambda^n = (\Lambda_{im}) = (\lambda_{ijlm} n_j n_l), \tag{15.20}$$

which allows us to put (15.19) in the direct form

$$(\Lambda - \lambda)\,\boldsymbol{u} = 0, \quad \lambda = v^2. \tag{15.21}$$

This shows that **u** is an eigenvector of tensor Λ, while v^2 is an eigenvalue of this (see section 13). Hence v^2 is a root of the characteristic equation

$$|\Lambda - v^2| = 0. \tag{15.22}$$

The problem thus reduces to that of solving for the eigenvectors and eigenvalues of the tensor Λ of (15.19) or (15.21). The vector equation (Christoffel's equation) is basic to the entire theory of elastic waves in crystals; (15.21) resembles any linear homogeneous equation in defining **u** only apart from an arbitrary factor. That is, solving for a given Λ, we can find only the directions of the displacement vector and the corresponding phase velocities of the waves that can propagate in the crystal. Here we should bear in mind that (15.20) indicates that to each direction of the wave (phase) normal **n** there corresponds a particular tensor Λ. The usual formulation of elastic-wave theory is that of finding **u** and **v** for all plane waves having the same (arbitrary) direction for **n** given the elastic properties (λ_{ijlm}) of the crystal.

16. General Properties of the Λ Tensor and Forms of Plane Elastic Waves in Crystals

The Λ tensor is symmetric for any direction of **n**, for the symmetry properties of the λ_{ijlm} tensor of (6.2) enable us to permute the subscripts and alter the notation for the dummy ones to get

$$\Lambda_{im} = \lambda_{ijlm} n_j n_l = \lambda_{lmij} n_j n_l = \lambda_{mlji} n_j n_l = \lambda_{mjli} n_l n_j = \Lambda_{mi},$$

or

$$\tilde{\Lambda} = \Lambda. \tag{16.1}$$

Also, Λ is a positive definite tensor, which may be demonstrated by noting that this is so by definition for Λ if for any vector $\mathbf{u} \neq 0$

$$\boldsymbol{u}\Lambda\boldsymbol{u} > 0. \tag{16.2}$$

We write the left-hand side of this in terms of subscripts and use (15.20) to get

$$u_i n_j \lambda_{ijlm} n_l u_m > 0. \tag{16.3}$$

But in section 6 we showed that the elastic energy Φ is always positive, so (6.4) gives us that $\gamma_{ij} c_{ijlm} \gamma_{lm} > 0$ for any deformation, i.e., for any symmetric tensor γ_{ij}. Let $\tau = \mathbf{u} \cdot \mathbf{n} = \tau^s + \tau^a$, in which $\tau^s = (1/2)(\mathbf{u} \cdot \mathbf{n} + \mathbf{n} \cdot \mathbf{u})$ is the symmetric part of dyad τ and $\tau^a = (1/2)(\mathbf{u} \cdot \mathbf{n} - \mathbf{n} \cdot \mathbf{u})$ is the antisymmetric part. The arguments of section 10 that gave (10.56) here show that $\overset{a}{\tau_{ij}} \lambda_{ijlm} = 0$, since $\overset{a}{\tau_{ij}} = -\overset{a}{\tau_{ji}}$ and λ_{jilm}; hence (16.3) may be put as

$$\tau_{ij} \lambda_{ijlm} \tau_{lm} = \tau^s_{ij} \lambda_{ijlm} \tau^s_{lm} > 0, \tag{16.4}$$

whence it is clear that it is a particular case of the general condition that the elastic energy be positive, where the arbitrary symmetric tensor γ_{ij} equals τ^s_{ij}. Hence the condition on the elastic energy guarantees that Λ will be positively definite for any direction of $\mathbf{n}$; but this result implies that all the eigenvalues must be positive, for (see section 13) the eigenvalues are the diagonal elements of a tensor in the principal system of axes, and the condition above (section 6) implies that these must be positive. Hence, if some eigenvalue of Λ were negative, it could not correspond to a wave. The above argument shows that this is impossible.

Our real symmetric tensor Λ in the general case has three distinct eigenvalues $\lambda_0 = v_0^2$, $\lambda_1 = v_1^2$, $\lambda_2 = v_2^2$, each corresponding to an eigenvector defining the direction of displacement in the wave. Hence, for any given direction of the wave normal there are, in general, three waves with different phase velocities; these three waves with a common wave normal are termed isonormal, and their displacement vectors, as for the vectors of any real symmetric tensor (section 13), are mutually perpendicular. However, there can be certain special directions of $\mathbf{n}$ in which Λ is uniaxial (section 13), i.e., two eigenvalues coincide. Then two of the three waves have the same phase velocity, and their displacement vectors may take any direction in a plane perpendicular to the displacement vector of the third wave (Table II). Tensor Λ is uniaxial for any direction of $\mathbf{n}$ in an isotropic medium, for from (8.1) and (15.17)

$$\lambda_{iklm} = a_1 \delta_{ik} \delta_{lm} + a_2 (\delta_{il} \delta_{km} + \delta_{im} \delta_{kl}),$$

whence we have

$$\Lambda_{kl} = \lambda_{iklm} n_i n_m = a_2 \delta_{kl} + (a_1 + a_2) n_k n_l, \tag{16.5}$$

or

$$\Lambda = a_2 + (a_1 + a_2) \boldsymbol{n} \cdot \boldsymbol{n}. \tag{16.6}$$

We have seen (Table II) that the axis of Λ is **n**, which is natural, since **n** represents the unique differentiated direction for an isotropic medium. Table II shows that one of the eigenvectors of the uniaxial Λ of (16.6) coincides in direction with **n**, so the displacement vector **u** for the corresponding wave will vibrate along the wave normal. This is called a *longitudinal wave*. The other two linearly independent displacement vectors [the eigenvectors of (16.6)] may be chosen arbitrarily in a plane perpendicular to **n**; they are the displacements of the *transverse waves* with equal phase velocities.

A crystal differs from an isotropic medium in that one of the three waves with the common wave normal **n** is neither purely longitudinal nor purely transverse. The **u** for a purely longitudinal wave must satisfy

$$[\boldsymbol{un}] = 0, \tag{16.7}$$

while that for a purely transverse one must satisfy

$$\boldsymbol{un} = 0. \tag{16.8}$$

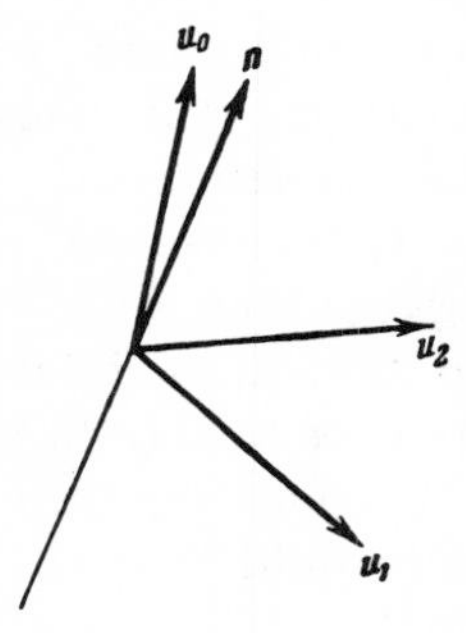

Fig. 3

A crystal has [**un**] ≠ 0 and **un** ≠ 0 for nearly all directions of **n**, but in any given case one of the three waves will have its displacement vector at the smaller angle to **n** than do the other two. Such a wave is called *quasilongitudinal*, while the others are called *quasitransverse*. Figure 3 shows an example of the quasilongitudinal ($\mathbf{u}_0$) and quasitransverse ($\mathbf{u}_1$ and $\mathbf{u}_2$) vectors corresponding to a given **n** in a crystal.

The failure of waves in a crystal to be purely longitudinal or purely transverse is one of the principal effects of the anisotropy in the elasticity.

In general, Λ has three distinct eigenvalues for a crystal, so there are three distinct eigenvalues for a crystal, so there are three eigenvectors with quite definite directions (section 13), which means that the displacement vectors for almost any direction of the wave normal will remain each parallel to a definite straight line while varying in magnitude in accordance with (15.5). We may say that plane elastic waves in a crystal are, as a rule, linearly polarized, the only exceptions being for those directions of **n** for which Λ becomes uniaxial, along which the linearly polarized waves may be accompanied by a wave of a different polarization. This topic is considered in more detail in section 22.

Consider now the properties of the waves propagating along directions with a definite relation to the symmetry elements. It is readily seen that an L^k axis of any order $(k \neq 1)$ is a *longitudinal normal* in any crystal, such a normal being a direction along which a purely longitudinal wave may propagate with that direction as its wave normal, as well as two purely transverse waves. Consider first an L^2 axis and assume that the **n** parallel to L^2 corresponds to an elastic wave whose **u** forms an angle with L^2 other than 0 or $\pi/2$ (Fig. 4). Now $\mathbf{n} \parallel L^2$, so differentiation of this direction for the wave normal does not alter the symmetry of the crystal-wave system relative to that of the crystal itself; hence the system should be unaltered by rotation through π around L^2. But this is possible only if **u** is either parallel or perpendicular to **n** (Fig. 4), since otherwise the displacement vector would be transferred to a different position **u'**, i.e., the crystal-wave system would not coincide with itself after the rotation. This means that only one purely longitudinal wave and two purely transverse waves can propagate along a twofold axis. This is true also if the wave normal is perpendicular to a symmetry plane, because the latter is equivalent to a twofold axis as regards its symmetry properties (section 8). These arguments may be transferred almost unaltered to

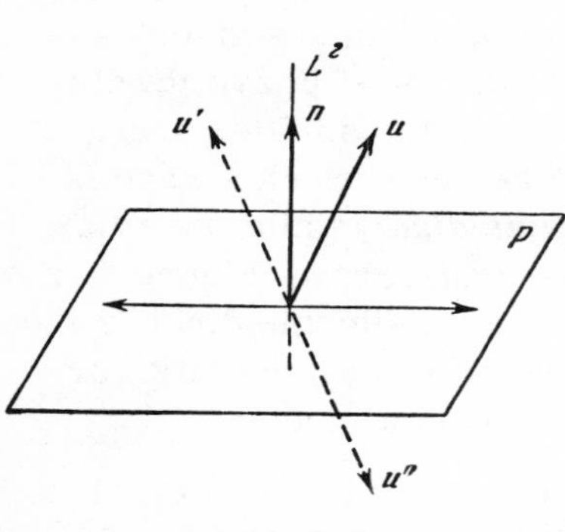

Fig. 4

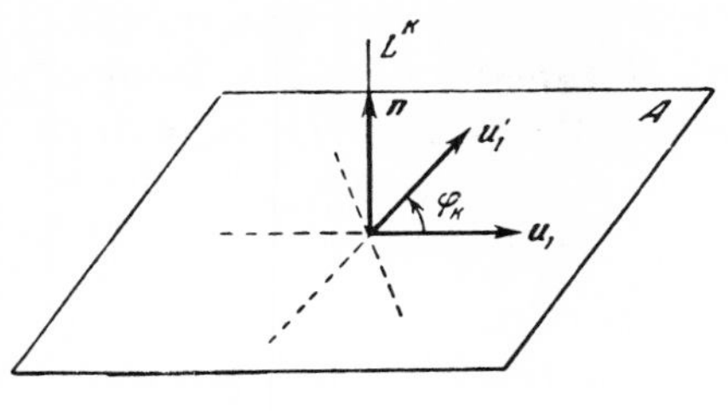

Fig. 5

any higher axis, along which purely longitudinal waves must propagate; but there are certain differences as regards the transverse waves. Consider a displacement vector $\mathbf{u}_1$ perpendicular to $\mathbf{n}$, which is parallel to an L^k axis ($k > 2$); then (Fig. 5) a rotation around L^k through an angle $\varphi_k = (2\pi/k) < \pi$ transfers $\mathbf{u}_1$ to $\mathbf{u}_1'$. The crystal-wave system is unaltered, so $\mathbf{u}_1'$ must also be an eigenvector of Λ having the same eigenvalue λ, i.e., from (15.21) we must have

$$\Lambda \boldsymbol{u}_1 = \lambda \boldsymbol{u}_1, \quad \Lambda \boldsymbol{u}_1' = \lambda \boldsymbol{u}_1'. \tag{16.9}$$

But φ_k differs from π, so $\mathbf{u}_1$ and $\mathbf{u}_1'$ are linearly independent, and hence any vector $\mathbf{u}$ lying in plane A may be put as a linear combination:

$$\boldsymbol{u} = \xi \boldsymbol{u}_1 + \eta \boldsymbol{u}_1'. \tag{16.10}$$

Then from (16.9)

$$\Lambda \boldsymbol{u} = \xi \Lambda \boldsymbol{u}_1 + \eta \Lambda \boldsymbol{u}_1' = \xi \lambda \boldsymbol{u}_1 + \eta \lambda \boldsymbol{u}_1' = \lambda \boldsymbol{u}, \tag{16.11}$$

i.e., any vector perpendicular to $\mathbf{n} \| L^k$ will be an eigenvalue of Λ for the given eigenvalue λ. Physically this means that all transverse waves propagating along such a symmetry axis will have the same velocity, and that their displacement vectors may be arbitrarily oriented in a plane perpendicular to L^k.

Consider also waves whose wave normal lies in a plane perpendicular to a symmetry axis of even order, or in a symmetry plane; we take the latter case for convenience (Fig. 4). If the wave normal lies in plane P, the crystal-wave system should be unaffected by reflection in that plane. This means that the displacement vector of each of the three isonormal waves in this case must lie either in plane P or normal to it, otherwise an inclined vector $\mathbf{u}$ would be transferred to another position $\mathbf{u}''$ (Fig. 4), and the symmetry would be violated.

The displacement vectors are mutually orthogonal, so one will be perpendicular to P, the other two lying in P. The symmetry properties in no way require that any of the displacement vectors should coincide with **n**; so in this case, generally speaking, there will not be a purely longitudinal wave, but one of the waves will be purely transverse.

A twofold axis is equivalent to a symmetry plane, and any even axis is simultaneously a twofold one, so this property occurs in any even axis.

Hence, one of the three isonormal waves will be purely transverse and will have its displacement perpendicular to the plane if the wave normal lies in a symmetry plane or in a plane perpendicular to a symmetry axis of even order.

In what follows we shall use a zero subscript to denote quantities relating to a quasilongitudinal wave (displacement $\mathbf{u}_0$, velocity $\mathbf{v}_0$), while subscripts 1 and 2 refer to the quasitransverse waves, 1 denoting the purely transverse wave, if there is one.

17. Special Directions for Elastic Waves in Crystals

The elastic waves in an isotropic medium have the following properties no matter what the direction of the wave normal:

1. One of the three isonormal waves is always purely longitudinal; we term longitudinal normals the directions of **n** along which such a wave can propagate. Any wave normal is therefore such a normal for an isotropic medium.

2. Purely transverse waves are associated with any wave normal.

3. Two waves with the same velocity propagate along any wave normal; any such normal is therefore an *acoustic axis*, which is defined as a direction of the wave normal such that the phase velocities of two isonormal waves coincide [22].

In general, none of these features occurs in a crystal; none of the orthonormal waves is purely longitudinal or purely transverse, and all three waves have different velocities. However, there are always directions of **n** such as to give the waves some of the above properties typical of isotropic media. For example, in section 16 we saw that there is always one purely transverse wave among the three propagating along a normal lying in a symmetry plane or in a plane perpendicular to a symmetry axis of even order. A symmetry axis of order higher than two is always an acoustic axis.

A special direction is one for which we find at least one of the properties 1, 2, and 3 [35]. These directions are important because the basic equation (15.21) becomes very much simpler for these and may be solved without difficulty. These directions thus play a special part in the theory and in practical applications, and hence their name.

We have seen that all directions parallel or perpendicular to symmetry planes or axes are special directions, except for directions perpendicular to threefold (odd) axes.

However, it should not be thought that all special directions are necessarily related to the symmetry elements of crystals in the above sense, for we have seen that even a triclinic crystal entirely lacking symmetry elements significant to the elastic properties yet can have longitudinal wave normals (section 18).

Here we consider the general conditions necessary and sufficient for a given direction of **n** to be a special direction. First we note that features 1–3 are not mutually independent; the presence of 1 or 3 necessarily implies the presence of 2. This is obvious for 1, since the other two waves must be purely transverse if one is purely longitudinal. In the case of 3, the Λ tensor must be uniaxial (section 13). Experiment shows that the velocity of the quasilongitudinal wave is always greater than those of the quasitransverse ones, so only the latter two can coincide. Hence $\mathbf{u}_0$ will be the axis of Λ, while any vector in a plane perpendicular to this may be considered as the displacement of a quasitransverse wave to this may be considered as the displacement of a quasitransverse wave (Fig. 6).

In plane M perpendicular to $\mathbf{u}_0$ there is always a direction perpendicular to **n**, so 3 implies the presence of a purely transverse wave, i.e., property 2. Figure 6 shows that the transverse displacement $\mathbf{u}_1$ is defined as to direction by

$$\boldsymbol{u}_1 \,\|\, [\boldsymbol{n}\boldsymbol{u}_0]. \tag{17.1}$$

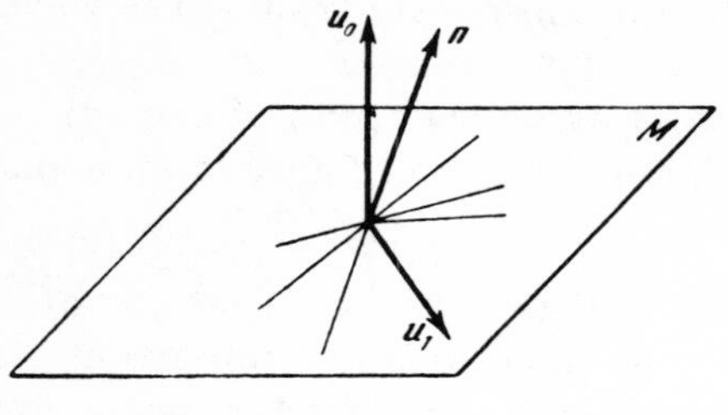

Fig. 6

We may say that property 2 occurs for any special direction; the presence of a purely transverse wave is a general condition for any special direction, so we must consider the conditions that the direction of the wave normal must satisfy in order that a purely transverse elastic wave can propagate along it. It will be clear from the above that this will be a condition for all special directions in a crystal generally.

From (15.21) we have

$$\Lambda \boldsymbol{u} = \lambda \boldsymbol{u}, \quad \lambda = v^2. \tag{17.2}$$

Let the wave normal $\mathbf{n}$ in $\Lambda = \Lambda^{\mathrm{n}}$ be such that

$$\boldsymbol{un} = 0, \tag{17.3}$$

i.e., the displacement vector $\mathbf{u} = \mathbf{u}_1$ corresponds to a purely transverse wave. This condition gives us, by multiplying (17.2) by $\mathbf{n}$, that

$$\boldsymbol{n}\Lambda\boldsymbol{u} = 0. \tag{17.4}$$

If now we multiply (17.2) by the vector $\Lambda\mathbf{n} = \mathbf{n}\Lambda$, we get

$$\boldsymbol{n}\Lambda^2\boldsymbol{u} = \lambda\boldsymbol{n}\Lambda\boldsymbol{u} = 0.$$

We could further derive analogous conditions by multiplying (17.2) by $\mathbf{n}\Lambda^{\mathrm{k}}$, which leads to the general condition $\mathbf{n}\Lambda^{\mathrm{k}+1}\mathbf{u} = \lambda\mathbf{n}\Lambda^{\mathrm{k}}\mathbf{u} = 0$ for any integer k;* but only three such conditions are independent, and we select the following as these:

$$\boldsymbol{nu} = \boldsymbol{n}\Lambda\boldsymbol{u} = \boldsymbol{n}\Lambda^2\boldsymbol{u} = 0. \tag{17.5}$$

---

* Note that k can be negative, which corresponds to the reciprocal tensor $\boldsymbol{\Lambda}^{-\mathrm{k}} = (\Lambda^{-1})^{\mathrm{k}}$; as for ordinary numbers, we have $\Lambda^0 = 1$, where by 1 we understand unit tensor.

The Hamilton– Cayley theorem of (13.35) enables us to put

$$\Lambda^3 = \Lambda_t \Lambda^2 - \bar{\Lambda}_t \Lambda + |\Lambda|. \tag{17.6}$$

Thus we have the identity

$$n\Lambda^3 u = n(\Lambda_t \Lambda^2 - \bar{\Lambda}_t \Lambda + |\Lambda|)u = \Lambda_t n\Lambda^2 u - \bar{\Lambda}_t n\Lambda u + |\Lambda| nu$$

and the equation $\mathbf{n}\Lambda^3\mathbf{u} = 0$ is a consequence of (17.5) and (17.6). Similarly it is easily shown that (17.6) implies that $\mathbf{n}\Lambda^k\mathbf{u} = 0$ will be a consequence of (17.5) for any positive or negative integer k. To eliminate **u** and obtain a condition containing only **n**, we use the fact that, from (17.5), the three vectors **n**, $\Lambda\mathbf{n}$, and $\Lambda^2\mathbf{n}$ are all perpendicular to **u**, so they are coplanar. The coplanarity condition for three vectors **A**, **B**, **C** is $\mathbf{A}[\mathbf{BC}] = 0$, so from (17.5), we have

$$n\Lambda^2[\Lambda n, n] = 0. \tag{17.7}$$

This is a direct consequence of (17.3), so it is a necessary condition for a transverse wave. From (17.3) and (17.4) we may determine the direction of this **u** for the purely transverse wave. This vector is perpendicular to the vector product of these:

$$u_1 = C[n, \Lambda n]. \tag{17.8}$$

We may say that vector **n** must satisfy (17.7) and the displacement must be given by (17.8) if a purely transverse wave can propagate with **n** as wave normal. We may show the converse, namely that, if (17.7) is obeyed, then along the corresponding **n** there can be a purely transverse wave whose displacement is the **u** of (17.8). To do this we must show that (17.7) makes $[\mathbf{n}, \Lambda\mathbf{n}]$ an eigenvector of tensor Λ, i.e., satisfies (17.2), which is equivalent to $[\mathbf{u}, \Lambda\mathbf{u}] = 0$. In other words, we must show that (17.7) implies

$$[\Lambda[n, \Lambda n], \quad [n, \Lambda n]] = 0. \tag{17.9}$$

Expanding the double vector product, we get

$$n \cdot (\Lambda n)\Lambda[n, \Lambda n] - \Lambda n \cdot n\Lambda[n, \Lambda n] = 0,$$

since $(\Lambda\mathbf{n})\Lambda[\mathbf{n}, \Lambda\mathbf{n}] = \mathbf{n}\Lambda^2[\mathbf{n}, \Lambda\mathbf{n}] = 0$ from (17.7) in the first term, while $\mathbf{n}\Lambda[\mathbf{n}, \Lambda\mathbf{n}] = \mathbf{n}[\Lambda\mathbf{n}, \Lambda\mathbf{n}] = 0$ in the second.

Thus (17.7) is a necessary and sufficient condition that must be satisfied by a wave normal in a crystal such that a purely transverse wave can propagate along it. This is simultaneously a necessary condition for all special directions. This condition was first derived in [15].

Condition (17.7) can be given a geometrical interpretation, because the definition of (15.20) shows that tensor $\Lambda = \Lambda^{n}$ is a homogeneous quadratic function of the components of $\mathbf{n}$: $\overset{n}{\Lambda}_{kl} = \lambda_{ikjl} n_i n_j$, so the left-hand side of (17.7) is a homogeneous function of 9th degree in the n_k. We multiply (17.7) by r^9 and consider $\mathbf{rn} = \mathbf{r}$ as the radius vector; (17.7) becomes

$$\boldsymbol{r}(\Lambda^r)^2[\Lambda^r \boldsymbol{r}, \boldsymbol{r}] = 0 \tag{17.10}$$

and is the equation of a surface of 9th degree; (17.10) is homogeneous in the coordinates, so this is a cone of 9th degree.

The set of all wave normals allowing purely transverse waves (all special directions) is thus a cone of 9th degree, which we call the cone of special directions, which may take a great variety of forms in accordance with the form of the λ_{ikjl} tensor. Condition (17.10) is satisfied identically by all directions in an isotropic or transversely isotropic medium (see Chapter 6), but for most crystals the cone of (17.10) splits up into a series of planes or cones of lower degree.

Condition (17.7) may be put in a somewhat different form if we use (13.34):

$$\Lambda^2 = \overline{\Lambda} - \overline{\Lambda}_t + \Lambda_t \Lambda, \tag{17.11}$$

which allows us to put

$$\boldsymbol{n}\Lambda^2 = \boldsymbol{n}\overline{\Lambda} - \overline{\Lambda}_t \boldsymbol{n} + \Lambda_t \boldsymbol{n}\Lambda.$$

The last two terms of this give identically zero when substituted into (17.7), so the latter becomes

$$\boldsymbol{n}\overline{\Lambda}[\Lambda\boldsymbol{n}, \boldsymbol{n}] = 0. \tag{17.12}$$

But $\overline{\Lambda} = |\Lambda|\,\Lambda^{-1}$, and dividing by $|\Lambda| \neq 0$, we get

$$n\Lambda^{-1}[\Lambda n, n] = 0. \tag{17.13}$$

Of course, (17.7), (17.12), and (17.13) are all entirely equivalent.

We assume that

$$[n, \Lambda n] \neq 0. \tag{17.14}$$

This case $[\mathbf{n}, \Lambda\mathbf{n}] = 0$ corresponds to a purely longitudinal wave (property 1) and will be considered below; from (17.2), knowing $\mathbf{u}$, we can find at once the corresponding velocity

$$\lambda = v^2 = \frac{u\Lambda u}{u^2}. \tag{17.15}$$

Substitution from (17.8) gives

$$v_1^2 = \frac{[n, \Lambda n]\,\Lambda\,[n, \Lambda n]}{[n, \Lambda n]^2}. \tag{17.16}$$

From (12.53) we can put that

$$\Lambda[n, \Lambda n] = |\Lambda|\left[\Lambda^{-1}n, \Lambda^{-1}\Lambda n\right] = [\bar{\Lambda} n, n]. \tag{17.17}$$

From (12.44) and (12.45) we have that

$$v_1^2 = \frac{[n, \Lambda n]\,[\bar{\Lambda} n, n]}{[n, \Lambda n]^2} = \frac{n\Lambda n \cdot n\bar{\Lambda} n - |\Lambda|}{n\Lambda^2 n - (n\Lambda n)^2}. \tag{17.18}$$

The square of the velocity of a purely transverse wave can be deduced in a different way. Converting (17.2) to absolute form, we have

$$v^2 = \frac{|\Lambda u|}{|u|}. \tag{17.19}$$

This gives us from (17.8) and (17.17) that

$$v_1^2 = \frac{|[n, \bar{\Lambda} n]|}{|[n, \Lambda n]|}. \tag{17.20}$$

The velocity and displacement of the transverse wave are obtained in the most general form when (17.7) is obeyed; the same quantities for the other two waves are then easily found. We have (see section 13) that

$$\Lambda_t = v_0^2 + v_1^2 + v_2^2. \tag{17.21}$$

$$\overline{\Lambda}_t = v_0^2 v_1^2 + v_1^2 v_2^2 + v_2^2 v_0^2. \tag{17.22}$$

$$|\Lambda| = v_0^2 v_1^2 v_2^2. \tag{17.23}$$

and so

$$v_0^2 + v_2^2 = \Lambda_t - v_1^2 = A. \tag{17.24}$$

$$v_0^2 v_2^2 = \overline{\Lambda}_t - v_1^2 (v_0^2 + v_2^2) = \overline{\Lambda}_t - A v_1^2 = \frac{|\Lambda|}{v_1^2} = B. \tag{17.25}$$

Then v_0^2 and v_2^2 are roots of

$$v^4 - Av^2 + B = 0. \tag{17.26}$$

From v_0^2 and v_2^2 we can find the displacements of the corresponding waves by the method presented at the beginning of section 13. Consider the tensor $\overline{\Lambda - v_0^2}$, inverse with respect to $\Lambda - v_0^2$; if $\overline{\Lambda - v_0^2} = 0$, then (see section 13) v_0^2 is a duplicated eigenvalue of Λ, so the velocities of both waves coincide. The quasilongitudinal wave has the highest of the three speeds in all known media, so this case is impossible; hence we assume $\overline{\Lambda - v_0^2} \neq 0$, in which case (see section 12) the tensor $\overline{\Lambda - v_0^2}$ is linear, i.e., is the simple dyad of (12.73). Also, $\overline{\Lambda - v_0^2}$ is a symmetric tensor, so

$$\overline{\Lambda - v_0^2} = C\boldsymbol{u}_0 \cdot \boldsymbol{u}_0. \tag{17.27}$$

Then $\mathbf{u}_0$ is proportional to a column (row) of tensor $\overline{\Lambda - v_0^2}$. Similarly, $\mathbf{u}_2$ is determined via $\overline{\Lambda - v_2^2}$.

We see that the velocities and displacements of the three isonormal waves are readily and easily found in general form for the special directions.

18. Longitudinal Normals and Acoustic Axes

Christoffel's equation $\Lambda\mathbf{u} = \lambda\mathbf{u}$ has the solution $\mathbf{u} = \mathbf{u}_0 = \mathbf{n}$ for a longitudinal normal, so

$$\Lambda \boldsymbol{n} = \lambda_0 \boldsymbol{n}, \quad \lambda_0 = v_0^2. \tag{18.1}$$

Here $\Lambda\mathbf{n} \| \mathbf{n}$, which may be put as

$$[\boldsymbol{n}, \Lambda \boldsymbol{n}] = 0. \tag{18.2}$$

Conversely, (18.1) follows from (18.2), so the latter is the necessary and sufficient condition for $\mathbf{n}$ to be a longitudinal normal; but then purely transverse waves will also propagate along $\mathbf{n}$, for (18.2) directly implies that (17.7) is satisfied.

We can replace (18.2) by the equivalent scalar relation simply by forming the square $[\mathbf{n}, \Lambda\mathbf{n}]^2 = 0$, or, from (12.45),

$$n\Lambda^2 n = (n\Lambda n)^2. \qquad (18.3)$$

It is clear that (18.2) and (18.3) are equivalent. We multiply (18.3) by r^8 and put $\mathbf{r} = r\mathbf{n}$ to get, by analogy with (17.10), that

$$r^2 \cdot r(\Lambda^r)^2 r = (r\Lambda^r r)^2. \qquad (18.4)$$

This homogeneous equation defines a cone of 8th degree, the cone of longitudinal normals.

It can be shown [10] that a crystal of any symmetry always has directions for **n** such as to allow purely longitudinal (and hence also purely transverse) waves. This is a direct consequence of the following general concept: the directions of **n** that permit purely longitudinal waves are those corresponding to extremal values of the invariant

$$K = \lambda_{iklm} n_i n_k n_l n_m = n\Lambda^n n. \qquad (18.5)$$

To prove this we find those points for K subject to $\mathbf{n}^2 = n_i^2 = 1$. Following the general rules for finding a local extremum, we take

$$f = K - 2\lambda(n_i^2 - 1),$$

in which -2λ is an undetermined Lagrange multiplier. The turning point corresponds to $\partial f/\partial n_i = 0$. In taking derivatives of K with respect to n_i we must remember that $\partial n_k/\partial n_i = \delta_{ik}$. The properties of δ_{ik} and the symmetry of tensor λ_{iklm} give us that

$$\frac{\partial f}{\partial n_i} = 4(\lambda_{iklm} n_k n_l n_m - \lambda n_i) = 0.$$

This is simply (18.1), so the expression in the K of (18.5) corresponds to a purely longitudinal wave. Multiplication of (18.1) by **n** gives

$$\lambda_0 = v_0^2 = n\Lambda n = K, \qquad (18.6)$$

or the square of the phase velocity for the purely longitudinal wave is equal to the extremal K; but K is a continuous function of the n_i, which are given within a bounded closed region (namely, the surface of unit sphere, $\mathbf{n}^2 = 1$), so K must have a maximum and a minimum within this region (section 13). Thus any crystal has at least two distinct directions permitting purely longitudinal waves, one corresponding to the absolute maximum in K and the other to an absolute minimum. There will be more than two such normals if there are also relative extrema in K. Finally, if the absolute maximum and minimum coincide, K becomes independent of **n**, and any direction is a longitudinal normal; the last is an isotropic medium, because (16.5) shows that here $K = \mathbf{n}\Lambda\mathbf{n} = a_1 + 2a_2 = \text{const}$.

This extremal property indicates that the cone of (18.4) should generally split up into a set of separate directions; in fact, we shall see that (18.4) corresponds to a continuous circular cone only for hexagonal crystals, which are transversely isotropic. For all other anisotropic media (18.3) defines some set of isolated directions.

We have from (18.1) and (18.6) very simple expressions for $\mathbf{u}_0$ and v_0^2 in the case of a longitudinal normal:

$$\boldsymbol{u}_0 = \boldsymbol{n}, \quad v_0^2 = \boldsymbol{n}\Lambda\boldsymbol{n}. \tag{18.7}$$

The corresponding quantities for the two purely transverse waves may be found as in (17.21)-(17.26); v_1^2 and v_2^2 are given by

$$v^4 + Av^2 + B = 0, \tag{18.8}$$

in which

$$\left.\begin{aligned} A &= v_1^2 + v_2^2 = \Lambda_t - v_0^2 = \Lambda_t - \boldsymbol{n}\Lambda\boldsymbol{n}, \\ B &= v_1^2 v_2^2 = \overline{\Lambda}_t - v_0^2 A = \frac{|\Lambda|}{\boldsymbol{n}\Lambda\boldsymbol{n}}. \end{aligned}\right\} \tag{18.9}$$

These velocities give $\mathbf{u}_1$ and $\mathbf{u}_2$ via equations such as (17.27), on the assumption that $\overline{\Lambda - v_1^2} \neq 0$, otherwise v_1^2 will be a duplicated root, i.e., **n** will be an acoustic axis.

Next we turn to the acoustic axes. Let $v_1^2 = v_2^2$; then (17.21)-(17.23) become

$$\Lambda_t = v_0^2 + 2v_1^2, \qquad \overline{\Lambda}_t = 2v_0^2 v_1^2 + v_1^4, \tag{18.10}$$

$$|\Lambda| = v_0^2 v_1^4. \tag{18.11}$$

We eliminate v_0^2 from (18.10) to get

$$v_1^4 - \frac{2}{3}\Lambda_t v_1^2 + \frac{1}{3}\overline{\Lambda}_t = 0. \tag{18.12}$$

Then

$$v_1^2 = \frac{1}{3}\left(\Lambda_t \pm \sqrt{(\Lambda_t)^2 - 3\bar{\Lambda}_t}\right),$$

$$v_0^2 = \frac{1}{3}\left(\Lambda_t \mp 2\sqrt{(\Lambda_t)^2 - 3\bar{\Lambda}_t}\right). \tag{18.13}$$

We can also get (18.12) from $\frac{d}{d(v^2)}|v^2 - \Lambda| = 0$, which is the condition for a repeated root [16]; experiment (see above) enables us to say that only the quasitransverse waves can coincide in velocity, and that we have $v_0^2 > v_1^2$, which implies that the lower signs should be taken in (18.13). For waves along an acoustic axis in any crystal we should have

$$v_0^2 = \frac{1}{3}\left(\Lambda_t + 2\sqrt{(\Lambda_t)^2 - 3\bar{\Lambda}_t}\right),$$

$$v_1^2 = \frac{1}{3}\left(\Lambda_t - \sqrt{(\Lambda_t)^2 - 3\bar{\Lambda}_t}\right), \tag{18.14}$$

in which v_0 is for the quasilongitudinal wave and v_1 is for the quasitransverse one. We substitute (18.14) into (18.11) to get

$$9\Lambda_t\bar{\Lambda}_t - 2(\Lambda_t)^3 + 2((\Lambda_t)^2 - 3\bar{\Lambda}_t)\sqrt{(\Lambda_t)^2 - 3\bar{\Lambda}_t} = 27|\Lambda|.$$

Eliminating the root, we have

$$4((\Lambda_t)^2 - 3\bar{\Lambda}_t)^3 = (27|\Lambda| - 9\Lambda_t\bar{\Lambda}_t + 2(\Lambda_t)^3)^2, \tag{18.15}$$

which may be put as

$$27|\Lambda|^2 + 4(\Lambda_t)^3|\Lambda| + 4(\bar{\Lambda}_t)^3 - (\Lambda_t\bar{\Lambda}_t)^2 - 18\Lambda_t\bar{\Lambda}_t|\Lambda| = 0.$$

This is simply the condition for a zero value for the determinant of the characteristic equation [16]

$$\lambda^3 - \Lambda_t\lambda^2 + \bar{\Lambda}_t\lambda - |\Lambda| = 0. \tag{18.16}$$

This condition is necessary and sufficient for two roots of (18.16) to coincide.

Section 13 shows that a real symmetric tensor Λ is uniaxial if it has two eigenvalues coincident, so it can be put in the form of (13.24):

$$\Lambda = \lambda_1 + (\lambda_0 - \lambda_1)\boldsymbol{u}_0 \cdot \boldsymbol{u}_0, \tag{18.17}$$

in which $\lambda_1 = v_1^2$ is the duplicated eigenvalue and $\lambda_0 = v_0^2$ is the single one, $\mathbf{u}_0$ being the eigenvector corresponding to λ_0 (quasilongitudinal wave). The Λ tensor for an acoustic axis can be put in the form of (18.17) because the velocities coincide in accordance with (18.15); but (18.17) itself implies the transverse condition (17.7), because substitution of the first into the second gives us identically zero, no matter what the direction of $\mathbf{u}_0$. As we saw in section 17, condition (17.7) is a direct consequence of condition (18.15) on the velocities.

The condition for the acoustic axes is readily given a geometrical interpretation. We multiply (18.15) by r^{12} and put $\mathbf{r} = r\mathbf{n}$ to get

$$4((\Lambda_t^r)^2 - 3\bar{\Lambda}_t^r)^3 - (27\,|\Lambda^r| - 9\bar{\Lambda}_t^r\Lambda_t^r + 2\,(\Lambda_t^r)^3)^2 = 0. \qquad (18.18)$$

This homogeneous equation defines a cone of 12th degree, the cone of the acoustic axes. As for the longitudinal normals, we may conclude that this cone will almost always split up into separate isolated directions, except for hexagonal crystals, where (17.7) is satisfied for all $\mathbf{n}$, whereupon (18.18) can give the continuous surface of a circular cone.

Condition (18.15) is very cumbrous and complicated, so it is best to discuss the possible acoustic axes in terms of the equivalent condition that tensor Λ can be put as in (18.17), a condition that has been considered for (12.69). To make the representation of (18.17) correct, it is necessary and sufficient that

$$\overline{\Lambda - v_1^2} = 0. \qquad (18.19)$$

This tensor condition is equivalent to the scalar one of (18.15), but it is very much simpler and splits up into a series of separate equations, which greatly facilitates the analysis, an approach that has already been utilized [18, 19]. We get an identity by expanding (18.19) via (13.33) and substituting for $\bar{\Lambda}$ in (17.12), so condition (17.7) is a consequence of condition (18.19) for the acoustic axes.

The speeds of all waves are given by (18.14) if this last condition is obeyed; $\mathbf{u}_0$ is given by (18.17) as

$$\boldsymbol{u}_0 \cdot \boldsymbol{u}_0 = \frac{1}{v_0^2 - v_1^2}(\Lambda - v_1^2). \qquad (18.20)$$

The displacement of the quasitransverse wave in this case is an arbitrary vector in a plane perpendicular to $\mathbf{u}_0$.

The most particular case of all is when all the properties characteristic of an isotropic medium (see start of section 17) occur together. This is the case of a longitudinal acoustic axis; in (18.17) we put $\mathbf{u}_0 = \mathbf{n}$, so for that $\mathbf{n}$ the Λ tensor takes the form for an isotropic medium:

$$\Lambda = a + b\boldsymbol{n} \cdot \boldsymbol{n}. \tag{18.21}$$

We know from section 16 that a symmetry axis of high order has this property; other directions can be longitudinal acoustic axes only if there are certain special relations between the elastic moduli. It is readily seen that such an axis can exist in a triclinic crystal, for example, only if there are additional relations between the parameters. In fact, we have from (17.2) with $\mathbf{u}^2 = 1$ that

$$v^2 = \boldsymbol{u}\Lambda^n\boldsymbol{u}.$$

Suppose that the x_3 axis coincides with the longitudinal acoustic axis, i.e., $\mathbf{u}_0 = \mathbf{n} = (0, 0, 1)$, while the displacements of the transverse waves are $\mathbf{u}_1 = (1, 0, 0)$, $\mathbf{u}_2 = (0, 1, 0)$; $v_1^2 = v_2^2$ for an acoustic axis, so we must have

$$\boldsymbol{u}_1\Lambda^n\boldsymbol{u}_1 = \boldsymbol{u}_2\Lambda^n\boldsymbol{u}_2$$

or

$$\lambda_{1331} = \lambda_{2332}, \quad \lambda_{55} = \lambda_{44}. \tag{18.22}$$

Hence a longitudinal acoustic axis in a triclinic crystal imposes (18.22) as additional condition, which is not obeyed in general by such crystals, so a triclinic crystal as a general rule cannot have such an axis.

19. Form of the Λ Tensor for the Various Crystal Systems

We have seen above that all the properties of plane elastic waves in a crystal are ultimately governed by the tensor

$$\Lambda = (\Lambda^n_{kl}) = (\lambda_{iklm} n_i n_m). \tag{19.1}$$

We must therefore establish the form of the Λ tensor for the various crystal systems in order to examine the laws of propagation. Here we use the matrices derived in sections 8 and 9 to define the independent elastic moduli for all systems. The problem reduces to computation of expressions of the form

$$\lambda_{iklm} n_i n_m = \lambda_{1kl1} n_1^2 + \lambda_{2kl2} n_2^2 + \lambda_{3kl3} n_3^2 +$$

$$+ (\lambda_{2kl3} + \lambda_{3kl2}) n_2 n_3 + (\lambda_{3kl1} + \lambda_{1kl3}) n_3 n_1 +$$

$$+ (\lambda_{1kl2} + \lambda_{2kl1}) n_1 n_2, \tag{19.2}$$

for k, l = (11), (22), (33), (23), (31), (12) with allowance for the condition of section 6:

$$\lambda_{iklm} = \lambda_{\alpha\beta} = \frac{1}{\rho} c_{\alpha\beta}. \tag{19.3}$$

One general remark must first be made. The diagonal element c_{66} is not independent in the generally accepted expressions for the matrices (8.2) (isotropic medium), (9.22) (hexagonal crystals), and (9.21) (trigonal crystals),* being expressed in terms of c_{11} and c_{12} via

$$c_{66} = \frac{1}{2}(c_{11} - c_{12}). \tag{19.4}$$

Of course, this may be used to express any of the three parameters in terms of the other two and so to eliminate that one. It appears [5] that it is better to deviate from the usual choice and to express the nondiagonal element c_{12} via

$$c_{12} = c_{11} - 2c_{66}, \tag{19.5}$$

leaving c_{11} and c_{66} (diagonal) as independent elements. This gives rather more convenient expressions for the components of (19.2) for the Λ tensor. Hence our choice of basic independent reduced elastic moduli is as follows: isotropic medium, λ_{11} and $\lambda_{44} = \lambda_{66}$; hexagonal crystal λ_{11}, λ_{33}, λ_{44}, λ_{66} and λ_{13}; and trigonal crystal λ_{11}, λ_{33}, λ_{44}, λ_{66}, λ_{13}, λ_{14}. Correspondingly, the reduced matrices for the elastic moduli will be taken as follows: (8.2) for an isotropic medium as

---

*All of these have a threefold axis parallel to the x_3 axis.

$$\begin{Vmatrix} \lambda_{11} & \lambda_{11}-2\lambda_{44} & \lambda_{11}-2\lambda_{44} & 0 & 0 & 0 \\ \lambda_{11}-2\lambda_{44} & \lambda_{11} & \lambda_{11}-2\lambda_{44} & 0 & 0 & 0 \\ \lambda_{11}-2\lambda_{44} & \lambda_{11}-2\lambda_{44} & \lambda_{11} & 0 & 0 & 0 \\ 0 & 0 & 0 & \lambda_{44} & 0 & 0 \\ 0 & 0 & 0 & 0 & \lambda_{44} & 0 \\ 0 & 0 & 0 & 0 & 0 & \lambda_{44} \end{Vmatrix}, \tag{19.6}$$

(9.22) for a hexagonal crystal as

$$\begin{Vmatrix} \lambda_{11} & \lambda_{11}-2\lambda_{66} & \lambda_{13} & 0 & 0 & 0 \\ \lambda_{11}-2\lambda_{66} & \lambda_{11} & \lambda_{13} & 0 & 0 & 0 \\ \lambda_{13} & \lambda_{13} & \lambda_{33} & 0 & 0 & 0 \\ 0 & 0 & 0 & \lambda_{44} & 0 & 0 \\ 0 & 0 & 0 & 0 & \lambda_{44} & 0 \\ 0 & 0 & 0 & 0 & 0 & \lambda_{66} \end{Vmatrix}, \tag{19.7}$$

and (9.21) for a trigonal crystal as

$$\begin{Vmatrix} \lambda_{11} & \lambda_{11}-2\lambda_{66} & \lambda_{13} & \lambda_{14} & 0 & 0 \\ \lambda_{11}-2\lambda_{66} & \lambda_{11} & \lambda_{13} & -\lambda_{14} & 0 & 0 \\ \lambda_{13} & \lambda_{13} & \lambda_{33} & 0 & 0 & 0 \\ \lambda_{14} & -\lambda_{14} & 0 & \lambda_{44} & 0 & 0 \\ 0 & 0 & 0 & 0 & \lambda_{44} & \lambda_{14} \\ 0 & 0 & 0 & 0 & \lambda_{14} & \lambda_{66} \end{Vmatrix}. \tag{19.8}$$

The following are the Λ tensors for the media in order of decreasing symmetry [10].

1. Isotropic medium. The choice of coordinate axes is arbitrary [see (19.6)] and

$$\Lambda_{kl}=a\delta_{kl}+bn_kn_l, \tag{19.9}$$

or

$$\Lambda=a+b\boldsymbol{n}\cdot\boldsymbol{n}, \tag{19.10}$$

with

$$a=\lambda_{44},\quad b=\lambda_{11}-\lambda_{44}. \tag{19.11}$$

2. Cubic system. This differs from an isotropic medium in that here (as for the other systems) we give the expression

for the Λ tensor not in the direct covariant* form analogous to (19.10) but in the particular coordinate system in which the tensor takes the simplest form. The coordinate axes coincide with twofold or fourfold axes [see Table I and (9.23)].

Then

$$\Lambda = c_1 + c_2 \boldsymbol{n} \cdot \boldsymbol{n} + c_3 \boldsymbol{\nu}, \tag{19.12}$$

in which

$$c_1 = \lambda_{44}, \quad c_2 = \lambda_{12} + \lambda_{44}, \quad c_3 = \lambda_{11} - (c_1 + c_2) = \lambda_{11} - \lambda_{12} - 2\lambda_{44} \tag{19.13}$$

and

$$\nu = \begin{vmatrix} n_1^2 & 0 & 0 \\ 0 & n_2^2 & 0 \\ 0 & 0 & n_3^2 \end{vmatrix}. \tag{19.14}$$

3. Hexagonal system. This represents a transversely isotropic medium. The x_3 axis coincides with the sixfold axes, while the x_1 and x_2 axes are set arbitrarily; (19.7) gives the components of the Λ tensor as having the form

$$\left.\begin{aligned} \Lambda_{11} &= \lambda_{11} n_1^2 + \lambda_{66} n_2^2 + \lambda_{44} n_3^2, & \Lambda_{23} &= (\lambda_{13} + \lambda_{44}) n_2 n_3, \\ \Lambda_{22} &= \lambda_{66} n_1^2 + \lambda_{11} n_2^2 + \lambda_{44} n_3^2, & \Lambda_{31} &= (\lambda_{13} + \lambda_{44}) n_3 n_1, \\ \Lambda_{33} &= \lambda_{44} (n_1^2 + n_2^2) + \lambda_{33} n_3^2, & \Lambda_{12} &= (\lambda_{11} - \lambda_{66}) n_1 n_2. \end{aligned}\right\} \tag{19.15}$$

4. Tetragonal system. The x_3 axis is placed along the fourfold axis; the x_1 axis for classes $\bar{4} \cdot m$, 4:2, $4 \cdot m$, and $m \cdot 4{:}m$ is taken as perpendicular to the symmetry plane or as parallel to the twofold axis, while for classes $\bar{4}$, 4, and 4:m it is taken to comply with condition (9.7). The physical feature that allows one to make this choice is that the direction defined by (9.7) is a longitudinal normal [12], as may be shown by considering (17.2) on the as-

---

* A relation is covariant if both parts are in a form from which it is obvious that they change identically in response to coordinate transformations.

sumption that when $\mathbf{n}$ lies along x_1 [$\mathbf{n}=(1, 0, 0)$] we have $\mathbf{u}=\mathbf{n}=(1, 0, 0)$, i.e., the displacement vector is parallel to the normal and we have a longitudinal wave. Then all the dummy subscripts are 1, and (17.2) becomes

$$\lambda_{i111} = v^2 n_i, \tag{19.16}$$

whence

$$\lambda_{2111} = \lambda_{16} = 0, \quad \lambda_{3111} = \lambda_{15} = 0. \tag{19.17}$$

But this is precisely the result obtained by setting the x_1 axis in accordance with (9.7), so we may (see section 9) consider the matrix of (9.3), which contains six independent parameters, as suitable for all the tetragonal classes. This gives us

$$\left.\begin{aligned} \Lambda_{11} &= \lambda_{11} n_1^2 + \lambda_{66} n_2^2 + \lambda_{11} n_3^2, & \Lambda_{23} &= (\lambda_{13} + \lambda_{44}) n_2 n_3, \\ \Lambda_{22} &= \lambda_{66} n_1^2 + \lambda_{11} n_2^2 + \lambda_{44} n_3^2, & \Lambda_{31} &= (\lambda_{13} + \lambda_{44}) n_3 n_1, \\ \Lambda_{33} &= \lambda_{44} (n_1^2 + n_2^2) + \lambda_{33} n_3^2, & \Lambda_{12} &= (\lambda_{12} + \lambda_{66}) n_1 n_2. \end{aligned}\right\} \tag{19.18}$$

This tensor differs from that of (19.15) for hexagonal crystals only in having a different expression for the component Λ_{12}.

5. <u>Trigonal system.</u> The x_3 axis is set along the threefold axis. Classes 3:2, 3 · m, and $\bar{6}$ · m have additional symmetry elements in the form of twofold axes perpendicular to L^3 or symmetry planes parallel to the latter, which allows us to determine the direction of x_1; but the less symmetrical classes 3 and $\bar{6}$ are shown by (9.20) to allow us to choose the x_1 axis to be such that the element $c_{15} = -c_{25}$ in matrix (9.19) becomes zero. Physically, this direction, as for tetragonal crystals, is distinguished because a purely longitudinal wave can propagate along it [12]. Then [see (19.17)] we have $\lambda_{15} = -\lambda_{25} = 0$, and matrix (9.19) coincides with (9.21). However, we will use the form (19.8), which gives

$$\Lambda^{\text{trig}} = \Lambda^{\text{hex}} + 2\lambda_{14} \begin{vmatrix} n_2 n_3 & n_1 n_3 & n_1 n_2 \\ n_1 n_3 & -n_2 n_3 & \frac{1}{2}(n_1^2 - n_2^2) \\ n_1 n_2 & \frac{1}{2}(n_1^2 - n_2^2) & 0 \end{vmatrix}. \tag{19.19}$$

Here Λ^{hex} is the Λ tensor for a hexagonal crystal as defined by (19.15).

6. Orthorhombic system. Here a natural basis for the coordinate system is available from the twofold axes and symmetry planes; (8.11) gives

$$\left.\begin{aligned} \Lambda_{11} &= \lambda_{11}n_1^2 + \lambda_{66}n_2^2 + \lambda_{55}n_3^2, & \Lambda_{23} &= (\lambda_{23} + \lambda_{44})\, n_2 n_3, \\ \Lambda_{22} &= \lambda_{66}n_1^2 + \lambda_{22}n_2^2 + \lambda_{44}n_3^2, & \Lambda_{31} &= (\lambda_{13} + \lambda_{55})\, n_3 n_1, \\ \Lambda_{33} &= \lambda_{55}n_1^2 + \lambda_{44}n_2^2 + \lambda_{33}n_3^2, & \Lambda_{12} &= (\lambda_{12} + \lambda_{66})\, n_1 n_2. \end{aligned}\right\} \qquad (19.20)$$

7. Monoclinic system. We set the x_3 axis along the twofold axis or perpendicular to the symmetry plane; the matrix for the elastic moduli then becomes as in (8.10). We have seen in section 16 that a purely longitudinal wave and two purely transverse ones can propagate along the x_3 axis; in this case it is [12] most convenient to take as the x_1 axis the direction of $\mathbf{u}_1$ for one of these transverse waves, whereupon x_2 coincides with $\mathbf{u}_2$, the displacement vector for the other transverse wave. Then $n_i = \delta_{3i}$, $u_{1j} = \delta_{1j}$; substituting these into (17.2), we get

$$\lambda_{i33j}u_{1j} = v_1^2 u_{1i}. \qquad (19.21)$$

We put i = 2 in (19.21) to get

$$\lambda_{2331} = \lambda_{45} = 0, \qquad (19.22)$$

so matrix (8.10) simplifies and contains only 12 independent parameters. This matrix is taken with (19.2), (19.3), and (19.20) to give

$$\Lambda^{\text{mon}} = \Lambda^{\text{rhombo}} + \begin{vmatrix} 2\lambda_{16}n_1n_2 & \lambda_{16}n_1^2 + \lambda_{26}n_2^2 & \lambda_{36}n_2n_3 \\ \lambda_{16}n_1^2 + \lambda_{26}n_2^2 & 2\lambda_{26}n_1n_2 & \lambda_{36}n_3n_1 \\ \lambda_{36}n_2n_3 & \lambda_{36}n_3n_1 & 0 \end{vmatrix}. \qquad (19.23)$$

8. Triclinic system. Here there are no symmetry elements to enable us to choose the basis, so we can use only the properties of the elastic waves. The general concept demonstrated in the previous section [10] indicates that any crystal, including a triclinic one, has at least two longitudinal normals. We take one of these as x_3 axis and set the x_1 and x_2 axes along the displacement vectors of the purely transverse waves [12]; here $n_i = \delta_{3i}$, and (17.2) becomes

$$\lambda_{i33j} u_j = \lambda u_i. \tag{19.24}$$

A purely longitudinal wave has $u_j = n_j = \delta_{3j}$, $u_1 = u_2 = 0$, so

$$\lambda_{1333} = \lambda_{35} = 0, \quad \lambda_{2333} = \lambda_{34} = 0. \tag{19.25}$$

Also, (19.22) gives $\lambda_{45} = 0$, so the 21 parameters of the general matrix of (6.14) reduce to 18 independent significant ones, and the Λ tensor takes the form

$$\Lambda^{\text{tri}} = \Lambda^{\text{mon}} + \Lambda', \tag{19.26}$$

in which

$$\left.\begin{aligned}
\Lambda'_{11} &= 2(\lambda_{56} n_2 n_3 + \lambda_{15} n_3 n_1), \\
\Lambda'_{23} &= \lambda_{56} n_1^2 + \lambda_{24} n_2^2 + (\lambda_{25} + \lambda_{46}) n_1 n_2, \\
\Lambda'_{22} &= 2(\lambda_{24} n_2 n_3 + \lambda_{16} n_3 n_1), \\
\Lambda'_{31} &= \lambda_{15} n_1^2 + \lambda_{46} n_2^2 + (\lambda_{14} + \lambda_{56}) n_1 n_2, \\
\Lambda'_{33} &= 0, \quad \Lambda'_{12} = (\lambda_{25} + \lambda_{46}) n_2 n_3 + (\lambda_{14} + \lambda_{56}) n_3 n_1.
\end{aligned}\right\} \tag{19.27}$$

These expressions for the Λ tensor indicate that those of (19.10), (19.12), (19.15), (19.18), and (19.20) may for any **n** be put as

$$\Lambda = \begin{pmatrix} A & pq & pr \\ qp & B & qr \\ rp & rq & C \end{pmatrix}. \tag{19.28}$$

However, this form is not suitable for all crystals; in fact, if one of the nondiagonal components is zero, e.g., $pq = 0$, then either $p = 0$ or $q = 0$, so one other component must be zero. Hence it cannot happen that only one nondiagonal element is zero in a tensor of the form of (19.28); whereas if we put $n_3/n_2 = \lambda_{66} - \lambda_{11}/2\lambda_{14}$ in the tensor of (19.19) for a trigonal crystal, which is always possible, we get Λ_{12} as zero, although Λ_{23} and Λ_{31} differ from zero. The same applies to monoclinic and triclinic crystals.*

---

* Musgrave [13] has used the form of (19.28) to construct a general theory of elastic waves in crystals, but this is inapplicable even to a crystal as important as quartz.

20. Convoluted Tensor for the Elastic Moduli

The λ_{iklm} tensor gives an exhaustive characterization of the elastic properties of a crystal, but much information about elastic waves in crystals may be obtained from the symmetric second-rank tensor μ derived from λ_{iklm} by convoluting with respect to a pair of internal (or external) subscripts:

$$\mu_{kl} = \mu_{lk} = \lambda_{kiil}. \tag{20.1}$$

To determine what physical properties are characterized by μ, we use it to produce the quadratic form from the components of the vector for the wave normal:

$$S = \boldsymbol{n}\mu\boldsymbol{n} = n_k\mu_{kl}n_l = n_k\lambda_{kiil}n_l. \tag{20.2}$$

Comparison with (15.20) shows that this quadratic form equals the trace of the Λ tensor:

$$S = \Lambda_{kk} = \Lambda_t. \tag{20.3}$$

But the trace of a tensor equals (see section 13) the sum of the eigenvalues, which here is the sum of the squares of the three waves corresponding to a given **n**. Thus

$$S = \Lambda_t = \boldsymbol{n}\mu\boldsymbol{n} = v_0^2 + v_1^2 + v_2^2. \tag{20.4}$$

In section 13 we showed that the extremal values of $\mathbf{n}\mu\mathbf{n}$ subject to $\mathbf{n}^2 = 1$ are the eigenvalues of μ, so the corresponding eigenvectors of μ define the directions of the wave normal that correspond to these extremal values* of the sum of the squares of the wave velocities.

We could use the matrices $c_{\alpha\beta} = c_{iklm}$ of sections 8 and 9 in order to obtain the explicit form of μ for the various crystal systems; but it is simpler to use the expressions for Λ derived in the previous section. The trace of Λ gives us as follows:

Isotropic medium

$$\Lambda_t = 3a + b = \lambda_{11} + 2\lambda_{44}, \tag{20.5}$$

---

* The components λ_{iklm} may be used in convolution to construct another second-rank tensor $\lambda'_{kl} = \lambda_{iikl}$ [7], but this does not have the importance of μ in relation to elastic waves.

Cubic system

$$\Lambda_t = 3c_1 + c_2 + c_3 = \lambda_{11} + 2\lambda_{44}. \qquad (20.6)$$

Hexagonal, tetragonal, and trigonal systems

$$\Lambda_t = (\lambda_{11} + \lambda_{44} + \lambda_{66})(n_1^2 + n_2^2) + (\lambda_{33} + 2\lambda_{44})\, n_3^2. \qquad (20.7)$$

Orthorhombic system

$$\Lambda_t = (\lambda_{11} + \lambda_{55} + \lambda_{66})\, n_1^2 + (\lambda_{22} + \lambda_{44} + \lambda_{66})\, n_2^2 + (\lambda_{33} + \lambda_{44} + \lambda_{55})\, n_3^2. \qquad (20.8)$$

Monoclinic system

$$\Lambda_t = \Lambda_t^{\text{rhomb}} + 2(\lambda_{16} + \lambda_{26})\, n_1 n_2. \qquad (20.9)$$

Triclinic system

$$\Lambda_t = \Lambda_t^{\text{mon}} + 2(\lambda_{24} + \lambda_{65})\, n_2 n_3 + 2(\lambda_{15} + \lambda_{46})\, n_3 n_1. \qquad (20.10)$$

From (20.4), $\Lambda_t = \mathbf{n}\mu\mathbf{n} = n_i \mu_{ik} n_k$, so the μ_{ik} define the factors to the products $n_i n_k$ in the expressions for Λ_t. The μ_{ik} equal half the factor to $n_i n_k$ if $i \neq k$, because the symmetry of μ causes each term to appear twice in $\mathbf{n}\mu\mathbf{n}$. For an isotropic medium or a cubic crystal we have

$$\Lambda_t = (\lambda_{11} + 2\lambda_{44})\, \delta_{ik} n_i n_k,$$

since

$$\delta_{ik} n_i n_k = n_i n_i = n^2 = 1.$$

Then

$$\mu^{\text{iso}} = \mu^{\text{cub}} = \begin{pmatrix} \lambda_{11} + 2\lambda_{44} & 0 & 0 \\ 0 & \lambda_{11} + 2\lambda_{44} & 0 \\ 0 & 0 & \lambda_{11} + 2\lambda_{44} \end{pmatrix}. \qquad (20.11)$$

Similarly we get

$$\mu^{\text{hex}} = \mu^{\text{tetr}} = \mu^{\text{trig}} = \begin{pmatrix} \lambda_{11} + \lambda_{44} + \lambda_{66} & 0 & 0 \\ 0 & \lambda_{11} + \lambda_{44} + \lambda_{66} & 0 \\ 0 & 0 & \lambda_{33} + 2\lambda_{44} \end{pmatrix}. \qquad (20.12)$$

$$\mu^{\text{rhomb}} = \begin{pmatrix} \mu_{11} & 0 & 0 \\ 0 & \mu_{22} & 0 \\ 0 & 0 & \mu_{33} \end{pmatrix}, \quad \begin{aligned} \mu_{11} &= \lambda_{11} + \lambda_{55} + \lambda_{66}, \\ \mu_{22} &= \lambda_{22} + \lambda_{44} + \lambda_{66}, \\ \mu_{33} &= \lambda_{33} + \lambda_{44} + \lambda_{55}, \end{aligned} \qquad (20.13)$$

$$\mu^{\text{mon}} = \begin{pmatrix} \mu_{11} & \mu_{12} & 0 \\ \mu_{12} & \mu_{11} & 0 \\ 0 & 0 & \mu_{33} \end{pmatrix}, \quad \mu_{12} = \lambda_{16} + \lambda_{26}, \qquad (20.14)$$

$$\mu^{\text{tri}} = \begin{pmatrix} \mu_{11} & \mu_{12} & \mu_{13} \\ \mu_{12} & \mu_{22} & \mu_{23} \\ \mu_{13} & \mu_{23} & \mu_{33} \end{pmatrix}, \quad \begin{aligned} \mu_{13} &= \lambda_{15} + \lambda_{46}, \\ \mu_{26} &= \lambda_{24} + \lambda_{56}. \end{aligned} \qquad (20.15)$$

All the μ_{ik} appearing in the last three formulas are expressed identically in terms of the $\lambda_{\alpha\beta}$.

Tensor μ is isotropic for a cubic crystal or an isotropic medium, whereas it is uniaxial (section 13) for the middle systems. The choice of coordinate axes used in section 19 means that the tensor μ for all crystals (other than monoclinic and triclinic) are diagonal; all the μ for the lowest systems are biaxial (section 13). The properties of tensor μ thus lead us to divide crystals into groups precisely as in optics: cubic crystals do not differ from isotropic media, crystals of the middle systems have rotational symmetry around high-order axes, and crystals of the lowest system have μ as a biaxial tensor.*

These relationships give rise to various conclusions: (20.4)-(20.6) show that cubic crystals and isotropic media have S (the sum of the squares of the velocities of the three waves) constant for all the directions of the wave normal [14]; (20.7) gives $\Lambda_t = \lambda_{11} + \lambda_{44} + \lambda_{66} + (\lambda_{33} + \lambda_{44} - \lambda_{11} - \lambda_{66}) n_3^2 = \text{const}$ for $n_3 = \text{constant}$, so in the case of the middle systems S is constant for all **n** lying in an arbitrary circular cone having its axis along a high-order axis. Also, (20.6)-(20.10) show that the symmetry axes (planes) of any crystal are the principal axes (planes) of tensor μ, the symmetry axes (or the directions perpendicular to symmetry planes) corresponding to extremal values of S [10] (section 13).

---

* These properties of tensor μ are common to all symmetric second-rank tensors and follow from the corresponding symmetry conditions.

Tensor μ, being symmetric, may (section 13) be put as

$$\mu = a_1 + a_2(c' \cdot c'' + c'' \cdot c'), \quad c'^2 = c''^2 = 1, \tag{20.16}$$

in which a_1 is the intermediate eigenvalue of μ. Then

$$S = n\mu n = a_1 + 2a_2 nc' \cdot nc''. \tag{20.17}$$

Thus S is constant for all **n** that satisfy

$$nc' \cdot nc'' = C = \text{const.} \tag{20.18}$$

To find the values that C may take, we consider the extremum of $f = \mathbf{nc}' \cdot \mathbf{nc}'' - \kappa/2(\mathbf{n}^2 - 1)$, in which $-\kappa/2$ is an undetermined Lagrange multiplier. We get

$$\frac{\partial f}{\partial n} = c' \cdot nc'' + nc' \cdot c'' - \varkappa n = 0,$$

or

$$(c' \cdot c'' + c'' \cdot c')\, n = \varkappa n, \quad \varkappa = 2nc' \cdot nc'' = 2C_{extr}. \tag{20.19}$$

Table II of section 13 shows that the solutions to this are $\mathbf{n} = [\mathbf{c}'\mathbf{c}'']$ and $\mathbf{n} = \mathbf{c}' \pm \mathbf{c}''$, with the corresponding values $\kappa = 0$ and $\kappa = \pm(1 \pm \mathbf{c}'\mathbf{c}'')$, so C can take values within the following limits:

$$-\frac{1 - c'c''}{2} \leqslant C \leqslant \frac{1 + c'c''}{2}. \tag{20.20}$$

A given C within these limits causes (20.18) to become the equation of a surface. We multiply (20.18) by r^2, put $r\mathbf{n} = \mathbf{r}$, and treat $\mathbf{r}$ as the radius vector to get

$$rc' \cdot rc'' - Cr^2 = 0. \tag{20.21}$$

This is a homogeneous equation defining a cone of second degree; symmetry considerations show that its axis is the bisector of the angle between $\mathbf{c}'$ and $\mathbf{c}''$. We take the angle $(\widehat{\mathbf{c}', \mathbf{c}''})$ as acute, which is always feasible, though it may require reversal of $\mathbf{c}'$ or $\mathbf{c}''$ with simultaneous change in the sign of a_2. Then for C positive the axis will be the internal bisector of angle $(\widehat{\mathbf{c}', \mathbf{c}''})$: $\mathbf{a} = \mathbf{c}' + \mathbf{c}''$, while for C negative it is the external bisector $\mathbf{b} = \mathbf{c}' - \mathbf{c}''$. The cone of (20.21) is elliptical; taking $C > 0$ and putting in (20.21) that

$$r = c' + c'' + r', \quad r'(c' + c'') = 0, \tag{20.22}$$

we direct vector $\mathbf{r}'$ lying in a plane perpendicular to axis $\mathbf{a}$ of the cone once $[\mathbf{c}', \mathbf{c}'']$ and a second time along $\mathbf{c}' - \mathbf{c}'' = \mathbf{b}$. We put in (20.21) that $\mathbf{r}_1' = k_1[\mathbf{c}'\mathbf{c}'']$, $\mathbf{r}_2 = k_2(\mathbf{c}' - \mathbf{c}'')$ to get, respectively,

$$Cr_1'^2 = (1 + c'c'')(1 + c'c'' - 2C),$$

$$Cr_2'^2 = \frac{(1 + c'c'')(1 + c'c'' - 2C)}{1 + (1 - c'c'')/2C}.$$

But $\mathbf{r}_1'^2 > \mathbf{r}_2'^2$, so the curve of the section of the cone by a plane is an ellipse, not a circle.

That is, the intersection between the cone of (20.21) and a plane perpendicular to the axis $\mathbf{a} = \mathbf{c}' + \mathbf{c}''$ is an ellipse, whose major axis is the vector $[\mathbf{c}'\mathbf{c}'']$ and whose minor axis is the vector $\mathbf{c}' - \mathbf{c}''$ (Fig. 7).

The cone of (20.21) degenerates to two planes for $C = 0$, the equations of these being $\mathbf{rc}' = 0$ and $\mathbf{rc}'' = 0$, i.e., these planes are perpendicular to $\mathbf{c}'$ and $\mathbf{c}''$; (20.17) and (20.18) give that here $S = a_1$.

Hence S is constant for all $\mathbf{n}$ lying on an elliptic cone of the form of (20.21) in the case of crystals of the lowest systems; in particular, any such crystal has two planes perpendicular to the $\mathbf{c}'$ and $\mathbf{c}''$ of (20.16) in which S is constant and equal to the middle eigenvalue of μ[10].

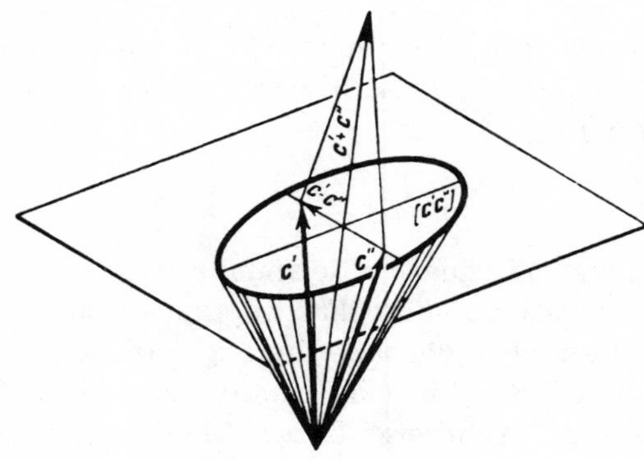

Fig. 7

Tensor μ also allows us to give a geometrical interpretation of the behavior of elastic waves in crystals. Consider the ellipsoid (section 13)

$$r\mu r = 1, \tag{20.23}$$

in which $\mathbf{r} = |\mathbf{r}|\,\mathbf{n}$. Then (20.4) gives

$$n\mu n = \frac{1}{r^2} = v_0^2 + v_1^2 + v_2^2. \tag{20.24}$$

Hence the inverse square of the radius vector of the ellipsoid of (20.23) equals the sum of the squares of the velocities of the three waves having their normal parallel to this radius vector [10].

Tensor μ also provides a natural choice of the coordinate axes for monoclinic and triclinic crystals (section 19). The x_3 axis may be set along the displacement of the purely longitudinal wave, while x_1 and x_2 may be set along the corresponding transverse displacements. The convenience of this arises from (19.22) and (19.25), but the choice is not unique for a triclinic crystal, because we have shown (section 18) that there is not a single direction in which a purely longitudinal wave can propagate. Instead of this we can choose as our coordinate axes the principal axes of tensor μ, which becomes diagonal [7]. This choice is shown by (8.10) to correspond to the following relation between the parameters in the case of a monoclinic crystal:*

$$\mu_{12} = \lambda_{45} + \lambda_{16} + \lambda_{26} = 0, \tag{20.25}$$

which allows us to eliminate λ_{45}, λ_{16}, or λ_{26}. From (6.14), this choice for a triclinic crystal leads to (20.25) together with two further relations:

$$\mu_{23} = \lambda_{24} + \lambda_{34} + \lambda_{56} = 0, \quad \mu_{13} = \lambda_{15} + \lambda_{35} + \lambda_{46} = 0, \tag{20.26}$$

which reduces the number of independent elastic constants to 18.

---

* Of course, (19.22) and (19.25) are now not obeyed, since the coordinate axes have been chosen in another way.

Chapter 4

Energy Flux and Wave Surfaces

21. The Energy-Flux Vector and the Ray Velocity

The propagation of an elastic wave in any medium is associated with the movement of energy. The laws governing this may be deduced from the time course of the kinetic and potential energies of a deformed elastic body enclosed in a volume V (section 5). From (5.8), (6.3), and (6.4) we have the densities of these forms of energy by

$$W = \frac{1}{2}\rho\dot{u}_i^2, \quad \Phi = \frac{1}{2}\sigma_{ik}\gamma_{ik} = \frac{1}{2}c_{iklm}\gamma_{ik}\gamma_{lm} \tag{21.1}$$

while for the total energy

$$E = \int_V (W + \Phi)\, dV. \tag{21.2}$$

It follows that

$$\frac{dE}{dt} = \int\left(\frac{\partial W}{\partial t} + \frac{\partial \Phi}{\partial t}\right) dV = \int\left(\rho\dot{u}_i\ddot{u}_i + \frac{\partial \Phi}{\partial \gamma_{ik}}\dot{\gamma}_{ik}\right) dV. \tag{21.3}$$

From (1.11), (4.3), and (5.18)

$$\frac{\partial \Phi}{\partial \gamma_{ik}} = \sigma_{ik} = c_{iklm}\gamma_{lm}, \quad \dot{\gamma}_{ik} = \frac{1}{2}\left(\frac{\partial \dot{u}_i}{\partial x_k} + \frac{\partial \dot{u}_k}{\partial x_i}\right). \tag{21.4}$$

Then the symmetry of σ_{ik} gives

$$\frac{\partial \Phi}{\partial t} = \frac{1}{2}\sigma_{ik}\left(\frac{\partial \dot{u}_i}{\partial x_k} + \frac{\partial \dot{u}_k}{\partial x_i}\right) = \sigma_{ik}\frac{\partial \dot{u}_i}{\partial x_k}.$$

Gauss's theorem allows us to transform the potential-energy term as follows:

$$\int \frac{\partial \Phi}{\partial t}\, dV = \int \sigma_{ik} \frac{\partial \dot{u}_i}{\partial x_k}\, dV = \int \left(\frac{\partial}{\partial x_k}(\sigma_{ik}\dot{u}_i) - \dot{u}_i \frac{\partial \sigma_{ik}}{\partial x_k}\right) dV =$$

$$= -\int \dot{u}_i \frac{\partial \sigma_{ik}}{\partial x_k}\, dV + \oint \sigma_{ik}\dot{u}_i\, dS_k.$$

Substitution in (21.3) gives

$$\frac{dE}{dt} = \int \dot{u}_i \left(\rho \ddot{u}_i - \frac{\partial \sigma_{ik}}{\partial x_k}\right) dV + \oint \sigma_{ik}\dot{u}_i\, dS_k.$$

The equations of equilibrium of (3.3) imply that the volume integral is zero for $\mathbf{g} = 0$, so

$$\frac{dE}{dt} + \oint \boldsymbol{P}\, d\boldsymbol{S} = 0, \tag{21.5}$$

in which

$$\boldsymbol{P} = -\sigma\dot{\boldsymbol{u}}, \quad P_i = -\sigma_{ik}\dot{u}_k. \tag{21.6}$$

The physical significance of (21.5) is that the change in the energy E enclosed in a fixed volume of an elastically deformed medium occurs from the flux of a vector **P** through the surface bounding the volume. This **P** is called the energy flux-density vector (Umov vector), and its direction indicates the direction of the energy flow at that point, the length being numerically equal to the amount of energy passing in unit time through unit area perpendicular to **P**. Dividing this vector by the energy density, we get a vector defining the magnitude and direction of the energy flow rate.

The formulas of (21.6) are completely general and apply for any deformation and stress in the crystal. We are interested in deformations of the elastic-wave type, so we assume that the displacement vector varies in accordance with (15.5); but the energy and energy flux are quadratic functions of the displacement vector, so we must consider only the real part of the complex expression of (15.5) (see section 15). We assume the elastic wave to be linearly polarized and put the displacement vector as

$$u_i = u_i^0 \cos\varphi, \quad \varphi = \boldsymbol{kr} - \omega t. \tag{21.7}$$

Here

$$\dot{u}_k = \omega u_k^0 \sin\varphi, \quad \frac{\partial u_k}{\partial x_l} = -k_l u_k^0 \sin\varphi,$$

$$\gamma_{ij} = \frac{1}{2}\left(\frac{\partial u_i}{\partial x_j} + \frac{\partial u_j}{\partial x_i}\right) = -\frac{1}{2}\left(k_i u_j^0 + k_j u_i^0\right)\sin\varphi$$

and substitution of (21.7) into (21.1) and (21.6) gives

$$W = \frac{1}{2}\rho\omega^2 \boldsymbol{u}^{02}\sin^2\varphi, \quad \Phi = \frac{1}{2}c_{ijlm}k_i k_l u_j^0 u_m^0 \sin^2\varphi, \tag{21.8}$$

$$P_i = -c_{ijlm}\gamma_{lm}\dot{u}_j = \omega c_{ijlm}k_l u_m^0 u_j^0 \sin^2\varphi. \tag{21.9}$$

Multiplying (15.16) by u_i, we get

$$c_{ijlm}k_j k_l u_i^0 u_m^0 = \rho\omega^2 u_i^{02}, \tag{21.10}$$

which means that a plane wave has its kinetic energy equal to its potential energy:

$$W = \Phi. \tag{21.11}$$

The total energy density is then

$$\mathscr{E} = 2W = 2\Phi = \rho\omega^2 \boldsymbol{u}^{02}\sin^2\varphi. \tag{21.12}$$

The time-averages of the density and flux are usually the quantities of practical interest:

$$\overline{\mathscr{E}} = \frac{1}{2}\rho\omega^2 \boldsymbol{u}^{02}, \quad \overline{P}_i = \frac{1}{2}\omega c_{ijlm}k_l u_j^0 u_m^0. \tag{21.13}$$

We divide $\overline{P}_i$ by $\overline{\mathscr{E}}$ to get the vector for the velocity of motion of the energy, which we denote by **s**. Now $c_{ijlm} = \rho\lambda_{ijlm}$ (15.17), $k_l = kn_l$ (15.8), $\omega/k = v$ (15.10), and taking **u** as unit displacement vector, we get

$$s_i = \frac{\overline{P}_i}{\overline{\mathscr{E}}} = \frac{\lambda_{ijlm}u_j u_m n_l}{v} = \frac{P_i}{\mathscr{E}}. \tag{21.14}$$

Vector $\mathbf{s} = (s_i)$ is the ray or group velocity, the latter name being given because the general theory [21] shows that the energy in a wave process propagates with the group velocity (not the phase velocity), that velocity being the speed of a wave packet, which is the result of superposition of numerous waves with similar frequencies The general theory shows that the group velocity of a wave of any type is given [21] by the derivative of the frequency with respect to the wave vector:

$$s_l = \frac{\partial \omega}{\partial k_l}, \quad \boldsymbol{s} = \frac{\partial \omega}{\partial \boldsymbol{k}}. \tag{21.15}$$

This gives (21.14) very simply; from (21.10), taking the displacement vector as unit, we have

$$\omega^2 = \lambda_{ijlm} k_j k_l u_i u_m. \tag{21.16}$$

Differentiation with respect to k_l gives $2\omega(\partial\omega/\partial k_l) = 2\lambda_{ijlm} k_j u_i u_m$, which implies that

$$s_l = \frac{\partial \omega}{\partial k_l} = \frac{\lambda_{ijlm} n_j u_i u_m}{\omega / k},$$

which coincides with (21.14).

In section 15 we introduced the tensor of (15.20): $\Lambda = \Lambda^n = (\lambda_{ijlm} n_j n_l)$. Similarly we define a second-rank tensor

$$\Lambda^u = (\Lambda^u_{il}) = (\lambda_{ijlm} u_j u_m). \tag{21.17}$$

Then (21.14) may be put in the form

$$\boldsymbol{s} = \frac{1}{v} \Lambda^u \boldsymbol{n}. \tag{21.18}$$

Now (15.20) gives

$$\boldsymbol{n} \Lambda^u \boldsymbol{n} = \boldsymbol{u} \Lambda^n \boldsymbol{u} = \lambda_{ijlm} n_i n_l u_j u_m = v^2, \tag{21.19}$$

so (21.18) gives the group and phase velocities as related by

$$\boldsymbol{s}\boldsymbol{n} = v. \tag{21.20}$$

The ray velocity therefore cannot be less than the phase one, and equals it only if the two coincide in direction. To a given direction of the wave normal in the crystal there correspond three waves with different phase velocities and mutually orthogonal displacements. Clearly, Λ^{u} has a different form for each of these; therefore, and also in view of the difference in v, the ray velocities of the three isonormal waves will, from (21.18), usually differ in magnitude and direction. For the quasilongitudinal and quasitransverse waves we may put, respectively, that

$$s_0 = \frac{1}{v_0}\Lambda^{u_0}n, \quad s_1 = \frac{1}{v_1}\Lambda^{u_1}n, \quad s_2 = \frac{1}{v_2}\Lambda^{u_2}n. \tag{21.21}$$

We note that $\mathbf{u}_0 = \mathbf{n}$ and

$$\Lambda^{n}\mathbf{u}_0 = v_0^2\mathbf{u}_0 = v_0^2\mathbf{n} = \lambda^{u_0}\mathbf{n}$$

for a longitudinal normal, so the first formula in (21.21) gives $\mathbf{s}_0 = \mathrm{v}_0\mathbf{n} = \mathbf{v}_0$; hence the vectors for the ray and phase velocities coincide in a purely longitudinal wave. This cannot in general be said of the transverse waves corresponding to the same normal.

We have from (21.21) that

$$v_0\mathbf{s}_0 + v_1\mathbf{s}_1 + v_2\mathbf{s}_2 = (\Lambda^{u_0} + \Lambda^{u_1} + \Lambda^{u_2})\mathbf{n};$$

but

$$\mathbf{u}_0 \cdot \mathbf{u}_0 + \mathbf{u}_1 \cdot \mathbf{u}_1 + \mathbf{u}_2 \cdot \mathbf{u}_2 = 1,$$

so (20.1) gives us that

$$\begin{aligned}\Lambda^{u_0}_{il} + \Lambda^{u_1}_{il} + \Lambda^{u_2}_{il} = \lambda_{ijkl}(\mathbf{u}_0 \cdot \mathbf{u}_0 + \mathbf{u}_1 \cdot \mathbf{u}_1 + \mathbf{u}_2 \cdot \mathbf{u}_2)_{jk} &= \lambda_{ijkl}\delta_{jk} \\ &= \lambda_{ijjl} \\ &= \mu_{il}.\end{aligned}$$

Hence

$$v_0\mathbf{s}_0 + v_1\mathbf{s}_1 + v_2\mathbf{s}_2 = \mu\mathbf{n}. \tag{21.21'}$$

Multiplying by $\mathbf{n}$ and using (21.20), we get (20.4).

A further very convenient expression may be derived for the ray-velocity vector; to find it we differentiate (21.19) with respect to n_k, taking all components of **n** as independent, i.e., neglecting the condition $\mathbf{n}^2 = n_k^2 = 1$:

$$\frac{\partial}{\partial n_k} v^2 = 2v \frac{\partial v}{\partial n_k} = \frac{\partial}{\partial n_k} (\boldsymbol{n}\Lambda^u \boldsymbol{n}) = \frac{\partial}{\partial n_k} (\boldsymbol{u}\Lambda^n \boldsymbol{u}). \tag{21.22}$$

The derivative on the right is found as follows:

$$\frac{\partial}{\partial n_k} (\boldsymbol{n}\Lambda^u \boldsymbol{n}) = 2 \frac{\partial \boldsymbol{n}}{\partial n_k} \Lambda^u \boldsymbol{n} + 2 \frac{\partial \boldsymbol{u}}{\partial n_k} \Lambda^n \boldsymbol{u}.$$

Here $\partial n_i/\partial n_k = \delta_{ik}$, so $(\partial \mathbf{n}/\partial n_k)\Lambda^u \mathbf{n} = (\Lambda^u \mathbf{n})_k$; further, **u** is unit vector, so $\mathbf{u}^2 = 1$ and $\partial \mathbf{u}^2/\partial n_k = 2\mathbf{u}(\partial \mathbf{u}/\partial n_k) = 0$. The equation $\Lambda^n \mathbf{u} = v^2 \mathbf{u}$ shows that $(\partial \mathbf{u}/\partial n_k)\Lambda^n \mathbf{u} = v^2(\partial \mathbf{u}/\partial n_k)\,\mathbf{u} = 0$, so (21.22) implies that

$$\frac{\partial v}{\partial n_k} = \frac{1}{v} (\Lambda^u \boldsymbol{n})_k,$$

or, omitting the subscripts,

$$\frac{\partial v}{\partial \boldsymbol{n}} = \frac{1}{v} \Lambda^u \boldsymbol{n}. \tag{21.23}$$

Comparison with (21.18) gives us the important relation

$$s_k = \frac{\partial v}{\partial n_k}, \quad \boldsymbol{s} = \frac{\partial v}{\partial \boldsymbol{n}}. \tag{21.24}$$

It is easily shown that (21.20) follows from this, for the phase velocity is a homogeneous function of first degree in the components of **n**, which follows from the definitive equation (15.22), $|v^2 - \Lambda^n| = 0$. We multiply this by k^6 to get $|(kv)^2 - \Lambda^{kn}| = 0$, which shows that multiplication of all components of **n** in $v = f(n_k)$ by a number k results in multiplication of v by the same number. But then we get from Euler's theorem for homogeneous functions that

$$n_k \frac{\partial v}{\partial n_k} = \boldsymbol{n} \frac{\partial v}{\partial \boldsymbol{n}} = v, \tag{21.25}$$

which (21.24) makes equivalent to (21.20).

From (21.24) and (21.20) we have

$$\frac{\partial}{\partial n_i} (n_k s_k) = s_i + n_k \frac{\partial s_k}{\partial n_i} = \frac{\partial v}{\partial n_i} = s_i,$$

$$n_k \frac{\partial s_k}{\partial n_l} = 0, \tag{21.26}$$

whose correctness may be verified directly by differentiating (21.18).

Relation (21.26) also shows that

$$\frac{\partial s_i}{\partial n_k} = \frac{\partial^2 v}{\partial n_k \, \partial n_i} = \frac{\partial^2 v}{\partial n_i \, \partial n_k} = \frac{\partial s_k}{\partial n_i}, \tag{21.27}$$

so $n_k(\partial s_k/\partial n_i) = n_k(\partial s_i/\partial n_k)$; but v is a homogeneous function of first degree in the n_k, so $s_k = \partial v/\partial n_k$ is a homogeneous function of the n_k of zero degree. Euler's theorem shows that the operator $n_k(\partial/\partial n_k)$ applied to a homogeneous function of the n_k of degree p is equivalent to multiplying the function by p; s_k has $p = 0$, so $n_k(\partial s_i/\partial n_k) = 0$.

We introduce the symmetric matrix

$$\beta = (\beta_{ik}) = \left(\frac{\partial s_i}{\partial n_k}\right) = \left(\frac{\partial s_k}{\partial n_i}\right) = \left(\frac{\partial^2 v}{\partial n_i \, \partial n_k}\right), \tag{21.28}$$

which allows us to put (21.26) as

$$\beta n = 0 \qquad (|\beta| = 0). \tag{21.29}$$

The importance of (21.24) leads us to derive it also from the definition of (21.15). Using (15.10) with $v = v(\mathbf{n}) = v(n_k)$, we may put that

$$s_l = \frac{\partial \omega}{\partial k_l} = \frac{\partial (kv)}{\partial k_l} = v \frac{\partial k}{\partial k_l} + k \frac{\partial v}{\partial k_l} = v \frac{\partial k}{\partial k_l} + k \frac{\partial v}{\partial n_i} \frac{\partial n_i}{\partial k_l}. \tag{21.29'}$$

But

$$\frac{\partial k}{\partial k_l} = \frac{\partial \sqrt{k_l^2}}{\partial k_l} = \frac{k_l}{k} = n_l, \tag{21.30}$$

$$\frac{\partial n_i}{\partial k_l} = \frac{\partial}{\partial k_l}\left(\frac{k_i}{k}\right) = \frac{\delta_{il}}{k} - \frac{n_i n_l}{k}. \tag{21.31}$$

Substitution into (21.29') gives

$$s_l = v n_l + \frac{\partial v}{\partial n_i}(\delta_{il} - n_i n_l) = \left(v - n_i \frac{\partial v}{\partial n_i}\right) n_l + \frac{\partial v}{\partial n_l}.$$

This by virtue of (21.25) reduces to (21.24)

We introduce the vector*

$$m = \frac{n}{v}, \tag{21.32}$$

having the direction of the wave normal and a magnitude equal to the reciprocal of the pahse velocity. This vector is entirely analogous to the important refraction vector of optics [17], so that name will be used here, the more so since the vector is especially important in the refraction and reflection of elastic waves (see Chapter 8). Refraction vector **m** allows us to put (21.18)-(21.20) as

$$s = \Lambda^{u} m \tag{21.33}$$

$$sm = m\Lambda^{u} m = u\Lambda^{m} u = 1. \tag{21.34}$$

We have $\Lambda^{u} = a + b\mathbf{u} \cdot \mathbf{u}$ for an isotropic medium, with $a + b = v_0^2$ (longitudinal wave) and $a = v_1^2$ (transverse wave). Also, for a longitudinal wave ($\mathbf{u} = \mathbf{n}$) we have $\mathbf{s} = (1/v_0)\Lambda^{u}\mathbf{n} = v_0\mathbf{n} = \mathbf{v}_0$, while for a transverse wave ($\mathbf{u} \perp \mathbf{n}$) we have $\mathbf{s} = (1/v_1)a\mathbf{n} = v_1\mathbf{n} = \mathbf{v}_1$. Hence the ray velocity for both types of wave in an isotropic medium coincides in magnitude and direction with the corresponding phase velocity.

Relation (21.24) does not conflict with this result, if we remember the reservation that all the n_k are to be taken as independent in differentiation with respect to **n** and that the phase velocity is a homogeneous function of first degree in the n_k. Therefore we must put, say, $v_1 = \sqrt{a n^2}$ for a transverse wave in an isotropic medium; applying (21.24) we get

$$s_1 = \frac{\partial v_1}{\partial n} = \sqrt{a}\,\frac{2n}{2\sqrt{n^2}} = \sqrt{a}\, n = v_1,$$

as should be the case.

---

* There is no accepted terminology for vector **m**; it has been called the reciprocal-velocity vector or the index vector [13, 25]. The term slowness vector is also used in the non-Russian literature [15].

Some important general relations follows from (21.33), which we solve for **m** to get

$$m = (\Lambda^u)^{-1} s. \tag{21.35}$$

Substitution into (21.34) gives

$$s (\Lambda^u)^{-1} s = 1. \tag{21.36}$$

This relation directly connects **u** with the corresponding ray velocity **s**; (21.36) for a given direction of **u** is the equation of an ellipsoid in ray-velocity space, the radius vector being $\mathbf{r} = \mathbf{s}$ (section 13). We conclude that the ends of the ray-velocity vectors for all elastic waves having a given **u** (with $\mathbf{u}^2 = 1$) must lie on the ellipsoid of (21.36). Of course, to each **u** there corresponds an ellipsoid distinct in shape, size, and orientation. As a rule, **s** is defined by one point (more precisely, two opposed points), or by several such pairs on the ellipsoid; but sometimes (see section 25) a given **u** will correspond to infinitely many **s**, whose ends form a line on the ellipsoid of (21.36).

We square (21.35) to get $\mathbf{m}^2 = ((\Lambda^u)^{-1}\mathbf{s})^2 = \mathbf{s}(\Lambda^u)^{-2}\mathbf{s}$, or, with (21.32) and (21.19),

$$n\Lambda^u n \cdot s (\Lambda^u)^{-2} s = 1. \tag{21.37}$$

This is equivalent to

$$v = \frac{1}{\sqrt{s (\Lambda^u)^{-2} s}}. \tag{21.38}$$

We multiply (21.35) scalarly by **n** to get

$$v = \frac{1}{n (\Lambda^u)^{-1} s}. \tag{21.39}$$

Also, (21.18) may be put in the form $\Lambda^u\mathbf{n} = \mathbf{s} \cdot \mathrm{v} = \mathbf{s} \cdot \mathbf{sn}$ of (21.20), or

$$(\Lambda^u - s \cdot s) n = 0. \tag{21.40}$$

This equation defines **n** by reference to a given **u** and **s**; for (21.40) to have a solution for **n**, it is necessary [see (12.25)] to comply with

$$|\Lambda^u - s\cdot s| = |\Lambda^u| - s\overline{\Lambda^u}s = 0. \tag{21.41}$$

This differs from (21.36) only in the factor $|\Lambda^u|$ [see (12.16)]. In addition to (21.40), another vector equation between **n**, **u**, and **s** is

$$(\Lambda^n - (ns)^2)\,u = 0, \tag{21.42}$$

i.e., Christoffel's equation of (15.21) and (21.20). Equations (21.40) and (21.42) do not follow one from the other; they are independent, since vector **s** is introduced via the definition of (21.18), which is equivalent to (21.40).

Suppose we are given the direction of **n** and have to find **s**. First of all we find **ns** = v from the condition for the existence of a solution to (21.42) for **u**:

$$|\Lambda^n - v^2| = 0. \tag{21.43}$$

In the general case this gives three distinct values of v: v_0, v_1, and v_2 (i.e., tensor Λ^n is biaxial, see section 13). Here $\overline{\Lambda^n - v^2} \neq 0$ and (13.8) gives

$$u\cdot u = \frac{\overline{\Lambda^n - v^2}}{(\overline{\Lambda^n - v^2})_t} = \frac{\overline{\Lambda^m - 1}}{(\overline{\Lambda^m - 1})_t}. \tag{21.44}$$

This defines **u** uniquely. The last relation in terms of subscripts is

$$u_j u_m = \frac{(\overline{\Lambda^n - v^2})_{jm}}{(\overline{\Lambda^n - v^2})_t}. \tag{21.45}$$

Then (21.17) for Λ^u becomes

$$\Lambda^u_{il} = \frac{\lambda_{ijlm}(\overline{\Lambda^n - v^2})_{jm}}{(\overline{\Lambda^n - v^2})_t}. \tag{21.46}$$

We substitute (21.46) with the corresponding root v^2 of (21.43) into

(21.18) to get the ray-velocity vector as

$$s_i = \frac{\lambda_{ijlk}\,(\overline{\Lambda^n - v^2})_{jk}\, n_l}{v\,(\overline{\Lambda^n - v^2})_t} = \frac{\lambda_{ijlk}\,(\overline{\Lambda^m - 1})_{jk}\, m_l}{(\overline{\Lambda^m - 1})_t}. \tag{21.47}$$

Formulas (21.43), (21.44), and (21.47) give a complete solution to the problem in the general case.

Since (21.44) with $v_0 \neq v_1 \neq v_2$ defines **u** uniquely, and hence the Λ^u of (21.46), we see from (21.47) that to each of the three isonormal waves there corresponds a definite unique **s**.

The situation is different when (21.43) has a duplicated root, which corresponds to an acoustic axis (section 18).

In this case the behavior of the ray-velocity is much more complicated and requires the special discussion presented in the next section.

22. Energy Vector with Acoustic Axes

Let tensor Λ^n have a duplicated eigenvalue $v_1^2 = v_2^2$ for a given direction of **n**; then (13.22) and (13.24) show that Λ^n is a uniaxial tensor whose axis is the displacement vector $\mathbf{u}_0$ of the quasilongitudinal wave. Table II gives us Λ^n for this case as

$$\Lambda^n = v_1^2 + (v_0^2 - v_1^2)\, u_0 \cdot u_0. \tag{22.1}$$

The displacement vector **u** of the quasitransverse wave may have any direction in a plane perpendicular to $\mathbf{u}_0$, i.e., it is restricted only by the one condition

$$uu_0 = 0. \tag{22.2}$$

Here (21.18) shows that **s** will vary in magnitude and direction with the direction of $\mathbf{u}(\mathbf{u}^2 = 1)$, but **s** will always satisfy (21.20). Taking **s** as radius vector, we have (21.20) for given **n** and $v = v_1$ as the equation of a plane P perpendicular to **n** and lying a distance v_1 from the origin. The end of **s** will always lie on that plane; it traces out a curve as **u** varies (Fig. 8), whose shape is found by introducing the two mutually orthogonal unit vectors

$$u' = \frac{[nu_0]}{|[nu_0]|}, \quad u'' = [u_0 u']. \tag{22.3}$$

In determining $\mathbf{u}'$ and $\mathbf{u}''$ we select the direction of $\mathbf{u}_0$ to be such as to form an acute angle with $\mathbf{n}$ (Fig. 8). We have $\mathbf{u}'\mathbf{u}_0 = \mathbf{u}''\mathbf{u}_0 = 0$, so the $\mathbf{u}$ of (22.2), the unit displacement vector of the quasitransverse wave, may be put as a linear combination:

$$u = u' \cos\theta + u'' \sin\theta, \quad u^2 = u'^2 = u''^2 = 1. \tag{22.4}$$

We assume here that $\mathbf{u}_0$ is not parallel to $\mathbf{n}$, i.e., that the acoustic axis is not longitudinal (section 18). Vectors $\mathbf{u}'$ and $\mathbf{u}''$ may be considered as loci in a plane perpendicular to $\mathbf{u}_0$; then θ is the angle between $\mathbf{u}$ and $\mathbf{u}'$. It is readily seen that tensor Λ^u now takes the form

$$\Lambda^u = \Lambda^{u'} \cos^2\theta + \Lambda^{u''} \sin^2\theta + \Lambda''' \cos\theta \sin\theta, \tag{22.5}$$

in which

$$\Lambda''' = \Lambda' + \Lambda'', \quad \Lambda'_{ik} = \lambda_{ijkl} u'_j u''_l, \tag{22.6}$$

$$\Lambda''_{ik} = \lambda_{ijkl} u''_j u'_l = \lambda_{ilkj} u''_l u'_j = \lambda_{kjil} u'_j u''_l = \Lambda'_{ki}. \tag{22.7}$$

Hence

$$\tilde{\Lambda}'' = \Lambda', \quad \tilde{\Lambda}''' = \Lambda'''. \tag{22.8}$$

Then we get (21.33) as the expression for $\mathbf{s}$:

$$s = \Lambda^u m_1 = \left(\cos^2\theta \Lambda^{u'} + \sin^2\theta \Lambda^{u''} + \sin\theta \cos\theta \Lambda'''\right) m_1, \tag{22.9}$$

in which $\mathbf{m}_1 = \mathbf{n}/v_1$; all possible orientations of the $\mathbf{u}$ of (22.4) are obtained by varying θ from 0 to 2π. This mean of s is

$$\langle s \rangle = \frac{1}{2\pi} \int_0^{2\pi} s\, d\theta = \frac{1}{2} \left(\Lambda^{u'} + \Lambda^{u''}\right) m_1 = s_0. \tag{22.10}$$

We can now put $\mathbf{s}$ in the form*

$$s = s_0 + s_1, \tag{22.11}$$

in which

$$s_1 = s - s_0 = \frac{1}{2} \cos 2\theta \left(\Lambda^{u'} - \Lambda^{u''}\right) m_1 + \frac{1}{2} \sin 2\theta \Lambda''' m_1. \tag{22.12}$$

---

* Subscripts 0 and 1 here differ in meaning from those of (21.21).

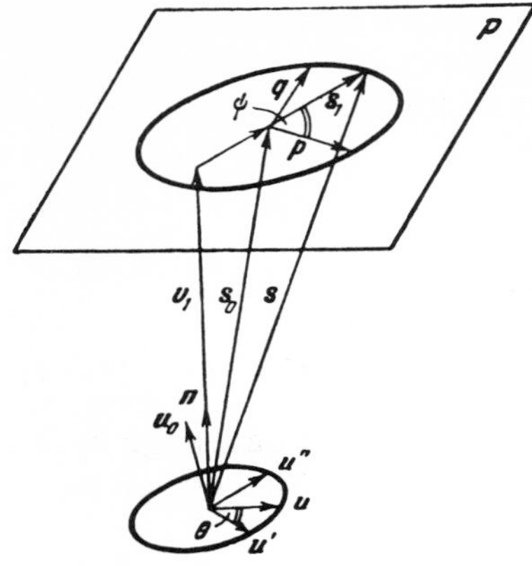

Fig. 8

But $\mathbf{n}\Lambda^{\mathbf{u}'}\mathbf{n} = \mathbf{n}\Lambda^{\mathbf{u}''}\mathbf{n} = \mathbf{u}'\Lambda^{\mathbf{n}}\mathbf{u}' = \mathbf{u}''\Lambda^{\mathbf{n}}\mathbf{u}'' = v_1^2$, $(\Lambda^{\mathbf{n}}\mathbf{u} = v_1^2\mathbf{u})$, so

$$ns_0 = v, \tag{22.13}$$

i.e., the end of $\mathbf{s}_0$ lies in the same plane as the ends of all the $\mathbf{s}$ vectors. On the other hand $\mathbf{n}\Lambda'''\mathbf{n} = 2\mathbf{u}'\Lambda^{\mathbf{n}}\mathbf{u}'' = 2v_1^2\mathbf{u}'\mathbf{u}'' = 0$, so $\mathbf{ns}_1 = 0$, i.e., $\mathbf{s}_1$ lies in plane P (Fig. 8); $\mathbf{s}_0$ is constant, so the shape of the curve described by the end of $\mathbf{s}$ is governed by the variation of $\mathbf{s}_1$. We get the equation of this curve by considering $\mathbf{s}_1$ as the radius vector in plane P and eliminating θ from

$$r = s_1 = p\cos 2\theta + q\sin 2\theta, \tag{22.14}$$

in which

$$p = \frac{1}{2}(\Lambda^{u'} - \Lambda^{u''})m_1, \quad q = \frac{1}{2}\Lambda'''m_1, \quad (pn = qn = 0). \tag{22.15}$$

It is simplest to eliminate θ by taking the vector product of (22.14) once by $\mathbf{p}$ and again by $\mathbf{q}$, squaring the results, and adding [17]. This gives

$$[rp]^2 + [rq]^2 = [pq]^2. \tag{22.16}$$

This is the equation of a curve of second degree, which cannot be other than an ellipse, because (22.14) shows that the length of $\mathbf{r}$ is bounded. The size and orientation of the axes of the ellipse of (22.16) can be found if we first find the θ at which $\mathbf{r}^2 = \mathbf{s}_1^2$ has its extreme values. For this purpose it is convenient to put (22.14) as

$$\boldsymbol{r} = \alpha \boldsymbol{p} + \beta \boldsymbol{q}, \quad \alpha^2 + \beta^2 = 1. \tag{22.17}$$

Then we have to find the maximum of

$$f(\alpha, \beta) = \boldsymbol{r}^2 - \lambda(\alpha^2 + \beta^2 - 1), \tag{22.18}$$

in which λ is an undetermined Lagrange multiplier. The condition for a turning point, $\partial f/\partial\alpha = \partial f/\partial\beta = 0$, gives

$$\left.\begin{aligned} (\boldsymbol{p}^2 - \lambda)\alpha + \boldsymbol{pq}\beta = 0, \\ \boldsymbol{pq}\alpha + (\boldsymbol{q}^2 - \lambda)\beta = 0. \end{aligned}\right\} \tag{22.19}$$

We multiply these, respectively, by α and β, and add the results, to get

$$\lambda = \alpha^2\boldsymbol{p}^2 + \beta^2\boldsymbol{q}^2 + 2\alpha\beta\boldsymbol{pq} = \boldsymbol{r}^2_{\text{exp}}. \tag{22.20}$$

On the other hand, the linear homogeneous equations (22.19) give us the following, because α and β are not simultaneously zero:

$$\begin{vmatrix} \boldsymbol{p}^2 - \lambda & \boldsymbol{pq} \\ \boldsymbol{pq} & \boldsymbol{q}^2 - \lambda \end{vmatrix} = \lambda^2 - (\boldsymbol{p}^2 + \boldsymbol{q}^2)\lambda + [\boldsymbol{pq}]^2 = 0. \tag{22.21}$$

This gives us the square of the semiaxes for the ellipse described by the end of $\mathbf{s}$ in the plane $\mathbf{ns} = v_1$:

$$\left.\begin{aligned} \lambda_1 = \boldsymbol{r}^2_{\text{max}} = \boldsymbol{a}^2 = \frac{1}{2}\left(\boldsymbol{p}^2 + \boldsymbol{q}^2 + \sqrt{(\boldsymbol{p}^2 + \boldsymbol{q}^2)^2 - 4[\boldsymbol{pq}]^2}\right), \\ \lambda_2 = \boldsymbol{r}^2_{\text{min}} = \boldsymbol{b}^2 = \frac{1}{2}\left(\boldsymbol{p}^2 + \boldsymbol{q}^2 - \sqrt{(\boldsymbol{p}^2 + \boldsymbol{q}^2)^2 - 4[\boldsymbol{pq}]^2}\right). \end{aligned}\right\} \tag{22.22}$$

We find corresponding α/β from (22.19) and substitute into (22.17) to get the directions of the principal axes of the ellipse:

$$\left.\begin{aligned} \boldsymbol{a} \parallel [\boldsymbol{q}[\boldsymbol{pq}]] - \lambda_1\boldsymbol{p} \parallel [\boldsymbol{p}[\boldsymbol{qp}]] - \lambda_1\boldsymbol{q}, \\ \boldsymbol{b} \parallel [\boldsymbol{q}[\boldsymbol{pq}]] - \lambda_2\boldsymbol{p} \parallel [\boldsymbol{p}[\boldsymbol{qp}]] - \lambda_2\boldsymbol{q}. \end{aligned}\right\} \tag{22.23}$$

From (22.17) we have $\mathbf{r} = \alpha\mathbf{p} \pm \sqrt{1-\alpha^2}\,\mathbf{q}$, so to each α there correspond two points A and B on the ellipse, which are obtained if to

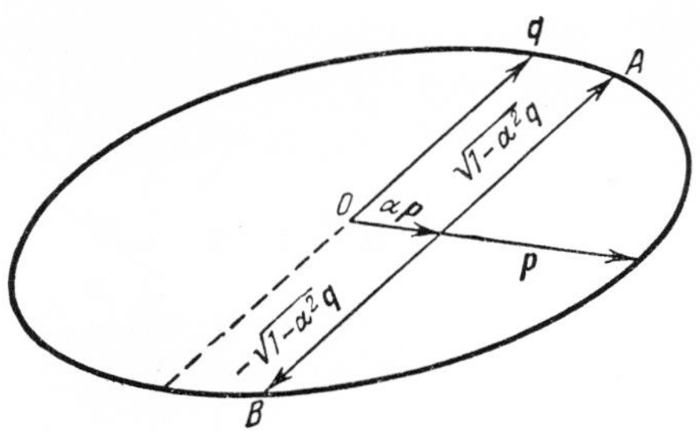

Fig. 9

vector $\alpha\mathbf{p}$ we add vector $\sqrt{1-\alpha^2}\,\mathbf{q}$ or the opposite vector $-\sqrt{1-\alpha^2}\,\mathbf{q}$ (Fig. 9). Thus chord AB parallel to $\mathbf{q}$ meets $\mathbf{p}$ at its middle. Similarly, $\mathbf{q}$ bisects all chords parallel to $\mathbf{p}$, so $\mathbf{p}$ and $\mathbf{q}$ are conjugate semidiameters of the ellipse of (22.14) and (22.16).

A change in the direction of the $\mathbf{u}$ of (22.4) due to increase in θ will cause a corresponding rotation of $\mathbf{s}_1$ in plane P, which (22.14) gives as twice the angle (Fig. 8). This rotation may be clockwise or counterclockwise looking along $\mathbf{n}$. The sense of the rotation may be determined by comparing the direction of the vector $[\mathbf{s}_1, d\mathbf{s}_1/d\theta]$ with that of $\mathbf{n}$. If $\mathbf{s}_1$ is taken as the radius vector of (22.14), then the vector $\mathbf{s}_1' = d\mathbf{s}_1/d\theta$, which also lies in plane P, will define the direction and speed of the motion of the end of $\mathbf{s}_1$ as θ varies. Vector $[\mathbf{s}_1\mathbf{s}_1']$ resembles $\mathbf{n}$ in being perpendicular to plane P. Figure 8 shows that a positive (negative) sign for $\mathbf{n}[\mathbf{s}_1\mathbf{s}_1']$ causes $\mathbf{s}_1$ to rotate with respect to $\mathbf{n}$ in the sense of a right-handed (left-handed) screw. From (22.14) we get the condition that for

$$\mathbf{n}[\mathbf{pq}] > 0 \qquad (22.24)$$

the sense of rotation of $\mathbf{s}_1$ as θ increases will be as for a right-handed screw with respect to $\mathbf{n}$, with the converse for $\mathbf{n}[\mathbf{pq}] < 0$.*

Thus a wave normal coincident with an acoustic axis corresponds to a cone of directions for the energy-flux vector of a quasi-transverse wave, each of which corresponds to a definite displacement vector. Hence the corresponding property is called, by

*See [22] for details of these properties of the energy-flux vector for the case of an acoustic axis.

analogy with crystal optics (see [23], for example) internal conical refraction. The equation for this cone is readily found. We take the vector product of **s** and (22.14), and use (22.11), to get

$$[\mathbf{s}_0\mathbf{s}] = [\mathbf{sp}]\cos 2\theta + [\mathbf{sq}]\sin 2\theta.$$

To eliminate θ we multiply this equation scalarly by **p** and by **q**, square the results, and add, getting

$$(\mathbf{s}[\mathbf{s}_0\mathbf{p}])^2 + (\mathbf{s}[\mathbf{s}_0\mathbf{q}])^2 = (\mathbf{s}[\mathbf{pq}])^2. \tag{22.25}$$

We introduce the symbols

$$\left.\begin{aligned} \mathbf{a} = [\mathbf{s}_0\mathbf{p}], \quad \mathbf{b} = [\mathbf{s}_0\mathbf{q}], \quad \mathbf{c} = [\mathbf{pq}] \parallel \mathbf{n}, \\ \gamma = \mathbf{a}\cdot\mathbf{a} + \mathbf{b}\cdot\mathbf{b} - \mathbf{c}\cdot\mathbf{c} \end{aligned}\right\} \tag{22.26}$$

and consider **s** as the radius vector **r** to put (22.25) as

$$\mathbf{r}\gamma\mathbf{r} = 0. \tag{22.27}$$

This equation is homogeneous in **r**, so it is the equation of a cone of second degree. The cone will not degenerate at $\mathbf{r} = 0$ provided that the symmetric tensor γ is not positively (negatively) definite, i.e., its eigenvalues must necessarily include positive and negative ones. Thus two of the three eigenvalues of γ (e.g., γ_1 and γ_2) will have the same sign, while the third (γ_3) will have the opposite sign. The eigenvector of γ corresponding to γ_3 will then be the axis of the cone, which will be circular if, and only if, $\gamma_1 = \gamma_2$, i.e., if tensor γ is uniaxial (section 13). Tensor γ is biaxial in the general case, and the cone is then elliptical (the section by a plane perpendicular to the axis is an ellipse). These properties lead us to conclude from (22.10), (22.15), and (22.26) that the cone of rays of (22.27) in the general case of an acoustic axis is elliptical and that its axis does not coincide with vector $\mathbf{s}_0$, because the latter is not an eigenvector of the tensor γ of (22.26).

The above general relations simplify somewhat in the particular case where the acoustic axis lies in a symmetry plane or in a plane equivalent to this that is normal to a fourfold axis (section 16). It is readily seen that (22.15) then shows that **p** lies in the symmetry plane, while **q** is perpendicular to it; in this case **n**, $\mathbf{u}_0$, and **u**" lie in the symmetry plane, while **u**' is perpendicular to it [see section 16 and (22.3)]. Reflection in the symmetry plane does not alter tensor λ_{iklm} and vectors **n**, $\mathbf{u}_0$, and **u**", while the direction of **u**' is reversed. Tensors $\Lambda^{u'}$, $\Lambda^{u''}$ and vector $\mathbf{m}_1 = \mathbf{n}/v_1$ are unaffected, while tensor Λ''' reverses in sign, because the components of **u**' appear in it to the first power [see (22.6) and (22.7)]. Thus **p** is unaltered by reflection, while **q** reverses in direction [see (22.15)], which implies that **p** lies in the symmetry plane and **q** is perpendicular to it. In the same way we show that the $\mathbf{s}_0$ of (22.10) also lies in the symmetry plane. It is clear that the ellipse of (22.16) lies symmetrically with respect to the symmetry plane, while the mutually perpendicular vectors **p** and **q** will in magnitude and direction coincide with the principal semiaxes of this ellipse. It is obvious that the axis of the ray cone of (22.27) will also lie in the symmetry plane. Figure 10 illustrates the relative disposition of the various vectors lying in the symmetry plane. Direction OD is the axis of the ray cone and coincides with the bisector of angle AOC; it is clear that this will always lie between $\mathbf{s}_0$ (as median) and v (the height of the triangle OAC). OD may coincide with $\mathbf{s}_0$ only if $\mathbf{s}_0 \parallel \mathbf{n}$.

A more special case is that of a longitudinal acoustic axis (section 18) with $\mathbf{u}_0 \parallel \mathbf{n}$. Such directions can, in principle, exist in crystals of the lower systems subject to special relations between the parameters (section 18), but it is usual for the properties of such an axis to occur only for the higher symmetry axes in the middle systems. In the latter case the cone of (22.27) becomes circular, and its axis coincides with the axis of high order, because the ellipse of (22.16) should not alter on rotation around **n** through an angle $\varphi < \pi$ when there is an axis of high order, which is possible only if the ellipse is a circle.

23. Elliptical Polarization in Elastic Waves and the Instantaneous Energy-Flux Vector

So far we have considered only linearly polarized plane elastic waves, which can (section 15) propagate along all directions in the crystal.* However, Λ^n becomes uniaxial along the acoustic axes, and here a wave of more general type can occur, namely one with elliptical polarization. Such waves can propagate in all directions in an isotropic medium.

---

*It is assumed that the crystal does not absorb elastic waves.

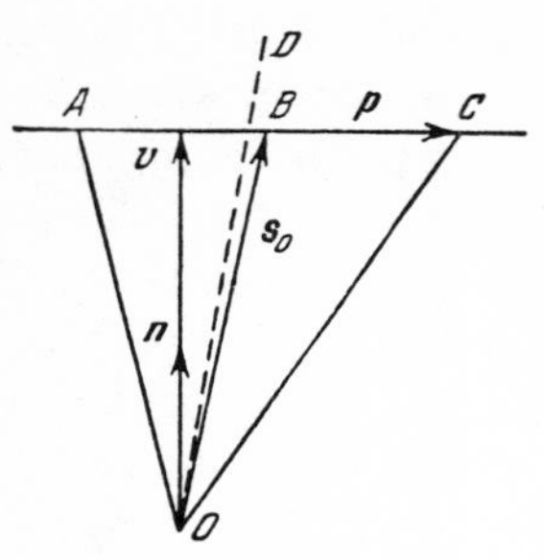

Fig. 10

In section 15 we showed that the displacement vector of a plane monochromatic wave may be put in the form of (15.5):

$$u = u^0 e^{i\varphi}, \quad \varphi = kr - \omega t, \tag{23.1}$$

in which $\mathbf{u}^0$ is a constant vector amplitude, which is an eigenvector of tensor Λ^n: $(\Lambda^n - v^2)\mathbf{u}^0 = 0$. Λ^n is biaxial for a crystal (section 13) for nearly all directions of $\mathbf{n}$, which means that it has three eigenvectors of precisely defined direction, which correspond to three linearly polarized isonormal waves. Here $\mathbf{u}^0$ may be considered as a real vector. A direct physical meaning also attaches to the real part of the $\mathbf{u}$ of (23.1) (section 15):

$$u_r = \mathrm{Re}\, u = u^0 \cos\varphi. \tag{23.2}$$

This indicates that $\mathbf{u}_r$ varies periodically with limits $\mathbf{u}^0$ and $-\mathbf{u}^0$, while remaining collinear with a fixed direction.

But $\mathbf{u}^0$ may also be complex:

$$u^0 = u' + iu'', \tag{23.3}$$

in which $\mathbf{u}'$ and $\mathbf{u}''$ are real vectors that are not collinear;* if they were, we would simply have the case of real $\mathbf{u}^0$ again, for if $\mathbf{u}'' = C\mathbf{u}'$ then $\mathbf{u}^0 = (1 + iC)\mathbf{u}' = C_1\mathbf{u}'$. The equation $(\Lambda^n - v^2)\,\mathbf{u}^0 = 0$ is linear and homogeneous, so the scalar factor C_1 plays no part and may be discarded, and (23.3) leads to new results only when $\mathbf{u}'$ and $\mathbf{u}''$ are linearly independent. In that case the real part of (23.1) becomes

$$u = u' \cos\varphi - u'' \sin\varphi. \tag{23.4}$$

A difference from (23.2) is that the end of $\mathbf{u}$ does not describe a straight line in the general case, but instead describes an ellipse, which becomes clear if we compare (23.4) with (22.14). This is why we call elliptically polarized a plane wave whose displacement vector is defined by (23.1) and (23.3). A linearly polarized wave (for $\mathbf{u}' \parallel \mathbf{u}''$) is a particular case; a circularly polarized wave

* The vectors $\mathbf{u}'$ and $\mathbf{u}''$ in (23.3) should not be confused with the vectors in (22.3) of the preceding section.

is another. The latter has a displacement vector (23.4) of constant length that describes a circle, which leads to

$$\mathbf{u}^2 = \frac{1}{2}(\mathbf{u}'^2 + \mathbf{u}''^2) + \frac{1}{2}(\mathbf{u}'^2 - \mathbf{u}''^2)\cos 2\varphi - \mathbf{u}'\mathbf{u}''\sin 2\varphi = \text{const},$$

which gives us the condition for circular polarization in the form of an invariant (see [17]):

$$\mathbf{u}'^2 - \mathbf{u}''^2 = \mathbf{u}'\mathbf{u}'' = 0. \qquad (23.5)$$

The vector of (23.4) must be eigenvector of Λ^n corresponding to a fixed eigenvalue v^2 for all φ if an elliptically polarized plane wave is to propagate in a crystal. Specifying some point $\mathbf{r}_0$, the $\mathbf{u}$ of (23.4) will take all directions in a plane containing $\mathbf{u}'$ and $\mathbf{u}''$ as the phase $\varphi = \mathbf{k}\mathbf{r}_0 - \omega t$ varies with time; it will remain an eigenvector of Λ^n only if the latter is uniaxial, with its axis (section 13) perpendicular to the plane of $\mathbf{u}'$ and $\mathbf{u}''$. Hence elliptically polarized transverse waves are possible in a crystal only when the wave normal is an acoustic axis. Here we have a complete analogy with crystal optics, since a light wave in a transparent crystal is always linearly polarized except along the optic axes [17]. A light wave or a transverse elastic wave can have any polarization during propagation in any direction if the medium is isotropic, of course.

All the properties of the ellipse described by the vector of (23.4) may be derived from the formulas of the previous section if we replace $\mathbf{p}$ by $\mathbf{u}'$ and $\mathbf{q}$ by $\mathbf{u}''$ [compare (22.14) and (23.4)]. In particular, (22.22) gives the lengths of the semiaxes as

$$\left.\begin{matrix} a^2 \\ b^2 \end{matrix}\right\} = \frac{1}{2}\left(\mathbf{u}'^2 + \mathbf{u}''^2 \pm \sqrt{(\mathbf{u}'^2 + \mathbf{u}''^2)^2 - 4[\mathbf{u}'\mathbf{u}'']^2}\right) \qquad (23.6)$$

and the directions of these are given by (22.23) as

$$\left.\begin{aligned} \mathbf{a} &\parallel [\mathbf{u}''[\mathbf{u}'\mathbf{u}'']] - a^2\mathbf{u}' \parallel [\mathbf{u}'[\mathbf{u}''\mathbf{u}']] - a^2\mathbf{u}'', \\ \mathbf{b} &\parallel [\mathbf{u}''[\mathbf{u}'\mathbf{u}'']] - b^2\mathbf{u}' \parallel [\mathbf{u}'[\mathbf{u}''\mathbf{u}']] - b^2\mathbf{u}''. \end{aligned}\right\} \qquad (23.7)$$

Consider now the behavior of the energy-flux vector along an acoustic axis when the elastic wave is elliptically polarized. The general formulas (21.6) and (21.12) are

$$\mathbf{P} = -\sigma\dot{\mathbf{u}}, \quad \mathcal{E} = \rho\dot{\mathbf{u}}^2, \qquad (23.8)$$

and in these we substitute from (23.4) for $\mathbf{u}=\mathbf{u}'\cos\varphi-\mathbf{u}''\sin\varphi$. Performing the operations of section 21 on this vector, we get

$$\left.\begin{aligned}\dot{\boldsymbol{u}} &= \omega(\boldsymbol{u}'\sin\varphi+\boldsymbol{u}''\cos\varphi),\\ \frac{\partial u_l}{\partial x_j} &= -k_j(u_l'\sin\varphi+u_l''\cos\varphi),\end{aligned}\right\} \tag{23.9}$$

which gives after substitution in (23.8)

$$\left.\begin{aligned}P_i &= -c_{ijlm}\gamma_{lm}\dot{u}_j = \omega c_{ijlm}k_m(u_j'u_l'\sin^2\varphi+\\ &\quad + u_j''u_l''\cos^2\varphi+(u_j'u_l''+u_j''u_l')\sin\varphi\cos\varphi),\\ \mathcal{E} &= \rho\omega^2(u'^2\sin^2\varphi+u''^2\cos^2\varphi+u'u''\sin 2\varphi).\end{aligned}\right\} \tag{23.10}$$

Also, (23.9) indicates that $\mathbf{u}$ describes an ellipse; comparison with (23.4) shows that, apart from the factor ω, $\dot{\mathbf{u}}$ differs from $\mathbf{u}$ in having $\mathbf{u}'$ replaced by $\mathbf{u}''$ and $\mathbf{u}''$ by $-\mathbf{u}'$. Formulas (23.6) and (23.7) imply that neither the lengths nor the orientations of the axes are affected by this substitution, so the ellipse described by $\dot{\mathbf{u}}$ has the same shape and orientation as that described by $\mathbf{u}$, differing merely in size by virtue of the factor ω. But from (15.8) and (15.12) we have

$$\boldsymbol{k}=\frac{\omega}{v}\boldsymbol{n}=\omega\boldsymbol{m}, \tag{23.10'}$$

so (23.10) for the energy-flux vector may [see (22.9)] be put as

$$\boldsymbol{P}=\rho\omega^2(\Lambda^{u'}\sin^2\varphi+\Lambda^{u''}\cos^2\varphi+\Lambda'''\sin\varphi\cos\varphi)\,\boldsymbol{m}, \tag{23.11}$$

in which Λ''' is given by (22.6)-(22.8).

Now we compare (23.11) for the instantaneous flux density for this wave with the corresponding expression [see (21.9)]

$$\boldsymbol{P}=\rho\omega^2\sin^2\varphi\Lambda^{u^0}\boldsymbol{m} \tag{23.12}$$

for the same vector of a linearly polarized wave. The main difference is that the vector of (23.12) changes only in magnitude in response to φ (from zero to $\mathbf{P}_0=\rho\omega^2\Lambda u^0\mathbf{m}$), while retaining a fixed direction, whereas that of (23.11) varies in magnitude and direction. Apart from the factor $\rho\omega^2$, (23.11) for $\mathbf{P}$ is entirely analogous to (22.9) for $\mathbf{s}$; but there is a major difference in the content of these formulas. Firstly, the **u'** and **u"** of (22.3)-(22.9) were taken as unit mutually perpendicular vectors, whereas those of (23.4)-(23.11) are arbitrary in magnitude and lie mutually at any angle. Secondly, the θ of (22.9) has been specified and defines the direction of the displacement vector of (22.4) for a linearly polarized wave, whereas in (23.11) the phase φ is, from (23.1), a function of time and coordinates. Hence a time-averaged [see (21.13)] of the energy-flux vector for a linearly polarized wave does not alter the direction of that vector. Moreover, (21.9) and (21.12)-(21.14) shows that the vector for the mean ray velocity exactly coincides with the vector for the instantaneous ray velocity, i.e., the latter is constant in magnitude and direction for a linearly polarized wave.

The vector **P** of an elliptically polarized wave is shown by (23.11) to vary in magnitude and direction, so the mean energy-flux vector becomes meaningless, as does the mean ray-velocity vector. We can use (23.9) and (23.10') to put (23.11) in the compact form

$$\boldsymbol{P} = \rho \Lambda^{\dot{u}} \boldsymbol{m}. \tag{23.13}$$

Multiplication by **n** gives

$$\boldsymbol{Pn} = \frac{\rho}{v} \boldsymbol{n} \Lambda^{\dot{u}} \boldsymbol{n} = \frac{\rho}{v} \dot{\boldsymbol{u}} \Lambda^{n} \dot{\boldsymbol{u}}.$$

It follows from $\Lambda^{\mathbf{n}}\mathbf{u} = \mathrm{v}^2\mathbf{u}$ after differentiation with respect to time that $\Lambda^{\mathbf{n}}\dot{\mathbf{u}} = \mathrm{v}^2\dot{\mathbf{u}}$, so $\dot{\mathbf{u}}\Lambda^{\mathbf{n}}\dot{\mathbf{u}} = \mathrm{v}^2\dot{\mathbf{u}}^2$ and

$$\boldsymbol{Pn} = v\rho\dot{\boldsymbol{u}}^2 = v\mathfrak{E}. \tag{23.14}$$

Hence, introducing the instantaneous ray-velocity vector **s** via the definition of (21.14), we have

$$\boldsymbol{s} = \frac{\boldsymbol{P}}{\mathfrak{E}}, \tag{23.15}$$

which, as before, satisfies (21.20):

$$\boldsymbol{sn} = v, \quad \boldsymbol{sm} = 1.$$

Using (23.8), (23.13), and (23.15) we may put

$$\boldsymbol{s} = \frac{\Lambda^{\dot{u}} \boldsymbol{m}}{\dot{\boldsymbol{u}}^2}, \tag{23.16}$$

or, introducing the unit vector $\dot{\mathbf{u}}_e = \dot{\mathbf{u}}/|\dot{\mathbf{u}}|$,

$$s = \Lambda^{\dot{u}_e} m. \tag{23.17}$$

This last expression differs from (21.33) only in having $\dot{\mathbf{u}}_e$ in place of $\mathbf{u}$, which is expected, since $\mathbf{u} = \mathbf{u}^0 \cos \varphi$, for linear polarization, so $\dot{\mathbf{u}} = \omega \mathbf{u}^0 \sin \varphi$, differs from $\mathbf{u}$ only by a scalar factor: $\dot{\mathbf{u}} = \omega \tan \varphi \mathbf{u}$. This difference drops out for the unit vectors, so (23.17) becomes (21.33); but (23.17) is more general, since it is dependent on the phase variable φ in the case of elliptical polarization.

The unit vector

$$\dot{u}_e = \frac{\dot{u}}{|\dot{u}|} = \frac{u' \sin \varphi + u'' \cos \varphi}{\sqrt{u'^2 \sin^2 \varphi + u''^2 \cos^2 \varphi + u'u'' \sin 2\varphi}} \tag{23.18}$$

always may be put in the form of (22.4):

$$\dot{u}_e = u_1 \cos \theta + u_2 \sin \theta, \quad u_1^2 = u_2^2 = 1, \quad u_1 u_2 = 0, \tag{23.19}$$

in which the constant unit mutually orthogonal vectors $\mathbf{u}_1$ and $\mathbf{u}_2$ lie in the same plane as the $\mathbf{u}'$ and $\mathbf{u}''$ of (23.4), and the direction of one of them in that plane may be chosen arbitrarily (θ is the angle between $\dot{\mathbf{u}}_e$ and $\mathbf{u}_1$). We take the directions of (22.3) for $\mathbf{u}_1$ and $\mathbf{u}_2$, which causes (23.17) and (23.19) to coincide precisely in form with (22.9) and (22.4), so all the subsequent relationships of the previous section apply also to elliptical polarization. A major difference, however, is that the θ of section 22 is taken as constant, denoting the fixed orientation of the displacement vector for the quasitransverse linearly polarized wave. The θ of (23.19) depends on the phase variable φ and varies with the latter as a function of time and coordinates; (23.18) and (23.19) give the dependence of θ on φ as

$$\cos \theta = u_1 \dot{u}_e = \frac{u_1 \dot{u}}{|\dot{u}|}, \quad \sin \theta = \frac{u_2 \dot{u}}{|\dot{u}|}. \tag{23.20}$$

Then $\varphi = 0$ gives $\theta = \theta_0 = (\widehat{\mathbf{u}_1 \mathbf{u}''})$; variation of φ from 0 to 2π corresponds to change in θ from θ_0 to $\theta_0 + 2\pi$.

The **s** of (23.17) thus behaves as follows for an elliptically polarized wave propagating along an acoustic axis **n**: **s** varies in accordance with (22.9)-(22.11) and (22.14), in which θ is defined by (23.9) and (23.20), with $\varphi = \mathbf{kr} - \omega t$. The end of **s** then traces out an ellipse in the plane $\mathbf{sn} = v$, the center of the ellipse lying at the point defined by the $\mathbf{s}_0$ of (22.10) and the equation of the ellipse having the form of (22.16), the shape, orientation, size, and direction of traversal being defined by (22.22)-(22.24). Vector **s** itself describes the surface of an elliptic cone, whose form and properties are described by (22.25)-(22.27). Vector $\mathring{\mathbf{u}}_e$ for any linearly independent **u'** and **u"** of (23.4) may be reduced to the standard form of (23.19), so all the relationships of section 22 remain completely valid for all such **u'** and **u"**. Specification of **u'** and **u"** entirely defines the polarization ellipse, so the **P** of (23.13) describes the cone of (22.25)-(22.27), and the **s** of (23.17) describes the ellipse of (22.16), as the elliptically polarized wave propagates along the specified acoustic axis in that crystal. For the same conditions, but with the wave linearly polarized, **P** is directed along a generator of the cone of (22.25), while **s** is directed to a fixed point on the ellipse of (22.16), whose position is dependent on the orientation of **u**.

Here we get an apparent contradiction, because we can pass continuously to linear polarization from elliptical by reducing one of the semiaxes of the ellipse continuously to zero; whereas the above would leave **s** describing the same cone (ellipse) no matter how narrow the polarization ellipse, while **s** for linear polarization would have a constant direction and magnitude. It seems impossible to make the continuous transition (from elliptical to linear polarization) compatible with a jump in the behavior of **s** (from motion on a cone to constant coincidence with one of the generators of that cone). This paradox is actually very simple to resolve if we consider the kinematics of the ray-velocity vector instead of pure geometry. From (23.4) we have $\mathbf{u} = \mathbf{u}' \cos\varphi - \mathbf{u}'' \sin\varphi$, which may be considered as the parametric equation of the ellipse traced out by the end of **u** as φ varies. We define the orientation of **u** via the angle θ' formed with vector **u'**. By analogy with (23.20) we have

$$\cos\theta' = \frac{u'u}{\sqrt{u'^2 u^2}}. \tag{23.21}$$

Consider the derivative

$$\frac{d\theta'}{d\varphi} = -\frac{1}{\sin\theta'}\frac{d}{d\varphi}\frac{u'u}{\sqrt{u'^2u^2}} = -\sqrt{\frac{[u'u]^2}{u'^2u^2}}\frac{d}{d\varphi}\frac{u'u}{\sqrt{u'^2u^2}}.$$

Simple steps give

$$\frac{d\theta'}{d\varphi} = \frac{|[u'u'']|}{u^2}. \tag{23.22}$$

But $\varphi = \mathbf{kr} - \omega t$, so the absolute magnitude of the angular velocity of $\mathbf{u}$ is

$$|\dot{\theta}'| = \left|\frac{d\theta'}{d\varphi}\dot{\varphi}\right| = \omega\frac{|[u'u'']|}{u^2}. \tag{23.23}$$

The area velocity V_σ (area described by $\mathbf{u}$ in unit time) is given by (23.23) as

$$V_\sigma = \frac{1}{2}u^2\dot{\theta}' = \frac{\omega}{2}|[u'u'']| = \text{const.} \tag{23.24}$$

Thus the area velocity is constant for elliptical polarization of a plane harmonic wave.*

The paradox is now resolved easily. A highly elongated ellipse (Fig. 11a) has a constant sector velocity, so the angular velocity, given by (23.4) as $\dot{\theta}' = 2V_\sigma/\mathbf{u}^2$, will be very large near points B and D, because $|\mathbf{u}|$ is small, whereas near A and C (major axis) $\dot{\theta}'$ will be relatively small. Let point A on the displacement ellipse (Fig. 11a) correspond to A' on the ray-velocity ellipse (Fig. 11b). Vector $\mathbf{s}_1$ rotates in accordance with the motion of $\mathbf{u}$, but twice as fast, since the angle is doubled in (22.14). Hence motion of $\mathbf{u}$ through π (A to C) corresponds to a complete rotation of $\mathbf{s}_1$ in the ellipse of (22.14), with return to point A'. The ellipse of $\mathbf{s}_1$ does not alter in shape, orientation, or size as the eccentricity of the $\mathbf{u}$ ellipse increases, but it spends an increasing proportion of its time near A' and travels along the rest ever more rapidly. In the limit, ellipse $\mathbf{u}$ becomes **a** straight line

---

* This law of areas is also true for mechanical motion in a coulomb field and for elliptically polarized plane harmonic waves of any type, although the phenomena are completely different in nature.

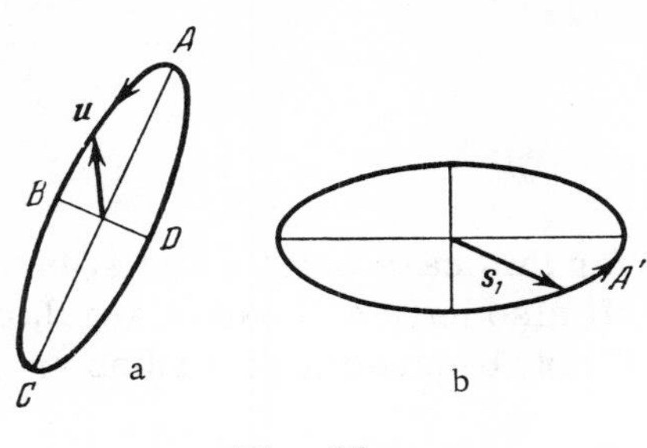

Fig. 11

while $\mathbf{s}_1$ spends all its time at A', which corresponds to no time spent running over the rest of the ellipse (infinite angular velocity). This is how we get a continuous transition from a state of elliptical polarization, in which the ray-velocity vector describes the entire cone of (22.25), to the state of linear polarization, in which it is directed always to one point on that cone. These arguments are easily confirmed by a simple calculation. Let ψ be the angle formed by $\mathbf{s}_1$ with the $\mathbf{p}$ of (22.14). Then the area velocity of $\mathbf{s}_1$ is

$$V_s = \frac{1}{2} s_1^2 \dot{\psi}. \tag{23.25}$$

Angle ψ for $\mathbf{s}_1$ is related to $2\theta'$ as is angle θ for $\mathbf{u}$ to φ; hence, by analogy with (23.21)-(23.24), we may put

$$\cos\psi = \mathbf{s}_1\mathbf{p}/\sqrt{\mathbf{s}_1^2\mathbf{p}^2}, \qquad d\psi/d(2\theta') = |[\mathbf{pq}]|/\mathbf{s}_1^2 \tag{23.26}$$

$$|\dot{\psi}| = |d\psi/d(2\theta')|2\dot{\theta}' = 2\omega(|[\mathbf{pq}]|/\mathbf{s}_1^2)\cdot(|\mathbf{u}'\mathbf{u}''|/\mathbf{u}^2). \tag{23.27}$$

Hence the section velocity V_S of $\mathbf{s}_1$ is

$$V_s = \omega\frac{|[\boldsymbol{pq}]|\cdot|[\boldsymbol{u}'\boldsymbol{u}'']|}{(\boldsymbol{u}'\cos\varphi - \boldsymbol{u}''\sin\varphi)^2}. \tag{23.28}$$

An elongated $\mathbf{u}$ ellipse is obtained by making $\mathbf{u}''$ small, for example. We see from (23.28) that for $\varphi = \pi/2$ the speed $V_s' = \omega|[\boldsymbol{pq}]|\dfrac{|[\boldsymbol{u}'\boldsymbol{u}'']|}{\boldsymbol{u}''^2}$ is large, while $\varphi = 0,\ \pi$, the speed $V_s'' = \omega|[\boldsymbol{pq}]|\dfrac{|[\boldsymbol{u}'\boldsymbol{u}'']|}{\boldsymbol{u}'^2}$ is small; $|\mathbf{u}''| \to 0$ corresponds to $V_S' \to \infty$, whereas $V_S'' \to 0$, in complete accordance with the above.

The general formula (23.28) allows us to establish the behavior of $\mathbf{s}_1$ for a circularly polarized wave also. Here (23.5) gives $\mathbf{u}'^2 - \mathbf{u}''^2 = \mathbf{u}'\mathbf{u}'' = 0$, so

$$(\mathbf{u}'\cos\varphi - \mathbf{u}''\sin\varphi)^2 = |[\mathbf{u}'\mathbf{u}'']| = \mathbf{u}'^2$$

and

$$V_s = \omega\,|[pq]| = \text{const.}$$

Thus for circular polarization (and for this case only) **u** and $\mathbf{s}_1$ both have constant sector velocities; **u** will also have a constant angular velocity, whereas $\mathbf{s}_1$, in general, will not, because it describes an ellipse.

The ray velocity coincides with the phase velocity for an isotropic medium (section 21); this property occurs for transverse waves with any direction of the displacement vector, so it persists also when the displacement vector rotates continuously, as in elliptical polarization. The vectors for the phase and ray velocities thus always coincide for an isotropic medium, no matter what the state of polarization.

Finally consider the directions in which **u** and $\mathbf{s}_1$ describe their ellipses. From (22.24) we see that the sense of motion of $\mathbf{s}_1$ is determined by the sign of the product $\mathbf{n}[\mathbf{s}_1(d\mathbf{s}_1/d\theta)]$ or $\mathbf{n}[\mathbf{s}_1\dot{\mathbf{s}}_1]$. The wave normal **n** always forms an acute angle with the displacement vector $\mathbf{u}_0 \| [\mathbf{u}'\mathbf{u}'']$ of the quasilongitudinal wave, so the direction of motion of **u** in (23.4) is characterized by the sign of $\mathbf{n}[\mathbf{u}\dot{\mathbf{u}}]$. Clearly, the two ellipses will be described in the same sense if $[\mathbf{u}\dot{\mathbf{u}}][\mathbf{s}_1\dot{\mathbf{s}}_1]$ is positive, but in opposite senses if it is negative.

24. Wave Surfaces

Various types of surface are used in crystal optics to characterize the laws of propagation of light [17, 23]. Similarly, the laws for elastic waves in crystals may also be illustrated by the use of surfaces. Here three main types are used, each being the locus of the end of one of the vectors characterizing the wave propagation. We shall term them all wave surfaces, although the term is commonly used to denote only one of them.

If we lay off from some point O the vectors for the phase velocities corresponding to all directions of the wave normal **n**, the surface formed by these ends is that of the phase velocity, or simply the velocity surface Here the radius vector is the phase velocity:

$$\boldsymbol{r} = \boldsymbol{v} = v\boldsymbol{n}. \tag{24.1}$$

The equation for this surface is obtained from the condition that there are nonzero solutions to $(\Lambda^n - v^2)\mathbf{u} = 0$; this condition [see (13.4) and (15.22)] takes the form

$$|v^2 - \Lambda^n| = v^6 - \Lambda^n_t v^4 + \overline{\Lambda}^n_t v^2 - |\Lambda^n| = 0. \tag{24.2}$$

From (15.20) we see that Λ^n is homogeneously quadratic in the components of unit vector $\mathbf{n}$, so

$$v^2\Lambda^n = \Lambda^v, \quad v^4\overline{\Lambda}^n = \overline{\Lambda}^v, \quad v^6|\Lambda^n| = |\Lambda^v|.$$

We multiply (24.2) by v^6 in order that it shall contain only components of vector $\mathbf{v}$, apart from the constant parameters entering into tensor λ_{iklm}. The result is

$$v^{12} - \Lambda^v_t v^8 + \overline{\Lambda}^v_t v^4 - |\Lambda^v| = 0. \tag{24.3}$$

This is the equation of the phase-velocity surface (surface V), which is an equation of degree twelve. To each direction of the wave normal there correspond three different phase velocities, which means that any line from O meets the surface of (24.3) three times, so the surface is one of three sheets. One of these, sheet L (longitudinal) corresponds to the quasilongitudinal waves; the other two, T_1 and T_2 (transverse), correspond to the quasitransverse waves. The speed of the former exceeds those of the latter two for all known media (for the same $\mathbf{n}$), so sheet L encloses T_1 and T_2, having no points in common with the latter. T_1 and T_2 may have common points, which lie on acoustic axes, since only along these directions do the two quasitransverse waves have the same velocity.

Surface V becomes two concentric spheres for an isotropic medium, the radius of the outer one corresponding to the speed of the longitudinal waves.

From (19.10) we have for an isotropic medium that $\Lambda^n = a + b\mathbf{n}\cdot\mathbf{n}$, $\Lambda^n_t = 3a + b$, $\overline{\Lambda}^n_t = 3a^2 + 2ab$, $|\Lambda^n| = a^2(a + b)$ (see Table II). Substituting this in (24.3) gives

$$v^6(v^2 - a - b)(v^2 - a)^2 = 0, \tag{24.4}$$

i.e., the left side splits up into rational factors linear in v^2; $v \neq 0$, so we have the equations of two spheres $\mathbf{r}^2 = v^2 = a + b = v_0^2$ and $\mathbf{r}^2 = a = v_1^2$. The only other case where (24.3) factorizes in this way is for hexagonal crystals (see section 32).

The refraction surface centered on O plays a more important part in the theory of elastic waves; this is formed by the ends of the refraction vectors. This surface does not have a universally accepted name; it has been called the reciprocal-velocity surface, the index surface, the inverse surface, or even the slowness surface [13, 24]. Its radius vector is the refraction vector

$$\boldsymbol{r} = \boldsymbol{m} = \frac{\boldsymbol{n}}{v}, \tag{24.5}$$

so it is natural to call it the refraction surface (surface M). I propose to use this term, the more so since this surface plays the principal part in problems of reflection and refraction of elastic waves (see Chapter 8). The equation of surface M is found from (24.2) by dividing the left side by v^6. We have $(1/v^2)\Lambda^n = \Lambda^m$, $(1/v^4)\bar{\Lambda}^n = \bar{\Lambda}^m$, $(1/v^6)|\Lambda^n| = |\Lambda^m|$, so the equation becomes

$$|\Lambda^m - 1| = |\Lambda^m| - \bar{\Lambda}_t^m + \Lambda_t^m - 1 = 0. \tag{24.6}$$

This is an equation of sixth degree in the components of the vector $\mathbf{m} = \mathbf{r}$; it resembles surface V in consisting of three separate sheets: L, T_1, and T_2, which corresponds, respectively, to the longitudinal and two transverse waves. Sheets T_1 and T_2 have common points along the acoustic axes. Equation (24.6) splits up into the equations of two spheres for an isotropic medium, for which

$$\Lambda^m = am^2 + b\boldsymbol{m} \cdot \boldsymbol{m}, \tag{24.7}$$

so from Table II

$$|\Lambda^m - 1| = |(am^2 - 1) + bm^2\boldsymbol{n} \cdot \boldsymbol{n}| = [(a+b)m^2 - 1](am^2 - 1)^2 = 0, \tag{24.7'}$$

whence we have

$$m^2 = \frac{1}{a+b} = \frac{1}{v_0^2}, \quad m^2 = \frac{1}{a} = \frac{1}{v_1^2}. \tag{24.8}$$

Equation (24.6) also splits up for hexagonal crystals (sections 31 and 32).

Surfaces V and M are related by the feature that the product of their radius vectors in a given direction is one: $|\mathbf{v}|\,|\mathbf{m}|=1$, so either surface may be derived from the other by inversion. The inversion transformation is specified in terms of point O (the inversion center) and a sphere of radius $\mathbf{r}_0$ drawn around it. A point on any surface lying at a distance $\mathbf{r}_1$ from O along a ray is related to a point on the same ray at a distance $\mathbf{r}_2$ from O, with $\mathbf{r}_1\mathbf{r}_2=\mathbf{r}_0^2$. In our case $\mathbf{r}_1=|\mathbf{v}|$, $\mathbf{r}_2=|\mathbf{m}|$, and $\mathbf{r}_0=1$, so the V and M surfaces are mutually inverse with respect to unit sphere having its inversion center at point O.

The special points are very important features of any surface or curve; surfaces V and M are best discussed from this viewpoint. A necessary condition for such points on a surface $f(\mathbf{r})=0$ is [28] that $\partial f/\partial \mathbf{r}=0$. Equation (24.6) is differentiated with respect to $\mathbf{m}$ and then multiplied by $\mathbf{m}$ to give for the special points of surface M that

$$\boldsymbol{m}\frac{\partial}{\partial \boldsymbol{m}}|\Lambda^m-1|=2\left(3\,|\Lambda^m|-2\bar{\Lambda}_t^m+\Lambda_t^m\right)=0. \tag{24.9}$$

This equation is obtained by applying Euler's theorem to the expressions $|\Lambda^m|$, $\bar{\Lambda}_t^m$, and Λ_t^m, which are, respectively, homogeneous functions of degrees six, four, and two. From (24.6) and (24.9) we have

$$\bar{\Lambda}_t^m-2\Lambda_t^m+3=0. \tag{24.10}$$

We multiply this by v^4 and use the fact that $v^2\Lambda_t^m=\Lambda_t^n$, $v^4\bar{\Lambda}_t^m=\bar{\Lambda}_t^n$, to get

$$3v^4-2v^2\Lambda_t^n+\bar{\Lambda}_t^n=0, \tag{24.11}$$

which is identical with (18.12), which corresponds to identity of the velocities, i.e., to an acoustic axis.

Then we may say that the refraction surface can have special points only along the acoustic axes. The same applies to the velocity surface of (24.3), since this differs from (24.6) only by inversion of the radius vector.

The third surface used to characterize elastic waves has the most direct physical meaning, being the locus of the points reached at time $t = 1$ by a wave disturbance arising at O at $t = 0$. This surface differs only as to scale from the actual wavefront propagating in all directions from a point source, so it is called the wave surface. This surface consists of the points reached by the energy of the wave disturbance at a given instant; the radius vector from the source to any point on this surface represents the distance traveled by the energy in unit time in that direction. But this distance is, by definition, equal to the ray velocity, so the wave surface may be defined as the envelope of the ends of the ray-velocity surface or the ray surface (surface S). It resembles the other surfaces in consisting of three sheets, which correspond to the quasilongitudinal wave and to the quasitransverse ones.

The definition of surface S shows that all points on it start to oscillate at the same instant; it is thus a surface of equal phase. Any small part of it at a large distance from the source may be considered as plane; this is how ordinary plane waves arise. The perpendicular **n** to this surface of equal phase will by definition be the phase (wave) normal of section 15. These concepts may be applied also to the case when the surface is considered at no great distance from the source; the area that may be treated as plane is naturally reduced, and the orientation of this area will be defined by that of the plane tangential to the wave surface at that point. This shows that the wave (phase) normal is the geometrical normal to the wave (ray) surface (Fig. 12, in which S is that surface). This demonstrates the close connection between surface S and surface M, whose radius vector is parallel to the wave normal. From (21.34), the angle between the radius vectors of surfaces S and M is defined by

$$\boldsymbol{sm} = 1. \qquad (24.12)$$

This, for a given **m**, may be considered as the equation of a plane perpendicular to **n** on which the end of vector **s** must lie. A plane of this type may be drawn for every **m** that satisfies (24.6), i.e., for each point on surface M. Each such plane P_m will pass through the corresponding point **s** on surface S and will touch the latter at that point, since P_m is perpendicular to **n**. Hence surface S is the envelope of the family of planes of (24.12) for which **m** satisfies (24.6).

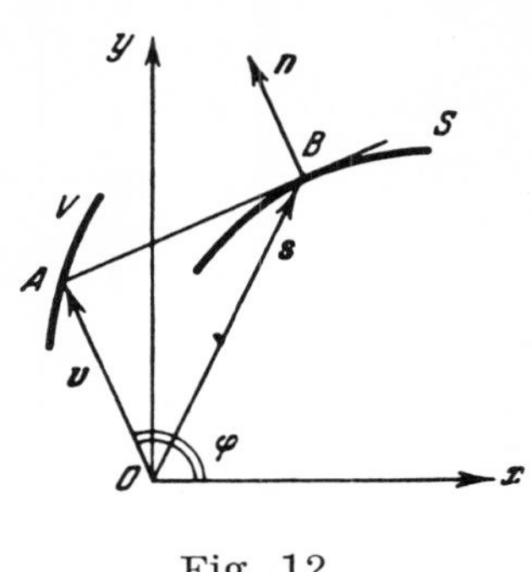

Fig. 12

Putting (24.12) as

$$sn = v \tag{24.13}$$

and assuming surface S as known, we can construct the phase-velocity vector $\mathbf{v} = v\mathbf{n}$ corresponding to a given $\mathbf{s}$. Through point $\mathbf{s}$ we draw the tangential plane and drop to it a perpendicular from O; (24.13) gives the length of this perpendicular as v and the direction as parallel to $\mathbf{n}$, so the foot of the perpendicular will be the end of vector $\mathbf{v}$, i.e., will lie on the phase-velocity surface of (24.3). Repeating this construction for all points on the ray surface, we find that surface V is the locus of the feet of the perpendiculars dropped from O to planes tangential to surface S (Fig. 12). Conversely, surface S is the antipode* of surface V.

The general equations for surfaces V and M, namely (24.3) and (24.6), are easily derived, but that for surface S is not. The radius vector of this surface is defined by (21.33) as

$$s = \Lambda^u m, \tag{24.14}$$

in which $\mathbf{m}$ satisfies (24.6) and $\mathbf{u}$ is unit displacement vector, which for a given $\mathbf{m}$ is defined by Christoffel's equation:

$$(\Lambda^m - 1) u = 0. \tag{24.15}$$

Now $\Lambda^u = a + b\mathbf{u}\cdot\mathbf{u}$ for isotropic media, with $\mathbf{u} = \mathbf{n}$, $v_0^2 = a + b$ for a longitudinal wave and $\mathbf{u} \perp \mathbf{n}$, $v_1^2 = a$ for a transverse one. Hence (24.14) gives $\mathbf{s}_0 = \mathbf{v}_0$ for a longitudinal wave and $\mathbf{s}_1 = \mathbf{v}_1$ for a transverse one; the phase and ray velocities coincide as do the corresponding surfaces, for an isotropic medium.

To find an explicit expression for $\mathbf{s}$, in which only the elastic constants λ_{iklm} appear in addition to the s_k, we have to express $\mathbf{u}$ in general form in terms of $\mathbf{m}$ from (24.15) and substitute into (24.14), which gives

* By definition, B is the antipode of A if surface A is the pode of B.

$$s = s(m_i, \lambda_{iklm}). \tag{24.16}$$

We now eliminate the three components of **m** from the four equations (24.6) and (24.16), which gives us that

$$F(s_i, \lambda_{iklm}) = 0, \tag{24.17}$$

which is the explicit equation for the wave surface. However, the same result may be obtained in another way. We use (21.40) and (21.42) as being equivalent to (24.14) and (24.15):

$$(\Lambda^u - s \cdot s)\, n = 0, \tag{24.18}$$

$$(\Lambda^n - (ns)^2)\, u = 0. \tag{24.19}$$

We eliminate **u** and **n**, with $\mathbf{u}^2 = \mathbf{n}^2 = 1$, to get again a relation of the type of (24.17). It is simple to eliminate either **u** or **n** in a very general way. Using (13.8) [see also (21.44)], we get from (24.18) that

$$n \cdot n = \frac{\overline{\Lambda^u - s \cdot s}}{(\overline{\Lambda^u - s \cdot s})_t} \tag{24.20}$$

and similarly from (24.19)

$$u \cdot u = \frac{\overline{\Lambda^n - (ns)^2}}{(\overline{\Lambda^n - (ns)^2})_t}. \tag{24.21}$$

Then from (21.17) and (21.45) we may put

$$\Lambda^n_{ik} = \frac{\lambda_{ijkl}(\overline{\Lambda^u - s \cdot s})_{jl}}{(\overline{\Lambda^u - s \cdot s})_t}, \tag{24.22}$$

$$\Lambda^u_{ik} = \frac{\lambda_{ijkl}(\overline{\Lambda^n - (ns)^2})_{jl}}{(\overline{\Lambda^n - (ns)^2})_t}. \tag{24.23}$$

We substitute these, respectively, into (24.19) and (24.18) to get

$$\left[\lambda_{ijkl}(\overline{\Lambda^u - s \cdot s})_{jl} - s(\overline{\Lambda^u - s \cdot s})\, s\delta_{ik}\right] u_k = 0,$$

$$\left[\lambda_{ijkl}(\overline{\Lambda^n - (ns)^2})_{jl} - (\overline{\Lambda^n - (ns)^2})_t s_i s_k\right] n_k = 0.$$

Here we have used the fact that $(\mathbf{ns})^2 = \mathbf{s}(\mathbf{n} \cdot \mathbf{n})\mathbf{s}$. These equations may also be put in the form

$$(\lambda_{ijkl} - \delta_{lk} s_j s_l)\left(\overline{\Lambda^u - s \cdot s}\right)_{jl} u_k = 0, \tag{24.24}$$

$$(\lambda_{ijkl} - s_i s_k \delta_{jl})\left(\overline{\Lambda^n - (ns)^2}\right)_{jl} n_k = 0. \tag{24.25}$$

The first contains only unit vector **u**, apart from **s** and λ_{ijkl}, while the second similarly contains only unit vector **n**. From (12.64) and (13.33) we get that

$$\overline{\Lambda^u - s \cdot s} = (\Lambda^u - \Lambda^u_t)(\Lambda^u + s^2) + \frac{1}{2}(\Lambda^u_t)^2 -$$

$$-\frac{1}{2}((\Lambda^u)^2)_t - (\Lambda^u s \cdot s + s \cdot s\Lambda^u) + \Lambda^u_t s \cdot s + s\Lambda^u s, \tag{24.26}$$

$$\overline{\Lambda^n - (ns)^2} = (\Lambda^n - \Lambda^n_t)(\Lambda^n + (ns)^2) + \frac{1}{2}(\Lambda^n_t)^2 - \frac{1}{2}((\Lambda^n)^2)_t + (ns)^4. \tag{24.27}$$

Since

$$\Lambda^u_{ik} = \lambda_{ijkl} u_j u_l, \qquad \Lambda^n_{ik} = \lambda_{ijkl} n_j n_l, \tag{24.28}$$

we have (24.27) as a homogeneous function of even degree in the components of **n**. We can also make (24.26) a homogeneous function of fourth degree in the components of **u** by introducing where necessary the factor $\mathbf{u}^2 = u_i u_k \delta_{ik} = 1$. Then substitution of (24.26)-(24.28) into (24.24) and (24.25) gives us two three-dimensional vector equations:

$$\beta_{ipqrsk} u_p u_q u_r u_s u_k = 0, \tag{24.29}$$

$$\gamma_{ipqrsk} n_p n_q n_r n_s n_k = 0, \tag{24.30}$$

whose left-hand sides are homogeneous functions of fifth degree in the u_k and n_k, respectively. The sixth-rank tensors β and γ are defined by

$$\beta_{ipqrsk} = (\lambda_{ijkl} - \delta_{ik} s_j s_l)\{(\lambda_{jpqg} - \lambda_{fpqf}\delta_{jg})(\lambda_{grsl} + s^2\delta_{rs}\delta_{gl}) +$$

$$+\frac{1}{2}(\lambda_{gpqg}\lambda_{frsf} - \lambda_{gpqf}\lambda_{frsg})\delta_{jl} +$$

$$+(\lambda_{gpqg} s_j s_l - \lambda_{jpqg} s_g s_l - \lambda_{lpqg} s_g s_j + \lambda_{gpqf} s_g s_f \delta_{jl})\delta_{rs}\}, \tag{24.31}$$

$$\gamma_{ipqrsk} = (\lambda_{ijkl} - s_i s_k \delta_{jl})\{(\lambda_{jpqg} - \lambda_{fpqf}\delta_{jg})(\lambda_{grsl} + s_r s_s \delta_{gl}) +$$

$$+\frac{1}{2}(\lambda_{gpqg}\lambda_{frsf} - \lambda_{gpqf}\lambda_{frsg})\delta_{jl} + s_p s_q s_r s_s \delta_{jl}\}. \tag{24.32}$$

The problem thus reduces to eliminating **u** from (24.29) and (24.31) or **n** from (24.30) and (24.32).

The general theory of elimination (see [26], section 82, for example) shows that k general homogeneous forms $f_1, f_2, f_3, \ldots, f_k$ of degree l in k variables $x_1, x_2, \ldots, x_k$ have a resultant R that is a polynomial in the coefficients of these forms, in particular, a homogeneous polynomial of degree l^{k-1} in the coefficients of each form, and hence a homogeneous polynomial of degree

$$p = kl^{k-1} \tag{24.33}$$

in the coefficients of all forms. The equation R = 0 is the result of eliminating the variables x_s from the system $f_1(x_1, \ldots, x_k) = 0, \ldots, f_k(x_1, \ldots, x_k) = 0$ $(s = 1, 2, \ldots, k)$. Applying these general deductions to (24.29) and (24.30), we see that in both cases R is a homogeneous polynomial of degree $p = 3 \times 5^2 = 75$ in the coefficient β_{ipqrsk} or γ_{ipqrsk}, respectively. This allows us to find the highest degree to which the components of **s** can appear in R and thus to determine the order of the wave surface. Here it is found that the terms in the β_{ipqrsk} of fourth order in **s** condense:

$$s_j s_l [(\lambda_{jpqg} - \lambda_{fpqf}\delta_{ig}) s^2 \delta_{gl} + \lambda_{gpqg} s_j s_l - \lambda_{jpqg} s_g s_l - \lambda_{lpqg} s_g s_j + \lambda_{gpqf} s_g s_f \delta_{jl}] \equiv 0.$$

Hence the β will be of second degree in **s**, so the degree of R and the order of the wave surface may be as high as 150.*

From these arguments we conclude formally that elimination of the n_k from (24.30) would give a resultant of degree $6 \times 75 = 450$ in s, since the γ of (24.32) contain uncondensed terms of sixth order in **s**: $3 s_i s_p s_q s_r s_s s_k$. However, the result of eliminating the six variables n_k and u_l from (24.18) and (24.19) cannot be dependent on the order in which the elimination is performed. As there is only one resultant, the degree of this as given by (24.30) cannot exceed that implied by (24.29). Hence the resultant of (24.30) must be such that terms of degree over 150 in **s** cancel out.

---

* The same conclusion is reached from the relation between the wave and refraction surfaces [13].

We cannot say that the order of surface S is absolutely obliged to be 150 even for a triclinic crystal; the actual degree may be much lower, as for hexagonal crystals* (see section 25). However, any detailed analysis of the general relations (24.29)-(24.32) is very difficult, in view of their complexity.

Only the general condition for special points will therefore be considered. If the surface is given in the parametric form

$$\boldsymbol{r}=\boldsymbol{r}(\alpha', \alpha''), \tag{24.34}$$

in which $\mathbf{r}$ is the radius vectors and α' and α'' are parameters, then the necessary condition [28] for special points is

$$\left[\frac{\partial \boldsymbol{r}}{\partial \alpha'}, \frac{\partial \boldsymbol{r}}{\partial \alpha''}\right]=0. \tag{24.35}$$

We have $\mathbf{r}=\mathbf{s}=\partial v/\partial \mathbf{n}$ from (21.24), in which unit vector $\mathbf{n}$ is a function of α' and α'' (e.g., the angle θ and φ in a spherical coordinate system, $n_1=\sin\theta\cos\varphi$, $n_2=\sin\theta\sin\varphi$, $n_3=\cos\theta$). Let

$$\frac{\partial \boldsymbol{n}}{\partial \alpha'}=\boldsymbol{n}', \quad \frac{\partial \boldsymbol{n}}{\partial \alpha''}=\boldsymbol{n}''. \tag{24.36}$$

Then, differentiating $\mathbf{n}^2=1$ with respect to α' and α'', we have

$$\boldsymbol{n}\boldsymbol{n}'=\boldsymbol{n}\boldsymbol{n}''=0. \tag{24.37}$$

Then

$$[\boldsymbol{n}'\boldsymbol{n}'']=C\boldsymbol{n}. \tag{24.38}$$

We assume $C\neq 0$ is unrelated to the form of the surface but is determined by the choice of the coordinate parameters α' and α''. For example, for $\alpha'=\theta$ and $\alpha''=\varphi$ we have $C=\sin\theta$, i.e., $C=0$ for $\theta=0$ or π. Hence $\theta=0$ is a special point of the coordinates but not of the surface. From (24.36)

$$\frac{\partial s_i}{\partial \alpha'}=\frac{\partial s_i}{\partial n_k}\frac{\partial n_k}{\partial \alpha'}=\frac{\partial s_i}{\partial n_k}n'_k. \tag{24.39}$$

---

*An example is the general equation (24.3) for surface V, which is actually of fourth degree in (24.4) for an isotropic medium, instead of degree twelve.

Introducing tensor β via (21.28), we get

$$\frac{\partial s}{\partial \alpha'} = \beta n', \quad \frac{\partial s}{\partial \alpha''} = \beta n'' \tag{24.40}$$

and condition (24.35) becomes

$$[\beta n', \beta n''] = \overline{\beta}[n'n''] = C\overline{\beta} n = 0. \tag{24.41}$$

Here (12.51) has been used. Putting $C \neq 0$ and using (13.34), $\overline{\beta} = \overline{\beta}_t + (\beta - \beta_t)\beta$, together with (21.29), we get $\overline{\beta}_t \mathbf{n} = 0$, or, since $\mathbf{n} \neq 0$,

$$\overline{\beta}_t = 0, \tag{24.42}$$

which may [see (13.37)] also be put as

$$(\beta_t)^2 = (\beta^2)_t. \tag{24.43}$$

This is the necessary condition for special points on the wave surface in a compact invariant form.

25. Sections of the Wave Surfaces by Symmetry Planes

The previous section shows that the surfaces can be of very high order, especially surface S, though even surface M (the simplest) is, from (24.6), clearly complicated and does not in the general case split up into rational equations of lower degree.

As is usual in geometry, a general discussion of these surfaces, and hence of the properties of the waves, may be based on the sections by various planes, especially symmetry planes.

In section 16 we saw that a wave normal in a symmetry plane P (or perpendicular to a symmetry axis of even order) implies that one of the waves is purely transverse, its displacement vector $\mathbf{u}_1$ being perpendicular to P. We set the x_3 axis perpendicular to P; then the components of $\mathbf{u}_1$ are defined by $u_{1l} = \delta_{3l}$. From $\Lambda^n \mathbf{u} = v^2 \mathbf{u}$, $(\mathbf{u}^2 = 1)$ we have that

$$v^2 = u \Lambda^n u. \tag{25.1}$$

Let $\mathbf{n}$ lie in P, with $\mathbf{u} = \mathbf{u}_1 \perp$ P. Then, introducing the symbols

$$\tau_{ab} = \lambda_{3ab3} = \lambda_{a33b} \, (a, \; b = 1, \; 2), \tag{25.2}$$

we get from (25.1) that

$$v^2 = \lambda_{3ab3} n_a n_b = \boldsymbol{n} \tau \boldsymbol{n}, \tag{25.3}$$

or, after multiplication by v^2,

$$v^4 - \boldsymbol{v} \tau \boldsymbol{v} = 0. \tag{25.4}$$

This equation defines in the plane P a curve whose radius vector is the phase velocity **v**; (25.4) is the equation for the section of the V of (24.3) by P. More precisely, V has three sheets, so (25.4) is the curve for the section of the sheets of T_1 or quasitransverse waves, which are purely transverse for normals lying in P.

Division of (25.3) by v^2 gives

$$\boldsymbol{m} \tau \boldsymbol{m} = 1. \tag{25.5}$$

This is the section by P of the T_1 sheet of surface M of (24.6). Taking as coordinate axes in P the directions of the eigenvectors of the symmetric positively definite tensor τ, we reduce the latter to diagonal form:

$$\tau = \begin{pmatrix} a^2 & 0 \\ 0 & b^2 \end{pmatrix}, \tag{25.6}$$

in which a^2 and b^2 are eigenvalues of τ. Then (25.4) becomes

$$(x^2 + y^2)^2 - (a^2 x^2 + b^2 y^2) = 0, \tag{25.7}$$

in which x and y are the components of **v** along these axes; (25.7) defines Booth's elliptical lemniscate, which is also the pode of the ellipse

$$\frac{x^2}{a^2} + \frac{y^2}{b^2} = 1, \tag{25.8}$$

relative to its center (see [27]). This curve is shown in Fig. 13a. The curve of (25.5) becomes, subject to (25.6) ($x = m_1$, $y = m_2$),

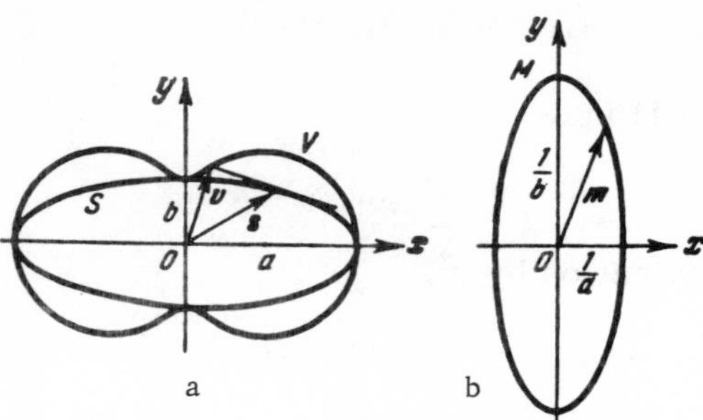

Fig. 13

$$a^2x^2 + b^2y^2 = 1, \tag{25.9}$$

which is an ellipse with semiaxes $1/a$ and $1/b$. This ellipse resembles that of (25.8) but is turned through 90° (Fig. 13b). The ellipse of (25.9) is shown by section 24 to be related to the curve of (25.7) by inversion.

The intersection between sheet T_1 and surface V or M in the symmetry planes P may be put in the explicit forms of (25.4) and (25.5), so the corresponding complete equations of (24.3) and (24.6) should split up in these planes. To show this we consider the tensor $\Lambda^n_{ij} = \lambda_{ikl j} n_k n_l$ in plane P; **n** lies in P, so only the components n_a differ from zero, in which $a = 1, 2$. In what follows we assign subscripts a, b, c, and d the values 1 and 2, whereas i, k, l, m, n, p, and q take the values 1 to 3. Then for **n** lying in P we may put

$$\Lambda^n_{ij} = \lambda_{iabj} n_a n_b. \tag{25.10}$$

We have seen in section 8 that a symmetry plane perpendicular to the x_3 axis makes the components of tensor $\lambda_{ikl j}$ zero if subscript 3 enters an odd number of times; a and b cannot take the value 3, so the only λ_{iabj} that differs from zero are λ_{cabd} and λ_{3ab3}. We accompany (25.2) by the symbols

$$\lambda^n = (\lambda^n_{cd}) = (\lambda_{cabd} n_a n_b). \tag{25.11}$$

Then tensor Λ^n in plane P becomes

$$\Lambda^n = \begin{pmatrix} \lambda_{1ab1}n_an_b & \lambda_{1ab2}n_an_b & 0 \\ \lambda_{2ab1}n_an_b & \lambda_{2ab2}n_an_b & 0 \\ 0 & 0 & \lambda_{3ab3}n_an_b \end{pmatrix} = \begin{pmatrix} \lambda_{11}^n & \lambda_{12}^n & 0 \\ \lambda_{21}^n & \lambda_{22}^n & 0 \\ 0 & 0 & \boldsymbol{n}\tau\boldsymbol{n} \end{pmatrix}, \qquad (25.12)$$

which may be put in the compact block form

$$\Lambda^n = \begin{pmatrix} \lambda^n & 0 \\ 0 & \boldsymbol{n}\tau\boldsymbol{n} \end{pmatrix}. \qquad (25.13)$$

The λ^n and τ defined by (25.2) and (25.11) are two-dimensional tensors given in P, so the subscripts to their components take only the values 1 and 2. We can use the relationships given for such tensors in section 14.

The principal invariants of Λ^n in (24.3) and (24.6) may be expressed in terms of those of the two-dimensional tensor λ^n and $\mathbf{n}\tau\mathbf{n}$ if the representation of (25.12) is correct. It is clear from (25.12) and (25.13) that

$$\Lambda_t^n = \lambda_t^n + \boldsymbol{n}\tau\boldsymbol{n}, \qquad |\Lambda^n| = |\lambda^n|\,\boldsymbol{n}\tau\boldsymbol{n}. \qquad (25.14)$$

An expression for $\bar{\Lambda}_t^n$ may be found from (13.37):

$$\bar{\Lambda}_t^n = \frac{1}{2}\left[(\Lambda_t^n)^2 - ((\Lambda^n)^2)_t\right]. \qquad (25.15)$$

From (25.13)

$$(\Lambda^n)^2 = \begin{pmatrix} (\lambda^n)^2 & 0 \\ 0 & (\boldsymbol{n}\tau\boldsymbol{n})^2 \end{pmatrix}, \quad ((\Lambda^n)^2)_t = ((\lambda^n)^2)_t + (\boldsymbol{n}\tau\boldsymbol{n})^2. \qquad (25.16)$$

From (25.14)-(25.16) we see that

$$\bar{\Lambda}_t^n = \boldsymbol{n}\tau\boldsymbol{n}\cdot\lambda_t^n + \frac{1}{2}\left[(\lambda_t^n)^2 - (\lambda^n)_t^2\right],$$

or, from (14.19)

$$\bar{\Lambda}_t^n = \boldsymbol{n}\tau\boldsymbol{n}\cdot\lambda_t^n + |\lambda^n|. \qquad (25.17)$$

It now remains to substitute from (25.14) and (25.17), with $\mathbf{n}$ replaced by $\mathbf{m}$, into (24.6), which becomes

$$(m\tau m - 1)\left(|\lambda^m| - \lambda_t^m + 1\right) = 0. \tag{25.18}$$

This equation splits up into (25.5) and the following one [see (14.25)]:

$$|\lambda^m - 1| = |\lambda^m| - \lambda_t^m + 1 = 0, \tag{25.19}$$

in which λ^m is a two-dimensional tensor defined in plane P by (25.11). The result of (25.18) could be derived directly from (25.13) and (24.6), because

$$|\Lambda^m - 1| = \begin{vmatrix} \lambda^m - 1 & 0 \\ 0 & m\tau m - 1 \end{vmatrix} = (m\tau m - 1)\,|\lambda^m - 1| = 0. \tag{25.20}$$

Equation (25.19) gives the curves for the intersection of P with sheets L and T_2 of surface M; (25.11) shows that λ^m is of fourth order in the components of **m**. An important point is that the method used in this book enables one to obtain this equation in a completely general compact invariant form suitable for all crystals. To examine detailed cases it merely remains to substitute in (25.19) for the λ_{abcd} for the particular crystal. In general, (25.19) splits up no further, although it describes two separate curves for sheets L and T_2, which do not have common points; curve L lies completely within curve T_2, because all known media have v_0 greater than the velocities of the quasitransverse waves and the length of the refraction vector $\mathbf{m} = \mathbf{n}/v$ is correspondingly less.

Now we have

$$v^4\lambda^m = \lambda^v, \tag{25.21}$$

and multiplying (25.19) by v^8 we at once get the equation for the corresponding sections of V by P:

$$v^8 - v^4\lambda_t^v + |\lambda^v| = 0. \tag{25.22}$$

This also describes two separate curves for sheets L and T_2, the first completely enclosing the second.

To calculate (25.19) and (25.22) from given elastic constants it is most convenient to use the dependence of the phase velocity on the direction of the wave normal. To find the latter we multiply (25.19) by v^4, the result being

$$v^4 - \lambda_t^n v^2 + |\lambda^n| = 0. \tag{25.23}$$

Then [see (14.26)] we get for v_0 in a symmetry plane that

$$v_0^2 = \frac{\lambda_t^n + \sqrt{2((\lambda^n)^2)_t - (\lambda_t^n)^2}}{2} \tag{25.24}$$

while for the quasitransverse waves ($v_2 < v_0$)

$$v_2^2 = \frac{\lambda_t^n - \sqrt{2((\lambda^n)^2)_t - (\lambda_t^n)^2}}{2}. \tag{25.25}$$

Also, from (25.5) we have for the purely transverse wave

$$v_1^2 = \boldsymbol{n}\tau\boldsymbol{n}. \tag{25.26}$$

Equations (25.24)-(25.26) may be considered as ones for the sections of all three sheets of the velocity surface in polar form, since v is the length of the radius vector $\mathbf{r} = \mathbf{v}$ and the components of the wave-normal vector are expressed in terms of the polar angle φ formed by $\mathbf{v}$ with x_1:

$$n_1 = \cos\varphi, \quad n_2 = \sin\varphi. \tag{25.27}$$

As in the previous section, we consider now the special points of (25.19) and (25.22). We differentiate (25.19) with respect to $\mathbf{m}$ and then multiply by $\mathbf{m}$ to get for these points that

$$\boldsymbol{m}\frac{\partial}{\partial \boldsymbol{m}}(|\lambda^m| - \lambda_t^m + 1) = 2(2|\lambda^m| - \lambda_t^m) = 0 \tag{25.28}$$

or

$$2|\lambda^n| = v^2\lambda_t^n. \tag{25.29}$$

Then from (25.23) we have

$$(\lambda_t^n)^2 - 4|\lambda^n| = 0. \tag{25.30}$$

But this means that the discriminant of (25.23) is zero, i.e., that the quasilongitudinal and quasitransverse waves have the same velocity. This does not occur for any known medium, so the curve of (25.19) cannot have special points. This is true also for (25.22).

Consider now the section of surface S by plane P. The results of the previous section show that this is a much more complicated problem than the surfaces M and V, although the section of sheet T_1 is easily found. Here for all directions of **n** in plane P the displacement vector $\mathbf{u} = \mathbf{u}_1 = (\delta_{3k})$ is constant. From (21.18), for this **u** we have $s_i = (1/v)\lambda_{i33k}n_k$ and if **n** lies in plane P, then **s** lies also in that plane, because we can put $s_a = (1/v_1)\lambda_{a33b}n_b$ or [see 25.2)]

$$\boldsymbol{s} = \frac{1}{v_1}\,\tau\boldsymbol{n} = \frac{\tau\boldsymbol{n}}{\sqrt{\boldsymbol{n}\tau\boldsymbol{n}}}. \tag{25.31}$$

The same result is obtained via (21.24), $\mathbf{s} = \partial v/\partial \mathbf{n}$, from (25.26); the section of T_1 is simply equation (21.36) written for P with $\mathbf{u} = \mathbf{u}_1 \perp$ P:

$$\boldsymbol{s}\tau^{-1}\boldsymbol{s} = 1. \tag{25.32}$$

In this case we do not have to eliminate **u** [compare the general three-dimensional case of (24.29)] simply because vector $\mathbf{u}_1 = (\delta_{3k})$ is constant for all purely transverse waves having normals lying in P. This equation is also obtained by eliminating **n** and v_1 from (25.31) and from $\mathbf{ns} = v_1$, $\mathbf{n}^2 = 1$, by analogy with (21.34)-(21.36). We take τ in the diagonal form of (25.6) and see that (25.32) coincides with the ellipse of (25.8), as should be the case, because section 24 shows that surface S is the antipode of surface V and the same relation applies for the curves representing sections of these surfaces by a symmetry plane.*

The ellipse of (25.32) is thus obtained from the section of sheet T_1 of the ray-velocity surface. The other two sheets of S present much greater difficulties. There are several ways of deriving the equations. Firstly, (21.24), (25.24), and (25.25) give

$$\boldsymbol{s} = \frac{\partial v}{\partial \boldsymbol{n}} = \frac{\dfrac{\partial}{\partial \boldsymbol{n}}\left(\lambda_{\rm t}^{n} \pm \sqrt{2((\lambda^{n})^2)_{\rm t} - (\lambda_{\rm t}^{n})^2}\right)}{2\sqrt{2\left(\lambda_{\rm t}^{n} \pm \sqrt{2((\lambda^{n})^2)_{\rm t} - (\lambda_{\rm t}^{n})^2}\right)}}. \tag{25.33}$$

---

* This relation between the curves applies only for a section of surfaces V and S by a symmetry plane or by the equivalent of this.

This two-dimensional vector equation may be considered as the parametric form of the equations for the curve, since it gives s_1 and s_2 as functions of **n**, i.e., effectively as functions of the one parameter φ of (25.27). We eliminate **n** from (25.33) and $\mathbf{n}^2 = 1$ to get the curve as $f(\mathbf{s}) = 0$.

Secondly, we can use the fact that the curve is the antipode (section 24) of curve (25.22), whose equation is best put in polar form [see (25.24) and (25.25)]:

$$v = v(n) = v(\varphi), \quad n_1 = \cos\varphi, \quad n_2 = \sin\varphi. \tag{25.34}$$

The points on curve S are derived from those on the V of (25.34) as follows. At point A of curve V (Fig. 12) we draw the straight line AB perpendicular to the radius vector **v**, which touches curve S at B. The equation of AB is

$$rn = v(\varphi), \quad n = (\cos\varphi, \sin\varphi). \tag{25.35}$$

Curve S is the envelope of the family of such lines drawn from all points of V, the parameter of the family of (25.35) being the polar angle φ. The equation of the envelope is [28] obtained by eliminating φ from (25.35) and from the derivative of this with respect to the parameter

$$r\frac{dn}{d\varphi} = \frac{dv}{d\varphi} = v', \quad \frac{dn}{d\varphi} = n' = (-\sin\varphi, \cos\varphi). \tag{25.36}$$

Equations (25.35) and (25.36) may be put as a system [$\mathbf{r} = \mathbf{s} = (x, y)$]:

$$\left.\begin{aligned} x\cos\varphi + y\sin\varphi &= v, \\ -x\sin\varphi + y\cos\varphi &= v', \end{aligned}\right\} \tag{25.37}$$

whence we have

$$\left.\begin{aligned} x &= v\cos\varphi - v'\sin\varphi, \\ y &= v\sin\varphi + v'\cos\varphi. \end{aligned}\right\} \tag{25.38}$$

These are the equations of the curve in parametric form, with $v(\varphi)$ given by (25.24) and (25.25). We eliminate φ from (25.38) to get the equation in the form $f(\mathbf{s}) = f(x, y) = 0$. It is readily shown that (25.38) is equivalent to the $\mathbf{s} = \partial v/\partial \mathbf{n}$ of (21.24) for the plane P. In fact,

$$\frac{dv}{d\varphi} = \frac{\partial v}{\partial n}\frac{dn}{d\varphi} = \frac{\partial v}{\partial n}n'; \tag{25.39}$$

but, on the other hand, (21.25) gives $\mathbf{v} = \mathbf{n}(\partial v/\partial \mathbf{n})$, so

$$x = \frac{\partial v}{\partial n}(\boldsymbol{n}\cos\varphi - \boldsymbol{n}'\sin\varphi), \quad y = \frac{\partial v}{\partial n}(\boldsymbol{n}\sin\varphi + \boldsymbol{n}'\cos\varphi).$$

From (25.35) and (25.36) we have the dependence of **n** and **n'** on φ, which gives $x = \partial v/\partial n_1$, $y = \partial v/\partial n_2$, which coincides with $\mathbf{s} = \partial v/\partial \mathbf{n}$.

We can also use the method of eliminating **u** and **n** from relations analogous to (24.29) and (24.30), which are derived via the representation of Λ^n of (25.13), which applies for plane P. Then equations (24.14), (24.18), and (24.19) for vectors **s**, **n**, and **u** lying in this plane become

$$\boldsymbol{s} = \lambda^u \boldsymbol{m}, \tag{25.40}$$

$$(\lambda^u - \boldsymbol{s}\cdot\boldsymbol{s})\boldsymbol{n} = 0, \tag{25.41}$$

$$(\lambda^n - (\boldsymbol{ns})^2)\boldsymbol{u} = 0. \tag{25.42}$$

It is clear [see (14.24)] that these imply

$$|\lambda^u - \boldsymbol{s}\cdot\boldsymbol{s}| = |\lambda^u| + \boldsymbol{s}\lambda^u\boldsymbol{s} - \lambda_t^u s^2 = 0, \tag{25.43}$$

$$|\lambda^n - (\boldsymbol{ns})^2| = |\lambda^n| - (\boldsymbol{ns})^2\lambda_t^n + (\boldsymbol{ns})^4 = 0. \tag{25.44}$$

As in (24.20) and (24.21), we get from (14.20) and (14.39) with $\alpha = \lambda^u - \mathbf{s}\cdot\mathbf{s}$ or $\alpha = \lambda^n - (\mathbf{ns})^2$ that

$$\boldsymbol{n}\cdot\boldsymbol{n} = \frac{\overline{\lambda^u - \boldsymbol{s}\cdot\boldsymbol{s}}}{\lambda_t^u - s^2} = \frac{\lambda_t^u - \lambda^u - s^2 + \boldsymbol{s}\cdot\boldsymbol{s}}{\lambda_t^u - s^2}, \tag{25.45}$$

$$\boldsymbol{u}\cdot\boldsymbol{u} = \frac{\overline{\lambda^n - (\boldsymbol{ns})^2}}{\lambda_t^n - 2(\boldsymbol{ns})^2} = \frac{\lambda_t^n - \lambda^n - (\boldsymbol{ns})^2}{\lambda_t^n - 2(\boldsymbol{ns})^2}. \tag{25.46}$$

We substitute these expressions into (25.42) and (25.41) to get two-dimensional equations analogous to (24.24) and (24.25):

$$(\lambda_{acbd} - \delta_{ab}s_c s_d)\left(\overline{\lambda^u - s \cdot s}\right)_{cd} u_b = 0, \tag{25.47}$$

$$(\lambda_{acbd} - s_a s_b \delta_{cd})\left(\overline{\lambda^n - (ns)^2}\right)_{cd} n_b = 0, \tag{25.48}$$

which take the form

$$\beta_{aba'b'} u_b u_{a'} u_{b'} = 0, \tag{25.49}$$

$$\gamma_{aba'b'} n_b n_{a'} n_{b'} = 0, \tag{25.50}$$

in which

$$\beta_{aba'b'} = \lambda^c_{ab}\lambda^c_{a'b'} - \lambda_{aefb}\lambda_{a'efb'} - s^2(\lambda^c_{ab}\delta_{a'b'} + \delta_{ab}\lambda^c_{a'b'}) + \delta_{ab}\lambda^s_{a'b'} + \lambda^s_{ab}\delta_{a'b'}, \tag{25.51}$$

$$\gamma_{aba'b'} = \lambda^c_{ab}\lambda^c_{a'b'} - \lambda_{aefb}\lambda_{a'efb'} - (\lambda^c_{ab}s_{a'}s_{b'} + s_a s_b \lambda^c_{a'b'}) + 2s_a s_b s_{a'} s_{b'}. \tag{25.52}$$

Here the symbols are

$$\lambda^c_{ab} = \lambda_{affb}, \quad \lambda^n_t = n\lambda^c n. \tag{25.53}$$

Tensor λ^c is the two-dimensional analog of tensor μ considered in section 20.

Note that (25.49) may be derived by differentiating (25.43), which is homogeneous in **u**, with respect to the latter:

$$(\partial/\partial \mathbf{u})|\lambda^u - \mathbf{u}^2 \mathbf{s}\cdot\mathbf{s}| = 0. \tag{25.49'}$$

Similarly, (25.50) may be derived by differentiating (25.44) with respect to **n**:

$$(\partial/\partial \mathbf{n})|\lambda^n - (\mathbf{ns})^2| = 0. \tag{25.50'}$$

Correspondingly, the three-dimensional equations (24.29) and (24.30) may be put in the form

$$(\partial/\partial \mathbf{u})|\Lambda^u - \mathbf{u}^2\mathbf{s}\cdot\mathbf{s}| = 0, \qquad (\partial/\partial \mathbf{n})|\Lambda^n - (\mathbf{ns})^2| = 0.$$

The problem of finding the section of surfaces S by plane P thus reduces to eliminating the u_a or n_a, respectively, from (25.49) and (25.51) or (25.50) and (25.52), which is very much simpler than that for S as a whole [compare (24.29)-(24.32)]. Relation (24.33) gives us that here the resultant is a homogeneous polynomial of degree $p = 2 \times 3 = 6$ in the components of tensor $\beta_{aba'b'}$ and $\gamma_{aba'b'}$.

Now the $\beta_{aba'b'}$ contain components of **s** (the radius vector of the curve) to the second degree, so the equation is not of degree higher than 12. Elimination via (25.50) and (25.52) would imply that terms of higher order cancel out (section 24).

We divide (25.49) by u_2^3 and introduce the variable $\xi = u_1/u_2$ to put these equations as

$$\left.\begin{aligned} A_1\xi^3 + B_1\xi^2 + C_1\xi + D_1 = 0, \\ A_2\xi^3 + B_2\xi^2 + C_2\xi + D_2 = 0. \end{aligned}\right\} \quad (25.54)$$

Here

$$\left.\begin{aligned} A_1 &= \beta_{1111}, \quad D_1 = \beta_{1222}, \\ B_1 &= \beta_{1112} + \beta_{1121} + \beta_{1211}, \\ C_1 &= \beta_{1122} + \beta_{1212} + \beta_{1221}, \end{aligned}\right\} \quad (25.55)$$

while the coefficients A_2, B_2, C_2, and D_2 differ only in that the first subscript to β is replaced by a two. The problem reduces to eliminating ξ from the two polynomials of third degree in (25.54). Using the usual method [16] of representing the resultant as a determinant, we get the equation for the curve as

$$R(s_a, \lambda_{aba'b'}) = \begin{vmatrix} A_1 & B_1 & C_1 & D_1 & 0 & 0 \\ A_2 & B_2 & C_2 & D_2 & 0 & 0 \\ 0 & A_1 & B_1 & C_1 & D_1 & 0 \\ 0 & A_2 & B_2 & C_2 & D_2 & 0 \\ 0 & 0 & A_1 & B_1 & C_1 & D_1 \\ 0 & 0 & A_2 & B_2 & C_2 & D_2 \end{vmatrix} = 0 \quad (25.56)$$

This again shows that the resultant is a polynomial of degree 6 in the $\beta_{aba'b'}$ and of degree 12 in the s_a. The determinant is best found from Laplace's formula [16] as the sum of the products of the second-order determinants composed of the elements from the first two rows and of the additional fourth-order minors. This gives us that

$$R=\begin{vmatrix} A_1 & B_1 \\ A_2 & B_2 \end{vmatrix} \cdot \begin{vmatrix} B_1 & C_1 & D_1 & 0 \\ B_2 & C_2 & D_2 & 0 \\ A_1 & B_1 & C_1 & D_1 \\ A_2 & B_2 & C_2 & D_2 \end{vmatrix} - \begin{vmatrix} A_1 & C_1 \\ A_2 & C_2 \end{vmatrix} \cdot \begin{vmatrix} A_1 & C_1 & D_1 & 0 \\ A_2 & C_2 & D_2 & 0 \\ 0 & B_1 & C_1 & D_1 \\ 0 & B_2 & C_2 & D_2 \end{vmatrix} +$$

$$+\begin{vmatrix} A_1 & D_1 \\ A_2 & D_2 \end{vmatrix} \cdot \begin{vmatrix} A_1 & B_1 & D_1 & 0 \\ A_2 & B_2 & D_2 & 0 \\ 0 & A_1 & C_1 & D_1 \\ 0 & A_2 & C_2 & D_2 \end{vmatrix} =$$

$$=(A, B)(C, D)[(B, C)+2(A, D)]+(A, C)(A, D)(B, D)-$$

$$-(A, B)(B, D)^2-(C, D)(A, C)^2-(A, D)^3; \qquad (25.57)$$

where for brevity we have

$$(A, B)=A_1B_2-A_2B_1=\begin{vmatrix} A_1 & A_2 \\ B_1 & B_2 \end{vmatrix}. \qquad (25.58)$$

From (25.51) and (25.55) it is clear that the equation is rather cumbrous in the general case; but the equation can be written in explicit form for general values of the elastic parameters for the more symmetrical crystals (see section 32).

We lack the explicit equation for (25.56), but analogy with section 24 allows us to find the condition for the special points. The necessary condition is [28] as follows when the curve is given via the parametric equations $\mathbf{r}=\mathbf{s}=\mathbf{s}(\varphi)$:

$$\frac{ds_a}{d\varphi}=0, \quad a=1, 2. \qquad (25.59)$$

Now $\mathbf{s}=\partial u/\partial \mathbf{n}$, $\mathbf{n}=\mathbf{n}(\varphi)=(\cos\varphi, \sin\varphi)$, so

$$\frac{ds_a}{d\varphi}=\frac{d}{d\varphi}\frac{\partial v}{\partial n_a}=\frac{\partial}{\partial n_b}\left(\frac{\partial v}{\partial n_a}\right)\frac{dn_b}{d\varphi}=\frac{\partial^2 v}{\partial n_a\,\partial n_b}n_b'=0, \qquad (25.60)$$

in which [see (25.36)]

$$\boldsymbol{n}'=(-\sin\varphi, \cos\varphi), \quad \boldsymbol{n}'\boldsymbol{n}=0, \quad \boldsymbol{n}'^2=1. \qquad (25.61)$$

We introduce the two-dimensional matrix

$$\beta=(\beta_{ab})=\left(\frac{\partial s_a}{\partial n_b}\right)=\left(\frac{\partial s_b}{\partial n_a}\right)=\left(\frac{\partial^2 v}{\partial n_a\,\partial n_b}\right). \qquad (25.62)$$

to get the condition of (25.60) as

$$\beta n' = 0. \tag{25.63}$$

By analogy with (21.29), we always have $|\beta| = |\partial s_a/\partial n_b| = 0$, so $\beta \mathbf{w} = 0$ always has a nonzero solution for **w**; but (25.61) implies that it is necessary for this solution to be a vector **n'** perpendicular to **n** in order to satisfy (25.63). We use (14.38) and (14.39) to get from (25.63) that

$$n' \cdot n' = \frac{\beta_c - \beta}{\beta_t}. \tag{25.64}$$

We multiply from the right by **n** to get $\beta_t \mathbf{n} - \beta \mathbf{n} = 0$; but (21.29) gives $\beta \mathbf{n} = 0$, so the condition for special points on this section of the ray surface by a symmetry plane takes the invariant form

$$\beta_t = \frac{\partial s_a}{\partial n_a} = \frac{\partial^2 v}{\partial n_a^2} = 0. \tag{25.65}$$

This may be considered as a particular case of (24.42), since $\bar{\beta}_t = \beta_t$ for a two-dimensional tensor [see (14.17)]. As an illustration we apply this condition to the section of sheet T_1 (purely transverse waves). From (25.31) we get for this case that

$$\beta = \frac{v_1^2 \tau - \tau n \cdot n \tau}{v_1^3}. \tag{25.66}$$

Condition (25.65) now becomes [see (14.18)] that

$$\beta_t = \frac{v_1^2 \tau_t - n \tau^2 n}{v_1^3} = \frac{|\tau|}{v_1^3} = 0. \tag{25.67}$$

This cannot be complied with, since τ (section 6) is a positively definite matrix, so the curve of (25.32) cannot have any special points, as should be the case, since it is an ellipse. However, the position is different for sections of the other sheets. Curves computed numerically for particular crystals [13, 25] do have special points in many cases, particularly when the anisotropy is strong, so (25.65) is often obeyed.

We note that

$$\frac{\partial^2 v^2}{\partial n_a^2} = 2\left[\left(\frac{\partial v}{\partial n_a}\right)^2 + v\frac{\partial^2 v}{\partial n_a^2}\right],$$

so that from (25.65) we get

$$2v^2\frac{\partial^2 v^2}{\partial n_a^2} - \left(\frac{\partial v^2}{\partial n_a}\right)^2 = 0,$$

which is more convenient for calculations, since Christoffel's equation gives v^2 directly, not v.

Chapter 5

General Theory of Elastic Waves in Crystals Based on Comparison with an Isotropic Medium

26. Mean Elastic Anisotropy of a Crystal

Anisotropy literally means deviation from isotropy, and it is this deviation that distinguishes a crystal from an isotropic body. It is entirely possible to conceive of an anisotropic body that differs by an arbitrarily small amount from an isotropic one; moreover, such a medium can actually be produced simply by subjecting an isotropic body to some directional action, e.g., compression or extension, electric or magnetic fields, a uniform temperature gradient, etc. The induced anisotropy may be as small as may be desired if the action is suitably weak; all the various features, including the laws of propagation for electromagnetic waves, will differ only slightly from those for an isotropic medium, which facilitates research on them. On the other hand, the properties will be far from those of an isotropic body if the anisotropy is large. Experiment shows that natural crystals differ from isotropic media to various extents, i.e., vary in anisotropy.

The degree of anisotropy is an important general characteristic of the properties of a crystal, but a strict quantitative definition is needed for this concept, whose qualitative meaning is intuitively clear. The object of this section is to give an exact quantitative characterization of the anisotropy of a crystal in respect of its elastic properties [15].

One of the basic concepts of crystal optics is that of optical anisotropy, which plays an important part in the general evaluation of the properties of a crystal. The optical parameters are determined by the dielectric-constant tensor $\varepsilon = (\varepsilon_{ik})$. The difference of the largest and smallest eigenvalues of ε is often used to characterize the anisotropy of the optical parameters;* another definition of the optical anisotropy is considered below. The problem becomes more difficult for the elastic properties, because here the tensor λ_{iklm} is of fourth rank. The convoluted tensor $\mu_{kl} = \lambda_{ikil}$ (or the $\mu'_{kl} = \lambda_{iikl}$ of section 20) is inadequate for this purpose, since this is the same as for an isotropic medium in the case of a cubic crystal, although the elastic properties of such a crystal can differ very substantially from those of an isotropic medium, which is not so for the optical properties.

It is clear that a crystal must be compared with an isotropic medium in order to characterize the anisotropy, so the problem becomes that of a suitable isotropic medium, which itself requires definition. We are concerned with elastic waves, so we will start from the basic second-rank tensor of (15.20):

$$\Lambda = (\Lambda_{kl}) = (\lambda_{ikls} n_i n_s), \tag{26.1}$$

in which λ_{ikls} is the elastic-modulus tensor for that crystal. Let the tensor for the isotropic medium be that of (19.10):

$$\Lambda_m = a_m + b_m \boldsymbol{n} \cdot \boldsymbol{n}. \tag{26.2}$$

The most natural definition of the isotropic medium most similar to the given crystal is that giving minimal difference between Λ_m and Λ, i.e., that causing all components of the tensor

$$\Lambda'_m = \Lambda - \Lambda_m \tag{26.3}$$

to be as small as possible in modulus. The condition will be obeyed if the sum of the squares of all components of Λ'_m is minimized; but from the symmetry of Λ and Λ_m we have

$$((\Lambda'_m)_{kl})^2 = (\Lambda'_m)_{kl} (\Lambda'_m)_{kl} = (\Lambda'_m)_{kl} (\Lambda'_m)_{lk} = (\Lambda'^2_m)_t. \tag{26.4}$$

---

* Use is also made of the difference of the corresponding principal refractive indices, which are the square roots of the eigenvalues of ε.

In addition, the components of Λ and Λ_m are dependent on the direction of the wave normal **n**; to eliminate this and obtain a relation dependent solely on the properties of the crystal, we must average (26.4) over all directions of **n**, i.e., integrate it over a complete solid angle and divide by 4π. The result of this averaging may be denoted by < >. Then for any quantity A dependent on unit vector n we have

$$\langle A\rangle = \frac{1}{4\pi}\int A(\mathbf{n})\,d\Omega = \frac{1}{4\pi}\int\limits_{\varphi=0}^{2\pi}\int\limits_{\theta=0}^{\pi} A(\theta,\varphi)\sin\theta\,d\theta\,d\varphi. \tag{26.5}$$

The problem then becomes one of minimizing the quantity*

$$\langle F_m\rangle = \langle \Lambda_m'^2\rangle_t = \langle(\Lambda-\Lambda_m)^2\rangle_t = \min. \tag{26.6}$$

We substitute from (26.2) and use the fact that the a_m and b_m are scalars independent of **n** to get that

$$\langle F_m\rangle = \langle\Lambda^2\rangle_t - 2a_m\langle\Lambda\rangle_t - 2b_m\langle \mathbf{n}\Lambda\mathbf{n}\rangle + 3a_m^2 + 2a_m b_m + b_m^2. \tag{26.7}$$

The determination of Λ_m amounts to finding the unknowns a_m and b_m that serve to minimize (26.6) and (26.7). The general conditions for a turning point show that

$$\frac{\partial\langle F_m\rangle}{\partial a_m} = \frac{\partial\langle F_m\rangle}{\partial b_m} = 0. \tag{26.8}$$

The a_m and b_m appear only in Λ_m, so (26.6) implies that

$$\frac{1}{2}\frac{\partial\langle F_m\rangle}{\partial a_m} = -\left\langle\frac{\partial\Lambda_m}{\partial a_m}(\Lambda-\Lambda_m)\right\rangle_t - \left\langle(\Lambda-\Lambda_m)\frac{\partial\Lambda_m}{\partial a_m}\right\rangle_t = 0; \tag{26.9}$$

an analogous relation applies for the derivative with respect to b_m. From (10.50), the trace of the product of two matrices is independent of the order of these, so (26.8) and (26.9) give that

$$\left\langle(\Lambda-\Lambda_m)\frac{\partial\Lambda_m}{\partial a_m}\right\rangle_t = \left\langle(\Lambda-\Lambda_m)\frac{\partial\Lambda_m}{\partial b_m}\right\rangle_t = 0, \tag{26.10}$$

---

* The operations of averaging and taking the trace may be interchanged: $\langle A\rangle_t = \langle A_t\rangle$.

or, from (26.2),

$$\langle \Lambda - \Lambda_m \rangle_t = \langle \boldsymbol{n} (\Lambda - \Lambda_m) \boldsymbol{n} \rangle = 0, \tag{26.11}$$

which leads to the equations

$$3a_m + b_m = \langle \Lambda \rangle_t, \quad a_m + b_m = \langle \boldsymbol{n} \Lambda \boldsymbol{n} \rangle, \tag{26.12}$$

whose solution is

$$a_m = \frac{1}{2} (\langle \Lambda \rangle_t - \langle \boldsymbol{n} \Lambda \boldsymbol{n} \rangle), \quad b_m = \frac{1}{2} (3 \langle \boldsymbol{n} \Lambda \boldsymbol{n} \rangle - \langle \Lambda \rangle_t). \tag{26.13}$$

This defines completely the Λ_m of (26.2) for the isotropic medium that an average is closest in elastic properties to the given crystal.

To find the a_m and b_m from (26.13) we have to calculate

$$\langle \Lambda_t \rangle = \langle \lambda_{ikli} n_k n_l \rangle = \lambda_{ikli} \langle n_k n_l \rangle, \tag{26.14}$$

$$\langle \boldsymbol{n} \Lambda \boldsymbol{n} \rangle = \langle \lambda_{ikls} n_i n_k n_l n_s \rangle = \lambda_{ikls} \langle n_i n_k n_l n_s \rangle. \tag{26.15}$$

We use the definition of (26.5) with $n_1 = \sin\theta \cos\varphi$, $n_2 = \sin\theta \sin\varphi$, $n_3 = \cos\theta$ to get by direct integration that

$$\langle n_k n_l \rangle = \frac{1}{3} \delta_{kl}, \tag{26.16}$$

$$\langle n_i n_k n_l n_s \rangle = \frac{1}{15} (\delta_{ik}\delta_{ls} + \delta_{il}\delta_{ks} + \delta_{is}\delta_{kl}). \tag{26.17}$$

Then

$$\langle \Lambda \rangle_t = \frac{1}{3} \lambda_{ikik}, \quad \langle \boldsymbol{n} \Lambda \boldsymbol{n} \rangle = \frac{1}{15} (\lambda_{iikk} + 2\lambda_{ikik}), \tag{26.18}$$

whence

$$a_m = \frac{1}{30} (3\lambda_{ikik} - \lambda_{iikk}), \quad b_m = \frac{1}{30} (3\lambda_{iikk} + \lambda_{ikik}). \tag{26.19}$$

We have $a_m(\partial\Lambda_m/\partial a_m) + b_m(\partial\Lambda_m/\partial b_m) = \Lambda_m$ from (26.2), so, multiplying (26.10) by a_m and b_m, and then adding, we get

$$\langle (\Lambda - \Lambda_m) \Lambda_m \rangle_t = \langle \Lambda \Lambda_m \rangle_t - \langle \Lambda_m^2 \rangle_t = 0. \tag{26.20}$$

Then (26.6) gives the minimum $\langle F_m \rangle$ as

$$\langle F_m \rangle_{min} = \langle \Lambda^2 \rangle_t - 2\langle \Lambda\Lambda_m \rangle_t + \langle \Lambda_m^2 \rangle_t = \langle \Lambda^2 \rangle_t - \langle \Lambda_m^2 \rangle_t, \qquad (26.21)$$

where (26.13) should be substituted into the Λ_m of (26.2). Clearly, none of the expressions $\langle \Lambda^2 \rangle_t$, $\langle \Lambda_m^2 \rangle_t$, $\langle (\Lambda - \Lambda_m)^2 \rangle_t$ can be negative, because each is the sum of nonnegative quantities. Moreover, $\langle \Lambda^2 \rangle_t$ and $\langle \Lambda_m^2 \rangle_t$ not only are nonnegative but also are always positive; this is so for $\langle \Lambda^2 \rangle_t$ because Λ is positively definite (section 16), while $\langle \Lambda_m^2 \rangle_t$ cannot equal zero, since then we would have $\Lambda_m = 0$, whence (26.2) and (26.3) would give $a_m = b_m = \langle \mathbf{n}\Lambda\mathbf{n} \rangle = \langle \Lambda_t \rangle = 0$, which conflicts with the nature of Λ. Thus we have from (26.2), (26.6), (26.13), and (26.21) that

$$\langle \Lambda^2 \rangle_t \geqslant \langle \Lambda_m^2 \rangle_t > 0\,, \qquad (26.22)$$

$$\langle \Lambda^2 \rangle_t > \langle F_m \rangle_{min} \geqslant 0. \qquad (26.23)$$

The $\langle F_m \rangle_{min}$ defined by (26.21) characterizes the deviation of the given crystal from the isotropic medium most similar as regards elastic properties. The meaning of this is best seen by applying analogous arguments of the optical parameters. Tensor ε reduces to a scalar ε_0 for an isotropic medium, and the condition for minimum difference between ε and ε_0 reduces, by analogy with (26.6), to the requirement that

$$F_\varepsilon = \langle (\varepsilon - \varepsilon_0)^2 \rangle_t \qquad (26.24)$$

be minimal. Now ε and ε_0 are independent of $\mathbf{n}$, so the need for the averaging of (26.6) drops out. The condition for a minimum states that

$$\frac{dF_\varepsilon}{d\varepsilon_0} = -2(\varepsilon - \varepsilon_0)_t = -2(\varepsilon_t - 3\varepsilon_0) = 0.$$

Thus in this case $\varepsilon_0 = \varepsilon_t/3$ and

$$(F_\varepsilon)_{min} = \left(\left(\varepsilon - \frac{1}{3}\varepsilon_t\right)^2\right)_t = \frac{1}{3}[(\varepsilon_1 - \varepsilon_2)^2 + (\varepsilon_2 - \varepsilon_3)^2 + (\varepsilon_3 - \varepsilon_1)^2], \qquad (26.25)$$

in which ε_1, ε_2, and ε_3 are the eigenvalues (principal values) of ε. Expression (26.25) occurs frequently in calculations [20] and characterizes the dielectric (optical) anisotropy of a crystal. Hence a derivation analogous to (26.6) and (26.21) gives the standard expression for the optical anisotropy. The quantity

$$\sqrt{(F_{\varepsilon})_{\min}} = \sqrt{\frac{1}{3}[(\varepsilon_1 - \varepsilon_2)^2 + (\varepsilon_2 - \varepsilon_3)^2 + (\varepsilon_3 - \varepsilon_1)^2]} \tag{26.26}$$

may be called the mean-square dielectric anisotropy, while $\sqrt{(F_{\varepsilon})_{\min}/(\varepsilon^2)_t}$ is the relative mean-square dielectric anisotropy. Similarly,

$$\sqrt{\langle F_m \rangle_{\min}} = \sqrt{\langle \Lambda^2 \rangle_t - \langle \Lambda_m^2 \rangle_t}, \tag{26.27}$$

is called the mean-square elastic anisotropy, Λ_m being given by (26.2) and (26.13). By analogy

$$\Delta_m = \sqrt{\frac{\langle F_m \rangle_{\min}}{\langle \Lambda^2 \rangle_t}} = \sqrt{\frac{\langle \Lambda^2 \rangle_t - \langle \Lambda_m^2 \rangle_t}{\langle \Lambda^2 \rangle_t}} \tag{26.28}$$

is the relative mean-square elastic anisotropy. These concepts are clearly extended automatically to any property characterized by a second-rank tensor. We have from (26.23) that Λ_m is always less than one:

$$0 \leqslant \Delta_m < 1. \tag{26.29}$$

From (26.2) and (26.13)

$$\langle \Lambda_m^2 \rangle_t = \frac{1}{2}(\langle \Lambda_t \rangle^2 + 3\langle \boldsymbol{n}\Lambda\boldsymbol{n} \rangle^2 - 2\langle \Lambda_t \rangle \langle \boldsymbol{n}\Lambda\boldsymbol{n} \rangle). \tag{26.30}$$

The above discussion leaves one point unelucidated. We defined the elastic anisotropy on the basis of tensor Λ, which characterizes the propagation of elastic waves in the crystal, but this does not make it obvious that Λ_m will characterize the deviation from isotropy in respect of all other elastic properties, for a complete characterization of all these properties is included in the elastic-modulus tensor λ_{iklm}. Therefore we pose the question as follows: find the tensor for the elastic moduli of an isotropic medium of (8.1),

$$\lambda_{iklm}^0 = c\delta_{ik}\delta_{lm} + a(\delta_{il}\delta_{km} + \delta_{im}\delta_{kl}). \tag{26.31}$$

which differs least from tensor λ_{iklm} for the given crystal. The condition for this, by analogy with (26.6), we take in the form

$$F_1 = \left(\lambda_{iklm} - \lambda^0_{iklm}\right)^2 = \text{min}. \tag{26.32}$$

We substitute (26.31) and obtain after simple steps involving the properties of Kronecker's symbol and the symmetry of the tensor λ_{iklm},

$$F_1 = (\lambda_{iklm})^2 - 2\left(c\lambda_{llkk} + 2a\lambda_{lklk}\right) + 3\left(3c^2 + 4ca + 8a^2\right).$$

The conditions for a turning point, $\partial F_1/\partial c = \partial F_1/\partial a = 0$, we have

$$3c + 2a = \frac{1}{3}\lambda_{llkk}, \quad c + 4a = \frac{1}{3}\lambda_{lklk},$$

and so

$$c = \frac{1}{15}\left(2\lambda_{llkk} - \lambda_{lklk}\right), \quad a = \frac{1}{30}\left(3\lambda_{lklk} - \lambda_{llkk}\right). \tag{26.33}$$

From (26.31), passing to the tensor $\Lambda^0 = (\lambda^0_{iklm} n_i n_m)$, we have

$$\Lambda^0_{kl} = \lambda^0_{iklm} n_i n_m = a\delta_{kl} + (c + a)\, n_k n_l \tag{26.34}$$

and comparison with (26.19) gives $a_m = a$, $b_m = a + c$, and $\Lambda^0 = \Lambda_m$. Hence the elastic constants of the isotropic medium most closely resembling the given crystal may be found from (26.6) or from the more general condition of (26.32), both sets being the same. Thus the isotropic medium defined by (26.2) and (26.13) is most similar to the given crystal not only as regards the propagation of elastic waves but also as regards elastic properties generally [15, 42].

27. Comparison with an Isotropic Medium

In section 26 we derived expression (26.13) for the elastic constants of some fictitious isotropic medium closest to the given crystal in its properties on average. This approach may be extended by seeking the parameters a and b that appear in the tensor

$$\Lambda^0 = a + bn \cdot n, \tag{27.1}$$

via a condition analogous to (26.6)

$$F = (\Lambda'^2)_c = \min, \tag{27.2}$$

in which

$$\Lambda' = \Lambda - \Lambda^0, \tag{27.3}$$

and $\Lambda = (\Lambda_{kl}) = (\lambda_{iklm} n_i n_m)$ is the tensor characterizing the given crystal [15].

Condition (27.2) differs from (26.6) only in that there is no averaging over the directions of vector **n**. Then (26.7) shows that in

$$F = (\Lambda^2)_t - 2a\Lambda_t - 2b\boldsymbol{n}\Lambda\boldsymbol{n} + 3a^2 + 2ab + b^2 \tag{27.4}$$

$(\Lambda^2)_t$, Λ_t, and $\mathbf{n}\Lambda\mathbf{n}$ are functions of **n**. Taking the direction of **n** as fixed, we can find a and b via calculations analogous to those of section 26. Only the results will be quoted, which are analogous to (26.11)-(26.13) and (26.20)-(26.23):

$$\Lambda'_t = \boldsymbol{n}\Lambda'\boldsymbol{n} = 0, \tag{27.5}$$

$$3a + b = \Lambda_t, \quad a + b = \boldsymbol{n}\Lambda\boldsymbol{n}, \tag{27.6}$$

$$a = \frac{1}{2}(\Lambda_t - \boldsymbol{n}\Lambda\boldsymbol{n}), \quad b = \frac{1}{2}(3\boldsymbol{n}\Lambda\boldsymbol{n} - \Lambda_t), \tag{27.7}$$

$$(\Lambda'\Lambda^0)_t = (\Lambda\Lambda^0)_t - (\Lambda^{0^2})_t = 0, \tag{27.8}$$

$$F_{\min} = (\Lambda^2)_t - (\Lambda^{0^2})_t, \tag{27.9}$$

$$(\Lambda^2)_t \geqslant (\Lambda^{0^2})_t > 0, \quad (\Lambda^2)_t > F_{\min} \geqslant 0. \tag{27.10}$$

These formulas differ from those of section 26 in that a, b, and F are functions of the direction of **n**, not constants; to each **n** there corresponds a Λ^0 of (27.1) for which $\Lambda' = \Lambda - \Lambda^0$ is least in the sense of (27.2).

It will be clear that Λ^0 may be chosen separately for each **n** from (27.5)-(27.7) to get for F (i.e., for Λ') lower values than those from the average choice used in section 26; this will be illustrated in Chapter 7 by reference to particular crystals. Of course, now we cannot compare Λ^0 with any isotropic medium (even a fictitious one), because a and b are functions of **n**; but it will become clear that the approach provides a very efficient and reasonably simple method of examining the elastic properties of crystals.

From (27.5) we have

$$\Lambda_t^0 = \Lambda_t. \tag{27.11}$$

From (20.4), this means that the sum of the squares of the speeds of the three isonormal waves with a given **n** will be the same for Λ and Λ^0.

From (27.3)

$$\Lambda = \Lambda^0 + \Lambda'; \tag{27.12}$$

while Christoffel's more general equation of (15.21) has the form

$$(\Lambda^0 + \Lambda')\, \boldsymbol{u} = \lambda \boldsymbol{u}, \quad \lambda = v^2. \tag{27.13}$$

The analogous equation for a medium described by the Λ^0 of (27.1) and (27.7) will be

$$\Lambda^0 \boldsymbol{u}_0 = (a + b\boldsymbol{n} \cdot \boldsymbol{n})\, \boldsymbol{u}_0 = \lambda_0 \boldsymbol{u}_0. \tag{27.14}$$

The solution to the latter equation (section 16) is

$$\boldsymbol{u}_0 = \boldsymbol{n}, \quad \lambda_0 = \boldsymbol{n}\Lambda^0\boldsymbol{n} = a + b. \tag{27.15}$$

From (27.3) and (27.5)

$$\boldsymbol{n}\Lambda\boldsymbol{n} = \boldsymbol{n}\Lambda^0\boldsymbol{n} = \lambda_0. \tag{27.16}$$

The last relation gives us a physical significance for $\mathbf{n}\Lambda\mathbf{n}$; (27.16) shows that this gives the square of the speed ($\lambda_0 = v_0^2$) of the longitudinal wave in the isotropic medium closest to the given crystal in the sense of (27.2). Analogy with the separation of Λ into two parts in (27.12) leads us to seek a solution to (27.13) in the form

$$\boldsymbol{u} = \boldsymbol{u}_0 + \boldsymbol{u}' = \boldsymbol{n} + \boldsymbol{u}', \quad \lambda = \lambda_0 + \lambda', \tag{27.17}$$

the additional vector **u'** being considered as perpendicular to **n**:

$$u'n = 0, \quad un = (n + u')\, n = 1. \tag{27.18}$$

Clearly, vector **u'** defines the deviation of the displacement vector **u** from the wave normal **n**.

Substitution of (27.17) into (27.13) gives

$$\Lambda^0 (n + u') + \Lambda' (n + u') = \lambda_0 (n + u') + \lambda' (n + u'). \tag{27.19}$$

We have $\Lambda^0\mathbf{n} = \mathbf{n}\Lambda^0 = \lambda_0\mathbf{n}$ from (27.14) and (27.15); this, with (27.5) and (27.18), gives, by multiplication of (27.19) by **n**, that

$$\lambda' = n\Lambda' u'. \tag{27.20}$$

We multiply (27.19) by $\mathbf{u} = \mathbf{n} + \mathbf{u}'$ and then by **n**, and substitute $\Lambda^0 = a + b\mathbf{n}\cdot\mathbf{n}$:

$$[n\,[(n + u'), \quad (a + bn \cdot n + \Lambda')(n + u')]\,] = 0. \tag{27.21}$$

We expand the double vector product via (12.38) to get

$$(n + u') \cdot n\,(a + bn \cdot n + \Lambda')(n + u') -$$

$$- (a + bn \cdot n + \Lambda')(n + u') \cdot n\,(n + u') = 0.$$

The above relationships allow-us to simplify this to

$$(b - \Lambda')\, u' + n \cdot n\Lambda' u' - \Lambda' n + n\Lambda' u' \cdot u' = 0. \tag{27.22}$$

We may transform these terms via (12.39):

$$n \cdot n\Lambda' u' - \Lambda' u' = (n \cdot n - 1)\, \Lambda' u' = n^{\times^2} \Lambda' u'.$$

Then (27.22) becomes

$$(b + n^{\times^2}\Lambda')\, u' = \Lambda' n - n\Lambda' u' \cdot u'. \tag{27.23}$$

We introduce a tensor α and a vector **h** via

$$\alpha = \frac{1}{b}\Lambda', \quad h = \alpha n. \tag{27.24}$$

From (27.5)

$$\alpha_t = n\alpha n = nh = 0 \tag{27.25}$$

and from (27.20)

$$\lambda' = bhu'. \tag{27.26}$$

Note that Λ and Λ^0 are symmetric tensors, and hence so are Λ' and α, since $\mathbf{n}\Lambda' = \Lambda'\mathbf{n}$. From (27.6) we have

$$h = \frac{1}{b}(\Lambda - a - bn \cdot n)\, n = \frac{1}{b}(\Lambda - n\Lambda n)\, n = \frac{1}{b}[n[\Lambda n, n]]. \tag{27.27}$$

From (27.24), we can put (27.23) in the form

$$(1 + n^{\times 2}\alpha)\, u' = h - hu' \cdot u'. \tag{27.28}$$

This equation plays a basic part in the entire theory [15]; it is absolutely exact, for nothing has been neglected in its deduction. Equation (27.28) is exactly equivalent to Christoffel's equation (27.13), although it has a very different character, for the linear equation $\Lambda\mathbf{u} = \lambda\mathbf{u}$ (with an unknown three-dimensional vector $\mathbf{u}$ and proper value λ) has been replaced by the nonlinear (27.28), in which the sole unknown is the vector $\mathbf{u}'$, which is restricted by the condition $\mathbf{u}'\mathbf{n} = 0$ to a plane perpendicular to the wave normal. Having found $\mathbf{u}'$, we at once get from (27.17) and (27.20) the corresponding values of the displacement vector $\mathbf{u} = \mathbf{n} + \mathbf{u}'$ and the velocity $v^2 = \lambda = \lambda_0 + \mathbf{n}\lambda'\mathbf{u}'$.

It is convenient also to transform the characteristic equation of (13.4) for Λ in conjunction with the conversion from (27.13) to (27.28):

$$|\lambda - \Lambda| = \lambda^3 - \Lambda_t\lambda^2 + \overline{\Lambda}_t\lambda - |\Lambda| = 0. \tag{27.29}$$

For this purpose we use (27.1) and (27.12), calculating the invariants Λ_t, $\overline{\Lambda}_t$, and $|\Lambda|$; (27.6) gives $\Lambda_t = 3a + b$, while (13.37) gives $\overline{\Lambda}_t$ as

$$\overline{\Lambda}_t = \frac{1}{2}((\Lambda_t)^2 - (\Lambda^2)_t) = \frac{1}{2}[(3a + b)^2 - ((\Lambda^0 + \Lambda')^2)_t].$$

From (27.8) and (27.24) we have

$$((\Lambda^0 + \Lambda')^2)_t = (\Lambda^{0^2})_t + 2(\Lambda^0\Lambda')_t + (\Lambda'^2)_t =$$

$$3a^2 + 2ab + b^2 + b^2(\alpha^2)_t.$$

whence we have

$$\bar{\Lambda}_t = a(3a + 2b) - b^2 g^2. \tag{27.30}$$

Here the symbol is

$$g^2 = \frac{1}{2}(\alpha^2)_t. \tag{27.31}$$

From this we may put

$$F_{\min} = b^2(\alpha^2)_t = 2b^2 g^2. \tag{27.31'}$$

Further, from (12.25)

$$|\Lambda| = |\Lambda^0 + \Lambda'| = |\Lambda^0| + (\bar{\Lambda}^0\Lambda')_t + (\Lambda^0\bar{\Lambda}')_t + |\Lambda'|. \tag{27.32}$$

From Table II we have

$$|\Lambda^0| = a^2(a+b), \quad \bar{\Lambda}^0 = a(a + b - b\boldsymbol{n}\cdot\boldsymbol{n}).$$

Then (27.5) gives

$$(\bar{\Lambda}^0\Lambda')_t = a(a+b)\Lambda'_t - ab\boldsymbol{n}\Lambda'\boldsymbol{n} = 0$$

and

$$(\Lambda^0\bar{\Lambda}')_t = a\bar{\Lambda}'_t + b\boldsymbol{n}\bar{\Lambda}'\boldsymbol{n}.$$

Passing to the α of (27.24) and (27.25), we get [see (12.21)] that

$$|\Lambda| = a^2(a+b) + ab^2\bar{\alpha}_t + b^3\boldsymbol{n}\bar{\alpha}\boldsymbol{n} + b^3|\alpha|.$$

But from (13.34), (13.37), and (27.25)

$$\bar{\alpha}_t = -\frac{1}{2}(\alpha^2)_t = -g^2, \tag{27.33}$$

$$\bar{\alpha} = \alpha^2 - g^2, \tag{27.34}$$

$$\boldsymbol{n}\bar{\alpha}\boldsymbol{n} = \boldsymbol{n}\alpha^2\boldsymbol{n} - g^2 = h^2 - g^2. \tag{27.35}$$

It can be shown that always

$$h^2 \leqslant g^2. \tag{27.36}$$

To do this we consider the following expression [see (12.45)]:

$$[n, \alpha h]^2 = n^2 (\alpha h)^2 - (n\alpha h)^2.$$

Now α and Λ' are symmetric tensors, so $\mathbf{n}\alpha = \alpha\mathbf{n} = \mathbf{h}$, $\alpha\mathbf{h} = \mathbf{h}\alpha$, because

$$[n, \alpha h]^2 = h\alpha^2 h - (h^2)^2 = n\alpha^4 n - (h^2)^2. \tag{27.37}$$

The Hamilton-Cayley theorem of (13.35) gives us, since $\alpha_t = 0$, that

$$\alpha^3 + \bar{\alpha}_t \alpha - |\alpha| = \alpha^3 - g^2\alpha - |\alpha| = 0. \tag{27.38}$$

On the other hand, multiplication of (27.38) by tensor α gives

$$\alpha^4 = |\alpha|\alpha + g^2\alpha^2. \tag{27.38'}$$

Scalar multiplication of (27.38) and (27.38') by $\mathbf{n}$ from right and left gives [see (27.25)]

$$n\alpha^3 n = h\alpha h = |\alpha|, \tag{27.39}$$

$$n\alpha^4 n = (\alpha h)^2 = h\alpha^2 h = g^2 n\alpha^2 n = g^2 h^2. \tag{27.40}$$

Then (27.37) becomes

$$[n, \alpha h]^2 = h^2 (g^2 - h^2). \tag{27.41}$$

Now we always have $[\mathbf{n}, \alpha\mathbf{h}]^2 \geq 0$ and $\mathbf{h}^2 \geq 0$, so (27.36) follows from this; the difference $g^2 - \mathbf{h}^2$ cannot become negative, so we denote it as follows:

$$g^2 - h^2 = \frac{1}{2} (\alpha^2)_t - h^2 = \varkappa^2, \tag{27.42}$$

in which κ is taken as positive.

From (27.33)-(27.35) and (27.42) we have

$$n\bar{\alpha}n = -\varkappa^2, \tag{27.43}$$

$$|\Lambda| = a^2(a+b) - ab^2g^2 + b^3(|\alpha| - \varkappa^2). \tag{27.44}$$

Then (27.29) takes the form

$$\lambda^3 - (3a+b)\lambda^2 + (3a^2 + 2ab - b^2g^2)\lambda -$$
$$a^2(a+b) + ab^2g^2 - b^3(|\alpha| - \varkappa^2) = 0.$$

We replace λ by the new unknown ξ via the formula

$$\xi = \frac{\lambda - a}{b}, \quad \lambda = a + b\xi. \tag{27.45}$$

Simple steps give us ξ as

$$\xi^3 - \xi^2 - g^2\xi + \varkappa^2 - |\alpha| = 0. \tag{27.46}$$

This, from (27.42), may be put also in the form

$$(\xi - 1)(\xi^2 - g^2) = h^2 + |\alpha|. \tag{27.47}$$

But a and b are always known from (27.7), so we at once get λ from (27.45) once ε has been found, and hence (27.46) is essentially the general characteristic equation for the phase velocities in any crystal. From (27.15), (27.17), and (27.45) we have

$$\lambda' = \lambda - (a+b) = b(\xi - 1). \tag{27.48}$$

Comparison with (27.26) gives

$$hu' = \xi - 1. \tag{27.49}$$

Then a known ξ gives us for **u'** from (27.28) the linear equation

$$(\xi + n^{\times 2}\alpha)\, u' = h, \tag{27.50}$$

whose solution for $|\xi + \mathbf{n}^{\times 2}\alpha| \neq 0$ is (see section 10) of the form

$$u' = (\xi + n^{\times 2}\alpha)^{-1} h. \tag{27.51}$$

Then a **u'** found from (27.28) gives us the displacement and velocity of the corresponding wave from (27.15), (27.17), and (27.26):

$$u = n + u', \quad v^2 = \lambda = a + b(1 + hu'). \tag{27.52}$$

If ξ is found from (27.46), we get the same quantities in the following form:

$$u = n + (\xi + n^{\times 2}\alpha)^{-1} h, \qquad v^2 = \lambda = a + b\xi. \tag{27.53}$$

We find the determinant in order to determine when tensor $(\xi + n^{\times 2}\alpha)$ allows inversion; from (12.25), remembering that ξ is a number, we have

$$|\xi + n^{\times 2}\alpha| = \xi^3 + \xi^2 (n^{\times 2}\alpha)_t + \xi (\overline{n^{\times 2}\alpha})_t,$$

since $|\mathbf{n}^{\times 2}\alpha| = |\mathbf{n}^{\times 2}|\,|\alpha| = 0$, because $|\mathbf{n}\times| = 0$. Further

$$(n^{\times 2}\alpha)_t = ((n \cdot n - 1)\alpha)_t = n\alpha n - \alpha_t = 0 \tag{27.54}$$

because of (27.25). Finally, from (12.23), (12.49), and (27.43) we have

$$(\overline{n^{\times 2}\alpha})_t = (\bar{\alpha} n \cdot n)_t = n\bar{\alpha} n = -\varkappa^2.$$

Thus

$$|\xi + n^{\times 2}\alpha| = \xi(\xi^2 - \varkappa^2). \tag{27.55}$$

The determinant then differs from zero for $\xi \neq 0, \pm\kappa$; to find the reciprocal tensor we use (12.16) and (13.33):

$$\overline{\xi + n^{\times 2}\alpha} = \xi^2 - \xi\,[n^{\times 2}\alpha - (n^{\times 2}\alpha)_t] + \overline{n^{\times 2}\alpha} =$$

$$= \xi^2 - \xi(n \cdot n\alpha - \alpha) + \bar{\alpha} n \cdot n = \xi^2 + \xi(\alpha - n \cdot h) + \bar{\alpha} n \cdot n, \tag{27.56}$$

$$(\xi + n^{\times 2}\alpha)^{-1} = \frac{\overline{\xi + n^{\times 2}\alpha}}{|\xi + n^{\times 2}\alpha|} = \frac{\xi^2 + \xi(\alpha - n \cdot h) + \bar{\alpha} n \cdot n}{\xi(\xi^2 - \varkappa^2)}. \tag{27.57}$$

Then (27.53) for **u** becomes

$$u = n + \frac{\xi h + \alpha h - h^2 \cdot n}{\xi^2 - \varkappa^2}. \tag{27.58}$$

The sole special case is related to

$$\xi = \pm \varkappa. \tag{27.59}$$

In all other cases we get **u** and v^2 at once from (27.53) and (27.58) when ξ is known. We substitute the ξ of (27.59) into (27.46) to get

$$|\alpha| = \mp \varkappa h^2. \tag{27.60}$$

We shall see in the next section that this is a different form of the general condition for special directions in a crystal (sections 17 and 18).

Equation (27.60) causes the second term in (27.58) to take the form of a 0/0 indeterminacy when $\xi = \pm \kappa$; here we calculate the square of the numerator in the second term in (27.58), and from (27.39), (27.40), and (27.42) we get that

$$(\xi \boldsymbol{h} + \alpha \boldsymbol{h} - \boldsymbol{h}^2 \cdot \boldsymbol{n})^2 = \boldsymbol{h}^2 (\xi^2 + \varkappa^2) + 2\xi |\alpha|. \tag{27.61}$$

Substitution from (27.59) and (27.60) gives zero; but a real vector itself is zero if its square is zero, so, subject to (27.59), we have

$$\alpha \boldsymbol{h} = \boldsymbol{h}^2 \cdot \boldsymbol{n} \mp \varkappa \boldsymbol{h}. \tag{27.62}$$

It is convenient to put (27.58) in several other forms in order to discuss the various particular cases. Firstly, it is readily verified that

$$\alpha \boldsymbol{h} - \boldsymbol{h}^2 \cdot \boldsymbol{n} = [\boldsymbol{n} [\alpha \boldsymbol{h}, \ \boldsymbol{n}]], \tag{27.63}$$

by expanding the double vector product. Moreover, **n**, **h**, and [**hn**] are mutually orthogonal, so any vector may be represented as a linear combination of these. In particular, the vector of (27.63), which is perpendicular to **n**, may be expanded as

$$[\boldsymbol{n} [\alpha \boldsymbol{h}, \ \boldsymbol{n}]] = a_1 \boldsymbol{h} + a_2 [\boldsymbol{h}\boldsymbol{n}]. \tag{27.64}$$

We multiply (27.64) by **h** and [**hn**] to get a_1 and a_2:

$$\left.\begin{aligned} a_1 &= \frac{1}{h^2} \boldsymbol{h} [\boldsymbol{n} [\alpha \boldsymbol{h}, \boldsymbol{n}]] = \frac{|\alpha|}{h^2}, \\ a_2 &= \frac{1}{h^2} [\boldsymbol{h}\boldsymbol{n}] [\boldsymbol{n} [\alpha \boldsymbol{h}, \ \boldsymbol{n}]] = \frac{[\boldsymbol{h}\boldsymbol{n}] \alpha \boldsymbol{h}}{h^2}. \end{aligned}\right\} \tag{27.65}$$

We thus get in place of (27.58) that

$$\boldsymbol{u} = \boldsymbol{n} + \frac{(h^2\xi + |\varkappa|)\,\boldsymbol{h} + [\boldsymbol{hn}]\,\alpha\boldsymbol{h}\cdot[\boldsymbol{hn}]}{h^2(\xi^2 - \varkappa^2)}. \qquad (27.66)$$

28. Special Directions

The general relationships derived in the previous section are based on comparing the tensor Λ for the crystal with the tensor Λ^0 for an isotropic medium as closely similar to it as possible for that **n**; this representation substantially simplified the analysis of many aspects of elastic waves in crystals. These relationships will now be applied to a discussion of special directions in crystals (see sections 17 and 18).

We showed in section 17 that a general necessary condition for any special direction (which is also a sufficient condition for purely transverse waves) is as in (17.7):

$$[\boldsymbol{n},\ \Lambda\boldsymbol{n}]\,\Lambda^2\boldsymbol{n} = 0. \qquad (28.1)$$

From (27.1), (27.12), and (27.24) we substitute

$$\Lambda = a + b(\boldsymbol{n}\cdot\boldsymbol{n} + \alpha),\quad \Lambda\boldsymbol{n} = (a+b)\,\boldsymbol{n} + b\boldsymbol{h}, \qquad (28.2)$$

to get after simple steps that

$$[\boldsymbol{n},\ \alpha\boldsymbol{n}]\,\alpha^2\boldsymbol{n} = [\boldsymbol{nh}]\,\alpha\boldsymbol{h} = 0. \qquad (28.3)$$

From (17.8), the displacement vector of the purely transverse wave is $\mathbf{u}_1 = C[\mathbf{n}, \Lambda\mathbf{n}]$. From (28.2) we get the equivalent expression

$$\boldsymbol{u}_1 = C\,[\boldsymbol{nh}]. \qquad (28.4)$$

It is readily seen that (28.3) and (28.4) may also be derived directly from (27.28) together with the requirement that the waves be transverse. From (27.17), $\mathbf{u} = \mathbf{n} + \mathbf{u}'$, with $\mathbf{u}' \perp \mathbf{n}$, so $\mathbf{un} = 0$ implies that $\mathbf{u}'$ must be an infinitely great vector; but then (27.28) can be satisfied only by $\mathbf{hu}' = 0$, otherwise the term in the second power of $\mathbf{u}'$ would not be balanced by the other terms. From $\mathbf{nu}' = \mathbf{hu}' = 0$ it follows that $\mathbf{u}' = C[\mathbf{nh}]$ for $C \to \infty$, which means that $\mathbf{u}_1 = C[\mathbf{nh}]$, which is (28.4). Then from (27.28) we get by multiplying by $\mathbf{h}$: $\mathbf{hn}^{\times 2}\alpha\mathbf{u} = \mathbf{h}^2$, that this [since $\mathbf{n}^{\times 2} = \mathbf{n}\cdot\mathbf{n} - 1$ and, from (27.25), $\mathbf{hn} = 0$] reduces to $\mathbf{h}\alpha\mathbf{u}' = -C\mathbf{h}\alpha[\mathbf{nh}] = \mathbf{h}^2$. Now $\mathbf{h}^2$ is a finite number, and $C \to \infty$, so the latter equation can be so only if (28.3) is obeyed.

We can give (28.3) a different form by using (12.45) to get

$$([\boldsymbol{nh}]\,\alpha\boldsymbol{h})^2 = (\boldsymbol{h}\,[\boldsymbol{n},\ \alpha\boldsymbol{h}]\,)^2 = \boldsymbol{h}^2\cdot[\boldsymbol{n},\ \alpha\boldsymbol{h}]^2 - [\boldsymbol{h}\,[\boldsymbol{n},\ \alpha\boldsymbol{h}]\,]^2. \tag{28.5}$$

Expansion of the double vector product and use of (27.39), (27.41), and (27.42) gives

$$([\boldsymbol{nh}]\,\alpha\boldsymbol{h})^2 = \boldsymbol{h}^4\varkappa^2 - |\alpha|^2. \tag{28.6}$$

Then condition (28.3) is equivalent to

$$|\alpha| = \pm\,\varkappa\boldsymbol{h}^2. \tag{28.7}$$

This is simply (27.60). The conclusion from (28.6) for the general case is

$$\boldsymbol{h}^4\varkappa^2 \geqslant |\alpha|^2. \tag{28.8}$$

Multiplication of (28.7) by b^3 and use of (27.24), (27.27), and (27.42) gives us that

$$|\Lambda'| = \pm\,[\boldsymbol{n},\ \Lambda\boldsymbol{n}]^2\sqrt{\frac{1}{2}(\Lambda'^2)_t - [\boldsymbol{n},\ \Lambda\boldsymbol{n}]^2}. \tag{28.9}$$

There are no irrational parts in $|\Lambda'|$ and $[\mathbf{n},\Lambda\mathbf{n}]$, so the square root in (28.9) should be easily extracted. Also, $|\Lambda'|$ is an integral function of third degree in the components of λ_{iklm}, while $[\mathbf{n},\Lambda\mathbf{n}]$ is of second degree, so $(1/2)(\Lambda'^2)_t-[\mathbf{n},\ \Lambda\mathbf{n}]^2$, subject to the transversality conditions, should be an exact square of some linear homogeneous function of the λ_{iklm}. From (27.7), (27.24), (27.27), and (27.42) we have that

$$\varkappa = \frac{\sqrt{\frac{1}{2}(\Lambda'^2)_t - [\boldsymbol{n},\ \Lambda\boldsymbol{n}]^2}}{\frac{1}{2}(3\boldsymbol{n}\Lambda\boldsymbol{n} - \Lambda_t)}, \tag{28.10}$$

should also be a rational function of the λ_{iklm} and n_k in this case. In particular, this property should occur unconditionally for all symmetry planes or planes perpendicular to symmetry axes of even order (section 17). It is readily shown from (12.25) and (27.1) that the n_k appear in $|\Lambda'| = |\Lambda - \Lambda^0|$ as a homogeneous function of degree 12, while $[\mathbf{n},\ \Lambda\mathbf{n}]^2$ appears as a homogeneous function of degree 8. Hence $\sqrt{(1/2)(\Lambda'^2)_t - [\mathbf{n},\ \Lambda\mathbf{n}]^2}$ must be a homogeneous function of degree 4 in the n_i in the case of purely transverse waves.

The main general relationships derived in the previous section may be summarized as follows:

$$\Lambda u = (\Lambda^0 + \Lambda')\, u = \lambda u, \quad \Lambda^0 = a + bn \cdot n, \quad \Lambda' = b\alpha; \tag{28.11}$$

$$\alpha n = h, \quad hn = \alpha_t = 0, \quad u = n + u', \quad u'n = 0; \tag{28.12}$$

$$\lambda = a + b\xi, \quad \xi^3 - \xi^2 - g^2\xi + \varkappa^2 - |\alpha| = 0; \tag{28.13}$$

$$(1 + n^{\times^2}\alpha)\, u' = h - hu' \cdot u', \quad hu' = \xi - 1; \tag{28.14}$$

$$u' = \frac{(h^2\xi + |\alpha|)\, h + [hn]\, \alpha h \cdot [hn]}{h^2\, (\xi^2 - \varkappa^2)}. \tag{28.15}$$

From (27.42) and (28.13) we get

$$\xi - 1 = \frac{h^2\xi + |\alpha|}{\xi^2 - \varkappa^2}. \tag{28.16}$$

On the other hand, from (28.6) and (28.13) we have

$$\frac{(\xi - 1)(h^2\xi - |\alpha|) - h^4}{[hn]\, \alpha h} = \frac{[hn]\, \alpha h}{\xi^2 - \varkappa^2}. \tag{28.17}$$

Then (28.15) can be replaced by

$$u' = \frac{\xi - 1}{h^2}\, h + \frac{(\xi - 1)(h^2\xi - |\alpha|) - h^4}{h^2 \cdot [hn]\, \alpha h}\, [hn]. \tag{28.18}$$

Further,

$$|\xi - \alpha| = \xi^3 - g^2\xi - |\alpha|. \tag{28.19}$$

Then it follows from (28.13) that

$$|\xi - \alpha| = \xi^2 - \varkappa^2. \tag{28.20}$$

Let (28.3) or (28.6) be obeyed:

$$|\alpha| \mp \varkappa h^2 = [hn]\, \alpha h = 0, \tag{28.21}$$

then **n** coincides with a special direction; here we assume $\mathbf{h} \neq 0$ (the case $\mathbf{h} = 0$ will be considered later). The general formulas (28.15) and (28.18) then become

$$u' = \frac{1}{\xi \mp \varkappa} h + \frac{[hn]\,\alpha h}{h^2(\xi^2 - \varkappa^2)}[hn], \tag{28.22}$$

$$u'_1 = \frac{\xi - 1}{h^2} h + \frac{(\xi - 1)(\xi \mp \varkappa) - h^2}{[hn]\,\alpha h}[hn]. \tag{28.23}$$

Substitution of (28.21) into (28.13) gives

$$\xi^3 - \xi^2 - (h^2 + \varkappa^2)\xi + \varkappa^2 \mp \varkappa h^2 \equiv (\xi \pm \varkappa)[(\xi - 1)(\xi \mp \varkappa) - h^2] = 0. \tag{28.24}$$

Hence in the case of (28.21) equation (28.13) splits up into the following:

$$\xi \pm \varkappa = 0, \tag{28.25}$$

$$(\xi - 1)(\xi \mp \varkappa) - h^2 = \xi^2 - (1 \pm \varkappa)\xi \pm \varkappa - h^2 = 0. \tag{28.26}$$

In case 1 the scalar coefficient to **h** in (28.22) equals $\mp 1/2\kappa$, so in cases 1 and 2 it remains finite, whereas the coefficient to [**hn**] becomes (0/0). But (28.17) implies that this indeterminacy equals ∞, so in case 1 it is more convenient to use (28.23), from which it follows that $\mathbf{u}' = C[\mathbf{hn}]$, $C \to \infty$, and so that $\mathbf{u} = \mathbf{n} + \mathbf{u}' \parallel [\mathbf{hn}]$.

In case 2 we have an indeterminate coefficient to [**hn**] in (28.23), although (28.17) shows that this indeterminacy is zero, so here it is more convenient to use (28.22), which implies directly that

$$u' = \frac{1}{\xi \mp \varkappa} h, \tag{28.27}$$

in which ξ is a root of (28.26).

Thus (28.25) corresponds to a purely transverse wave with $\mathbf{u}_1 = [\mathbf{hn}]$, while (28.26) corresponds to two waves whose displacements are

$$u = n + \frac{h}{\xi \mp \varkappa} = n + \frac{\xi - 1}{h^2} h, \tag{28.28}$$

in which ξ is a root of (28.26). We multiply together the two vectors $\mathbf{u}_0$ and $\mathbf{u}_2$ of (28.28), which correspond to the two roots ξ_0 and ξ_2 of (28.26), to get, as we should, that

$$u_0 u_2 = 1 + \frac{(\xi_0 - 1)(\xi_2 - 1)}{h^2} = 0,$$

if we remember that, from (28.26), $\xi_0 \xi_2 = \pm \kappa - \mathbf{h}^2$, $\xi_0 + \xi_2 = 1 \pm \kappa$.

The solutions to (28.26) are

$$\xi_0 = \frac{1}{2}\left(1 \pm \varkappa + \sqrt{(1 \mp \varkappa)^2 + 4h^2}\right), \qquad (28.29)$$

$$\xi_2 = \frac{1}{2}\left(1 \pm \varkappa - \sqrt{(1 \mp \varkappa)^2 + 4h^2}\right). \qquad (28.30)$$

The coefficients of $\mathbf{h}$ in (28.28) are correspondingly

$$\frac{\xi_0 - 1}{h^2} = \frac{1}{2h^2}\left(\sqrt{(1 \mp \varkappa)^2 + 4h^2} - (1 \mp \varkappa)\right), \qquad (28.31)$$

$$\frac{\xi_2 - 1}{h^2} = \frac{-1}{2h^2}\left(\sqrt{(1 \mp \varkappa)^2 + 4h^2} + (1 \mp \varkappa)\right). \qquad (28.32)$$

If $\kappa < 1$, which is usually the case in practice, we have

$$|\xi_0 - 1| < |\xi_2 - 1|.$$

The second term in (28.28) defines the deviation of the displacement vector from $\mathbf{n}$ and is less for the wave of (28.29), which therefore is quasilongitudinal, while the wave of (28.30) is quasitransverse.

In the $v^2 = a + b\xi$ of (28.13) we have $a = (1/2)(\Lambda_t - \mathbf{n}\Lambda\mathbf{n})$ as always positive, because Λ_t is the sum of all three eigenvalues of Λ, and $\mathbf{n}\Lambda\mathbf{n}$ does not exceed the largest eigenvalue (see section 13). It cannot be shown that b is positive in general, although $b > 0$ practically always. The isonormal wave having the highest speed is that corresponding to the root ξ of (28.13) that is largest in the algebraic sense.

We have $\xi_1 = \mp \kappa$ from (28.25) for the purely transverse wave in the present case of a special direction. Since $\sqrt{(1 \mp \kappa)^2 + 4h^2} \geq 1|\mp \kappa$, we have

$$\xi_0 \geqslant \frac{1}{2}(1 \pm \varkappa + 1 \mp \varkappa) = 1, \quad \xi_2 \lessgtr \pm \varkappa. \tag{28.33}$$

Hence $\xi_0 > \xi_1$ and $\xi_0 > \xi_2$, so the quasilongitudinal wave has the highest speed of the three. Experimental and numerical calculations show that this is so for all directions in all known media, but so far it has been possible to prove it in general only for isotropic bodies, for which it follows because the elastic energy is positively determinate. The velocity of the longitudinal wave in an isotropic medium is $v_0^2 = a + b = \lambda_{11}$, while the transverse one has $v_1^2 = a = \lambda_{44}$ [see (19.11)], so $v_0^2 - v_1^2 = \lambda_{11} - \lambda_{44} > 0$ [see (6.16) and (19.6)].

Consider now the acoustic axes. We have seen in sections 17 and 18 that for these the transversality condition must be obeyed, so (28.21)-(28.23), (28.25), and (28.26) must apply; but to these we add the condition that the velocities of two waves must coincide. Only the transverse and quasitransverse wave can coincide in speed, which means that the solution $\xi = \mp\kappa$ of (28.25) must also be a solution of (28.26), which leads [15] to $\mathbf{h}^2 = \pm\, 2\kappa$. This and $|\alpha| = \pm\, \kappa\mathbf{h}^2$ serve to define the acoustic axes, but the lower sign should be discarded, since for $0 < \kappa < 1$ the condition $\mathbf{h}^2 = -\, 2\kappa(1-\kappa)$ cannot be complied with, because the left side is positive whereas the right side is negative. Then the necessary and sufficient conditions for the acoustic axes are

$$|\alpha| = \varkappa \boldsymbol{h}^2, \qquad \boldsymbol{h}^2 = 2\varkappa(1 + \varkappa). \tag{28.34}$$

The velocity of the quasitransverse waves is

$$\xi_1 = -\varkappa, \quad v_1^2 = a - b\varkappa. \tag{28.35}$$

Equations (28.26) and (28.34) give us the speed of the quasilongitudinal wave as

$$(\xi - 1)(\xi - \varkappa) - 2\varkappa(1 + \varkappa) = (\xi + \varkappa)[\xi - (1 + 2\varkappa)] = 0.$$

Thus

$$\xi_0 = 1 + 2\varkappa, \qquad v_0^2 = a + b(1 + 2\varkappa). \tag{28.36}$$

For the quasilongitudinal wave we have from (28.28), (28.34), and (28.36) that

$$\boldsymbol{u}_0 = \boldsymbol{n} + \frac{\boldsymbol{h}}{1+\varkappa}. \tag{28.37}$$

Expressions (28.22) and (28.33) both give an indeterminate result for the case of (28.35), as should occur, because the displacement vectors of the quasitransverse waves can have any direction in a plane perpendicular to the $\mathbf{u}_0$ of (28.37).

Consider now the case of a longitudinal normal (section 17); here $\mathbf{u}=\mathbf{n}$, so $\mathbf{u}'=0$, and (28.14) gives us that

$$\boldsymbol{h} = \alpha \boldsymbol{n} = 0. \tag{28.38}$$

Conversely, if $\mathbf{h}=0$, then $(1+\mathbf{n}^{\times 2}\alpha)\mathbf{u}'=0$, which implies $\mathbf{u}'=0$, because $|1+\mathbf{n}^{\times 2}\alpha| = 1-\kappa^2 \neq 0$ from (27.55). Thus (28.38) is a necessary and sufficient condition for a longitudinal normal; from (27.27), it coincides with (18.2). This case has somewhat of a special place, since $\mathbf{h}$ appears in most of the above formulas. In particular, it is readily seen that, if (28.38) is obeyed, the tensors $\Lambda = \boldsymbol{a}+b(\mathbf{n}\cdot\mathbf{n}+\alpha)$ and α commutate:

$$\Lambda\alpha = \alpha\Lambda. \tag{28.39}$$

Conversely, the commutation of α with Λ implies $\mathbf{n}\cdot\mathbf{h}=\mathbf{h}\cdot\mathbf{n}$, which gives (28.38) after multiplication by $\mathbf{n}$. The condition for a longitudinal normal is thus equivalent to the commutation condition of (28.39), which also may be expressed via (27.3) and (27.24) as

$$\Lambda\Lambda' - \Lambda'\Lambda = \Lambda\Lambda^0 - \Lambda^0\Lambda = 0. \tag{28.40}$$

Thus tensor Λ commutates with Λ^0 when, and only when, $\mathbf{n}$ is a longitudinal normal.

(28.6) and (28.38) imply $|\alpha|=0$; from (28.14) we have for a longitudinal wave ($\mathbf{u}'=0$) that

$$\xi_0 = 1, \quad v_0^2 = a + b = \lambda_0. \tag{28.41}$$

Equation (27.47) becomes

$$(\xi - 1)(\xi^2 - g^2) = 0,$$

so for transverse waves

$$\xi_1 = g = \varkappa, \quad v_1^2 = a + bg = a + b\varkappa, \tag{28.42}$$

$$\xi_2 = -\varkappa, \quad v_2^2 = a - b\varkappa. \tag{28.43}$$

It should be borne in mind that here $g = \kappa$, because $\mathbf{h} = 0$ from (27.42). It remains to find the displacement vectors of the transverse waves, which is most simply done by the method presented in sections 13 and 17. From (17.27) we may put $\overline{\Lambda - v_1^2} = \mathbf{C}\mathbf{u}_1 \cdot \mathbf{u}_1$; using (28.11) and (28.42), we get

$$\overline{\alpha + \boldsymbol{n} \cdot \boldsymbol{n} - g} = C_1 \boldsymbol{u}_1 \cdot \boldsymbol{u}_1. \tag{28.44}$$

From (12.64) and Table II we get after simple steps that

$$\overline{\alpha + (-g + \boldsymbol{n} \cdot \boldsymbol{n})} = (\alpha - 1)(\alpha + g) + g\boldsymbol{n} \cdot \boldsymbol{n} = C_1 \boldsymbol{u}_1 \cdot \boldsymbol{u}_1. \tag{28.45}$$

Multiplication of this tensor equation by any vector $\mathbf{p}$ gives, because $C_1\mathbf{u}_1 \cdot \mathbf{u}_1\mathbf{p} \| \mathbf{u}_1$, that

$$\boldsymbol{u}_1 \| (\alpha - 1)(\alpha + g)\boldsymbol{p} + g\boldsymbol{n}\boldsymbol{p} \cdot \boldsymbol{n}. \tag{28.46}$$

Reversal of the sign of g gives us an analogous expression for the displacement $\mathbf{u}_2$ of the second transverse wave. It is readily verified that the $\mathbf{u}_1$ and $\mathbf{u}_2$ defined in this way are perpendicular to $\mathbf{n}(\mathbf{n}\mathbf{u}_1 = \mathbf{n}\mathbf{u}_2 = 0)$.

Multiplication of (28.45) by $(\alpha + \mathbf{n} \cdot \mathbf{n} - g)$ gives $|\alpha + \mathbf{n} \cdot \mathbf{n} - g| = 0$ on the left and $(\alpha^2 + g^2\mathbf{n}^{\times 2})$ on the right, because now $\alpha^3 = g^2\alpha$ from (27.38). Then, subject to condition (28.38), we have

$$\alpha^2 = -g^2\boldsymbol{n}^{\times 2} = g^2(1 - \boldsymbol{n} \cdot \boldsymbol{n}). \tag{28.47}$$

We always have from (27.34) that $\overline{\alpha} = \alpha^2 - g^2$, so in this case

$$\overline{\alpha} = -g^2\boldsymbol{n} \cdot \boldsymbol{n}. \tag{28.48}$$

Of course, (28.47) does not imply that we can extract the square root and put $\alpha = ig\mathbf{n}^{\times}$ $(i^2 = -1)$; this is clear from the fact that α is a symmetric tensor, whereas $\mathbf{n}^{\times}$ is an antisymmetric one, and also because α is real. We also cannot put $\alpha = \pm g\mathbf{n}^{\times 2}$, although this makes α symmetric and gives* $\alpha^2 = -g^2\mathbf{n}^{\times 2}$ in accordance with (28.47). In fact, if we did have $\alpha = \pm g\mathbf{n}^{\times 2}$, then (12.21), (12.23), and (12.49) would give $\alpha = g^2\mathbf{n} \cdot \mathbf{n}$, which contradicts (28.48). Moreover, $(\pm g\mathbf{n}^{\times 2})_t = \pm g(\mathbf{n} \cdot \mathbf{n} - 1)_t = \mp 2g$, which contradicts the basic condition

* This relation follows from (12.39): $\mathbf{n}^{\times 2} = \mathbf{n} \cdot \mathbf{n} - 1$, $\mathbf{n}^{\times 3} = -\mathbf{n}^{\times}$, $\mathbf{n}^{\times 4} = -\mathbf{n}^{\times 2}$.

$\alpha_t = 0$. Finally, $\alpha = \pm g n^{\times 2} = \mp g \pm g n \cdot n$ is a uniaxial tensor (section 13), which has two eigenvalues the same. All the same the condition $h^2 = |\alpha| = 0$ causes the characteristic equation for α to take the form of (28.19), $|\xi - \alpha| = \xi(\xi^2 - g^2) = 0$, which has three distinct roots.

Substitution of (28.47) into (28.46) gives a simple expression for the displacement of the transverse wave:

$$u_1 = C(\alpha - g n^{\times 2}) p, \tag{28.49}$$

in which p is an arbitrary vector whose choice is restricted only by the condition that $(\alpha - g n^{\times 2}) p \neq 0$. It is at once clear that $n u_1 = 0$ (since $n\alpha = h = 0$, $n n^{\times} = 0$), so u_1 actually is at right angles to n. Further, from (28.47) we have

$$\alpha u_1 = C(\alpha^2 - g\alpha n^{\times 2}) p = C(-g^2 n^{\times 2} + g\alpha) p = g u_1. \tag{28.50}$$

Then

$$\Lambda u_1 = (a + b n \cdot n + b\alpha) u_1 = (a + bg) u_1$$

in accordance with (28.42). The displacement of the second transverse wave differs only in the sign of g:

$$u_2 = C(\alpha + g n^{\times 2}) p, \tag{28.51}$$

with $n u_2 = 0$, while from (28.43) $\Lambda u_2 = v_2^2 u_2 = (a - bg) u_2$. It is easily verified that $u_1 u_2 = 0$.

The case of a longitudinal acoustic axis (section 18) is obtained if we require that v_1 and v_2 are the same; (28.42) and (28.43) show that this is possible only if $g^2 = (1/2)(\alpha^2)_t$. But α is a symmetric tensor, so $(\alpha^2)_t$ equals the sum of the squares of all components of α, and hence $g = 0$ is equivalent to condition (18.21):

$$\alpha = 0, \quad \Lambda = a + b n \cdot n = \Lambda^0. \tag{28.52}$$

It is clear that the properties of a crystal are as for an isotropic medium as regards elastic waves along an acoustic axis. The velocities are $v_0^2 = a + b = n\Lambda n$ (longitudinal wave) and $v_1^2 = a = (1/2)(\Lambda_t - n\Lambda n)$ (transverse wave).

There are also some special cases when certain quantities in the principal relationships become zero; these are related to the elastic parameters and to the direction of **n**, namely: $\mathbf{h}^2$, κ^2, $g^2 = \mathbf{h}^2 + \kappa^2$, $|\alpha|$. For $\mathbf{h}^2 = 0$ ($\mathbf{h} = 0$) and $g^2 = 0$ ($\alpha = 0$) we get cases already considered (longitudinal normal and longitudinal acoustic axis). Consider now the case

$$\varkappa = 0, \quad g^2 = \frac{1}{2}(\alpha^2)_t = \boldsymbol{h}^2. \tag{28.53}$$

From (27.41) and (27.42) we see that then $[\mathbf{n}, \alpha\mathbf{h}] = 0$, $\alpha\mathbf{h} = \mathbf{C}\mathbf{n}$; multiplication by **n** gives $\mathbf{C} = \mathbf{h}^2$, so (28.53) implies that

$$\alpha \boldsymbol{h} = \boldsymbol{h}^2 \cdot \boldsymbol{n}. \tag{28.54}$$

But then (28.3) gives $[\mathbf{hn}]\alpha\mathbf{h} = 0$, i.e., (28.53) is a particular case of the transversality condition, in which, from (28.7),

$$|\alpha| = 0. \tag{28.55}$$

Let $\mathbf{a} = \alpha[\mathbf{hn}]$; clearly, $\mathbf{na} = \mathbf{h}[\mathbf{hn}] = 0$, $\mathbf{ha} = \mathbf{h}\alpha[\mathbf{hn}] = 0$ from (28.54). Consider $[\mathbf{hn}]\mathbf{a} = [\mathbf{hn}]\alpha[\mathbf{hn}]$; from (12.47) and (12.48) we have

$$\boldsymbol{n} \cdot \boldsymbol{n} + \frac{\boldsymbol{h} \cdot \boldsymbol{h}}{h^2} + \frac{[\boldsymbol{hn}] \cdot [\boldsymbol{hn}]}{h^2} = 1, \tag{28.56}$$

because $\mathbf{n}$, $\mathbf{h}/|\mathbf{h}|$, $[\mathbf{hn}]/|\mathbf{h}|$ form a set of three mutually orthogonal unit vectors, and so

$$\alpha_t = (\alpha 1)_t = \boldsymbol{n}\alpha\boldsymbol{n} + \frac{\boldsymbol{h}\alpha\boldsymbol{h}}{h^2} + \frac{[\boldsymbol{hn}]\,\alpha\,[\boldsymbol{hn}]}{h^2} = 0. \tag{28.57}$$

In our case $\mathbf{n}\alpha\mathbf{n} = 0$, $\mathbf{h}\alpha\mathbf{h} = |\alpha| = 0$, which means that $[\mathbf{hn}]\alpha[\mathbf{hn}] = 0$; so $\mathbf{an} = \mathbf{ah} = \mathbf{a}[\mathbf{hn}] = 0$, $\mathbf{a} = 0$. We have the relations

$$\alpha \boldsymbol{n} = \boldsymbol{h}, \quad \alpha \boldsymbol{h} = \boldsymbol{h}^2 \cdot \boldsymbol{n}, \quad \alpha[\boldsymbol{nh}] = 0, \tag{28.58}$$

which define the result of multiplying tensor α by the three linearly independent vectors **n**, **h**, and $[\mathbf{nh}]$. We have seen in section 12 that (12.54), (12.56), and (12.58) then allow us to derive the following expression for the tensor α ($\bar{\mathbf{a}}_1 = \mathbf{n}$, $\bar{\mathbf{a}}_2 = \mathbf{h}$, $\bar{\mathbf{a}}_3 = [\mathbf{nh}]$; $\mathbf{b}_1 = \mathbf{h}$, $\mathbf{b}_2 = \mathbf{h}^2\mathbf{n}$, $\mathbf{b}_3 = 0$):

$$\alpha = \frac{\boldsymbol{h} \cdot [\boldsymbol{h}\,[\boldsymbol{nh}]\,] + h^2\boldsymbol{n} \cdot [\,[\boldsymbol{nh}]\,\boldsymbol{n}]}{\boldsymbol{n}\,[\boldsymbol{h}\,[\boldsymbol{nh}]\,]} = \boldsymbol{h} \cdot \boldsymbol{n} + \boldsymbol{n} \cdot \boldsymbol{h}. \tag{28.59}$$

Then for $\kappa = 0$ we may put tensor Λ in the form

$$\Lambda = a + b(n \cdot n + h \cdot n + n \cdot h). \tag{28.60}$$

This at once shows that $\mathbf{u}_1 = [\mathbf{hn}]$ is an eigenvector of this tensor (the displacement vector of the transverse wave), while $v_1^2 = a$ is the corresponding velocity. The characteristic equation of (28.13) takes the form $\xi(\xi^2 - \xi - h^2) = 0$, with $\xi_1 = 0$ corresponding to the purely transverse wave, $\xi_0 = (1/2)(1 + \sqrt{1 + 4\mathbf{h}^2})$ corresponding to the quasilongitudinal one and $\xi_2 = (1/2)(1 - \sqrt{1 + 4\mathbf{h}^2})$ to the quasitransverse one (if $b > 0$). These velocities and the corresponding displacements may be obtained as particular cases of (28.24)-(28.30) for $\kappa = 0$.

From (28.6) we have for $|\alpha| = 0$ that

$$[nh]\,\alpha h = \pm \varkappa h^2, \tag{28.61}$$

and (28.15) simplifies somewhat to

$$u' = \frac{\xi h \pm \varkappa\,[hn]}{\xi^2 - \varkappa^2}. \tag{28.62}$$

However, we still have to solve a cubic equation in order to find ξ:

$$\xi^3 - \xi^2 - g^2\xi + \varkappa^2 = 0. \tag{28.63}$$

We note finally that in the general case of transversality, with $\mathbf{h}\alpha[\mathbf{hn}] = 0$, $|\alpha| = \pm\kappa\mathbf{h}^2$, we also may find, by analogy with (28.58), the result of multiplying tensor α by the three linearly independent mutually orthogonal vectors **n**, **h**, and [**hn**]:

$$\alpha n = h, \quad \alpha h = h^2 n \pm \varkappa h, \quad \alpha\,[hn] = \mp \varkappa\,[hn]. \tag{28.64}$$

Then the general expression for α in the case of any special direction is

$$\alpha = h \cdot n + n \cdot h \pm \frac{\varkappa}{h^2}(h \cdot h - [hn] \cdot [hn]). \tag{28.65}$$

The signs in this formula correspond to the sign of $|\alpha|$.

The method used here is based on comparison with an isotropic medium [15] and provides a means of general discussion of the behavior of elastic waves in crystals. This gives exact general expressions for the velocities and displacements for the special directions in all crystals, which are of covariant form and independent of the choice of coordinate system. However, even greater advantages appear in the approximate solution of the problems considered in later sections.

29. Approximate Theory of Quasilongitudinal Waves

In this section we consider mainly equation (27.28):

$$(1 + \boldsymbol{n}^{\times^2}\alpha)\,\boldsymbol{u}' = \boldsymbol{h} - \boldsymbol{h}\boldsymbol{u}' \cdot \boldsymbol{u}'. \tag{29.1}$$

From (27.57) we introduce the tensor

$$\beta = (1 + \boldsymbol{n}^{\times^2}\alpha)^{-1} = \frac{1 + \alpha - \boldsymbol{n}\cdot\boldsymbol{h} + \overline{\alpha}\boldsymbol{n}\cdot\boldsymbol{n}}{1 - \varkappa^2}, \tag{29.2}$$

which allows us to transform (29.1) to

$$\boldsymbol{u}' = \beta\boldsymbol{h} - \boldsymbol{h}\boldsymbol{u}' \cdot \beta\boldsymbol{u}'. \tag{29.3}$$

Vector **u'** lies in a plane perpendicular to the wave normal **n** (section 28); this two-dimensional **u'** is found from the nonlinear equation (29.3) to get the displacement and velocity of the corresponding wave from

$$\boldsymbol{u} = \boldsymbol{n} + \boldsymbol{u}', \quad v^2 = a + b(1 + \boldsymbol{h}\boldsymbol{u}'). \tag{29.4}$$

Thus the determination of the **u'** that satisfies (29.3) completely solves a basic problem in the theory of elastic waves in crystals, but an exact solution can be obtained without difficulty only for particular directions of **n** (section 28). In the general case, the problem is one of finding the roots of (27.29) or (28.13), a characteristic equation of third degree. Application of Cardano's formula (see [16]) to (28.13),

$$\xi^3 - \xi^2 - g^2\xi + \varkappa^2 - |\alpha| = 0, \tag{29.5}$$

in which (28.14) gives $\xi = 1 + \mathbf{hu}'$, leads to

$$\xi = \frac{1}{3} + \sqrt[3]{\frac{B}{2} + \sqrt{D}} + \sqrt[3]{\frac{B}{2} - \sqrt{D}}\,. \tag{29.6}$$

Here

$$B = \frac{2}{27} + \frac{g^2}{3} - \varkappa^2 + |\alpha|, \tag{29.7}$$

and D is the discriminant of (29.5) [16]:

$$D = \frac{1}{4}\left\{(\varkappa^2 - |\alpha|)^2 - \frac{2}{27}(\varkappa^2 - |\alpha|)(2 + 9g^2) - \frac{g^4}{27}(1 + 4g^2)\right\}. \tag{29.8}$$

The roots must be such that their product is $(1+3g^2)/9$.

From ξ we determine (sections 27 and 28) the velocity as $v^2 = a + b\xi$ and the displacement as

$$u = n + \frac{(h^2\xi + |\alpha|)\, h + [hn]\varkappa h \cdot [hn]}{h^2(\xi^2 - \varkappa^2)}. \tag{29.9}$$

Formulas (29.6)-(29.8) show that the exact solution of (29.5) is complicated and largely unusable in a general investigation, though it can be used in numerical calculations with specified values for the λ_{iklm} and directions for **n**; but here the exact formulas are inconvenient because they are cumbrous, and they are not really essential because, as a rule, the elastic constants cannot presently be measured to better than 10^{-3}. This makes it clear that there is no real justification for using the exact formulas; hence we have the task of deriving reasonably simple and convenient approximate relations for the main features. Little success has been obtained in attempts to apply the general methods of ordinary perturbation theory [29], since the results are applicable only to directions of **n** lying close to symmetry axes of the crystal.

Here I present a very simple but effective method for approximate determination of the velocities and displacements for any crystal and any direction of the wave normal, which can be extended to give any desired accuracy [30]. The method is based on comparing a given crystal with the isotropic medium most similar in elastic properties (section 27), the parameters a and b of this fictitious medium being given as functions of the direction of the wave normal by (27.7). The procedure is extremely simple and amounts to applying the method of successive iteration to the nonlinear equation (29.3). We assume that **u'** (deviation of the displacement vector from **n**) is small, whereupon we may take as our zero approximation:*

$$u_0' = 0. \tag{29.10}$$

Substitution of this on the right in (29.3) gives us the first approximation for **u'** as

$$u_1' = \beta h. \tag{29.11}$$

In turn, the second approximation is obtained by substituting this on the right in (29.3):

$$u_2' = \beta h - h\beta h \cdot \beta^2 h. \tag{29.12}$$

Similarly, the third approximation is

$$u_3' = \beta h - h\beta h (1 - h\beta^2 h)(\beta^2 h - h\beta h \cdot \beta^3 h). \tag{29.13}$$

Clearly, any approximation is derived from the previous one via the recurrence formula

$$u_{k+1}' = \beta h - h u_k' \cdot \beta u_k', \quad u_0' = 0. \tag{29.14}$$

To a given **n** in the crystal there correspond three elastic waves with mutually orthogonal displacements, whereas the process of (29.10)-(29.14), if it converges, will give a single value for **u** = **n** + **u'**. It is readily seen that this is the displacement vector of the quasilongitudinal wave, for the successive approximations of (29.10), (29.11), etc., all tend to the displacement vector for which |**u'**| is largest, which is that of the quasilongitudinal wave.

To see this, we compare (29.10)-(29.14) with an exact solution, e.g., for a special direction (section 28). We have from (28.28) and (28.31) for the quasilongitudinal wave in this case that

$$u' = Ch, \quad C = \frac{1 \mp \varkappa}{2h^2}\left(\sqrt{1 + \frac{4h^2}{(1 \mp \varkappa)^2}} - 1\right). \tag{29.15}$$

On the other hand, for **u'** = **Ch** equation (29.3) becomes

*Here subscripts 0, 1, and 2 are used not to distinguish the quasilongitudinal and quasitransverse waves but to denote the order of the approximation.

$$Ch = (1 - C^2h^2)\,\beta h. \tag{29.16}$$

From (29.2) we get in the general case

$$\beta h = \frac{h + \alpha h - h^2 \cdot n}{1 - \varkappa^2} = \frac{h + [n\,[\alpha h,\, n]\,]}{1 - \varkappa^2}. \tag{29.17}$$

We use (27.64) and (27.65) to represent $\beta \mathbf{h}$ in the form

$$\beta h = \frac{(h^2 + |\alpha|)\,h + [hn]\,\alpha h \cdot [hn]}{h^2\,(1 - \varkappa^2)}. \tag{29.18}$$

Formulas (29.17) and (29.18) are completely general; for the particular case of a special direction [see (28.21)] we have

$$[hn]\,\alpha h = |\alpha| \mp \varkappa h^2 = 0, \tag{29.19}$$

so

$$\beta h = \frac{h}{1 \mp \varkappa}. \tag{29.20}$$

Then (29.16) becomes

$$C = \frac{1 - C^2h^2}{1 \mp \varkappa}. \tag{29.21}$$

Thus, since the direction of $\mathbf{u}'$ is known, to find $\mathbf{u}' = C\mathbf{h}$ we must perform an iteration on the scalar equation (29.21). Putting $C_0 = 0$, we get from successive substituting that

$$\left.\begin{aligned} C_1 &= \frac{1}{1 \mp \varkappa}, \quad C_2 = \frac{1}{1 \mp \varkappa}\left(1 - \frac{h^2}{(1 \mp \varkappa)^2}\right), \\ C_3 &= \frac{1}{1 \mp \varkappa}\left(1 - \frac{h^2}{(1 \mp \varkappa)^2} + \frac{2h^4}{(1 \mp \varkappa)^4} - \frac{h^6}{(1 \mp \varkappa)^6}\right). \end{aligned}\right\} \tag{29.22}$$

On the other hand, expanding the root in (29.15) via a binomial expansion, we get

$$C = \frac{1}{1 \mp \varkappa}\left(1 - \frac{h^2}{(1 \mp \varkappa)^2} + \frac{2h^4}{(1 \mp \varkappa)^4} - \frac{5h^6}{(1 \mp \varkappa)^6} + \ldots\right). \tag{29.23}$$

Comparison of (29.22) with (29.23) gives rise to two conclusions. Firstly, we see that the iteration of (29.14) and (29.22) selects the one of the two values for C in (28.31) and (28.32) that corresponds to the quasilongitudinal wave. Secondly, the successive approximations of (29.22) show that iteration k for C contains regular terms up to order $|\mathbf{h}|^{2(k-1)}$ inclusive. Usually $|\mathbf{h}|$ is small, and each iteration increases the order of approximation by $\mathbf{h}^2$, so the iteration process converges very rapidly. This is also clear from the form of (29.3), since the addition to the small term $\beta\mathbf{h}$ is of the third order of smallness, not the second. Finally, the accuracy of the elastic constants is not very high (see above), so it appears (see section 37) that even the first approximation is quite sufficient in many cases, while the second suffices for a great many problems.

The successive-approximation process can be given an even more convenient form if we use

$$\beta^2 \boldsymbol{h} = \frac{2\beta \boldsymbol{h} - \boldsymbol{h}}{1 - \varkappa^2}, \tag{29.24}$$

which is readily derived from (29.2) and (29.17). This implies that $\beta^k\mathbf{h}$ for any k must be expressed as a linear combination of the two vectors $\mathbf{h}$ and $\beta\mathbf{h}$. It is also clear from (29.11)-(29.14) that $\mathbf{u}'$ in any approximation is expressed via a linear combination of vectors of the form $\beta^k\mathbf{h}$ with different k, so (29.14) indicates that it is always possible to form the representation

$$\boldsymbol{u}'_k = \eta_k \boldsymbol{h} + \zeta_k \beta \boldsymbol{h}, \tag{29.25}$$

in which η_k and ζ_k are scalar coefficients. We substitute this expression in (29.14) to get

$$\eta_{k+1}\boldsymbol{h} + \zeta_{k+1}\beta\boldsymbol{h} = \beta\boldsymbol{h} - (\eta_k \boldsymbol{h}^2 + \zeta_k \boldsymbol{h}\beta\boldsymbol{h})(\eta_k \beta\boldsymbol{h} + \zeta_k \beta^2 \boldsymbol{h}).$$

Use of (29.24) and comparison of the coefficients of $\mathbf{h}$ on right and left gives us the following recurrence formulas:

$$\left.\begin{aligned} \eta_{k+1} &= \frac{\zeta_k}{1-\varkappa^2}(\eta_k \boldsymbol{h}^2 + \zeta_k \boldsymbol{h}\beta\boldsymbol{h}), \\ \zeta_{k+1} &= 1 - (\eta_k \boldsymbol{h}^2 + \zeta_k \boldsymbol{h}\beta\boldsymbol{h})\left(\eta_k + \frac{2\zeta_k}{1-\varkappa^2}\right). \end{aligned}\right\} \tag{29.26}$$

Here from (29.17) we have

$$\boldsymbol{h}\beta\boldsymbol{h} = \frac{\boldsymbol{h}^2 + |\alpha|}{1-\varkappa^2}. \tag{29.27}$$

We need know only η_k and ζ_k for the zero approximation in order to use (29.26). Since $\mathbf{u}'_0 = 0$, these are

$$\eta_0 = \zeta_0 = 0. \tag{29.28}$$

Then (29.26) gives $\eta_1 = 0$ and $\zeta_1 = 1$, which leads to (29.11) and so on. From η_k and ζ_k we find the correction of the velocity as

$$\lambda' = b\boldsymbol{h}\boldsymbol{u}' = b\,(\eta_k \boldsymbol{h}^2 + \zeta_k \boldsymbol{h}\beta\boldsymbol{h}). \tag{29.28'}$$

(29.14) or (29.25)-(29.28) will give the displacement vector of the quasilongitudinal wave as

$$\boldsymbol{u}_k = \boldsymbol{n} + \boldsymbol{u}'_k \tag{29.29}$$

for any crystal along any direction $\mathbf{n}$ in any approximation. The corresponding velocity is given by (29.4) as

$$v_k^2 = a + b\left(1 + \boldsymbol{h}\boldsymbol{u}'_k\right) = \boldsymbol{n}\Lambda\boldsymbol{n} + b\boldsymbol{h}\boldsymbol{u}'_k. \tag{29.30}$$

For a longitudinal normal, when $\mathbf{h} = 0$ from (28.38), the zero approximation gives the exact result:

$$\boldsymbol{u} = \boldsymbol{u}_0 = \boldsymbol{n}, \qquad v^2 = v_0^2 = \boldsymbol{n}\,\Lambda\boldsymbol{n}. \tag{29.31}$$

The explicit forms are given below for the velocity and displacement of the quasilongitudinal wave in the first two approximations. Here it should be borne in mind that the orders of the various terms should be evaluated from the total degree of the components of the tensor α appearing in them, since this tensor characterizes the relative deviation from an isotropic medium, which is used as the zero approximation. The order of each term is then equal to the sum of the powers of α, $|\mathbf{h}|$, κ, and g appearing in it. From (29.3) and (29.11), the first approximation for $\mathbf{u}'$ is characterized by the neglect of terms of third order, so it is exact up to terms of second order inclusive (although these do not appear in $\mathbf{u}'_1$). Correspondingly, the second approximation must retain terms up to the fourth order inclusive, those of fifth order being discarded. In general, the k-th approximation for $\mathbf{u}'$ retains terms up to order 2k inclusive.

As regards the velocity, we have (29.30) as exact, so the error in v_k^2 is determined by that in $\mathbf{u}_k'$; but $\mathbf{u}_k'$ in (29.30) is multiplied by **h**, so the order of the terms retained in v_k^2 is one higher than that in $\mathbf{u}_k'$. Hence the k-th approximation for v_k^2 contains terms up to order 2k+1 inclusive.

Thus in our first approximation we may consider the denominator* in (29.11) or (29.17) as equal to one, since this involves only neglect of terms of third order and above. Hence

$$\boldsymbol{u}_1 = \boldsymbol{n} + \boldsymbol{h} + [\boldsymbol{n}[\alpha\boldsymbol{h},\ \boldsymbol{n}]] = (1 - \boldsymbol{h}^2)\boldsymbol{n} + \boldsymbol{h} + \alpha\boldsymbol{h}. \tag{29.32}$$

From (29.30) we have the velocity in this approximation as

$$v_1^2 = a + b(1 + \boldsymbol{h}^2 + \boldsymbol{h}\alpha\boldsymbol{h}) = \boldsymbol{n}\,\Lambda\boldsymbol{n} + b(\boldsymbol{h}^2 + |\alpha|). \tag{29.33}$$

The second approximation is given by (29.12), (29.17), and (29.24) as follows, if we neglect terms of order 5 and above:

$$\boldsymbol{u}_2 = \boldsymbol{n} + (1 + \varkappa^2 - \boldsymbol{h}^2 - |\alpha|)\boldsymbol{h} + (1 + \varkappa^2 - 2\boldsymbol{h}^2)[\boldsymbol{n}[\alpha\boldsymbol{h},\ \boldsymbol{n}]], \tag{29.34}$$

$$v_2^2 = a + b[1 + \boldsymbol{h}^2(1 + \varkappa^2 - \boldsymbol{h}^2) + |\alpha|(1 + \varkappa^2 - 3\boldsymbol{h}^2)]. \tag{29.35}$$

Equation (29.33) is essentially that of sheet L of the phase-velocity surface in the form v = v(**n**) (see section 24). Discarding the term in $|\alpha|$ and writing (29.33) in expanded form via (27.7) and (27.27), we get

$$v^2(3\boldsymbol{n}\,\Lambda\boldsymbol{n} - \Lambda_t) + \Lambda_t\boldsymbol{n}\,\Lambda\boldsymbol{n} - (\boldsymbol{n}\,\Lambda\boldsymbol{n})^2 - 2\boldsymbol{n}\,\Lambda^2\boldsymbol{n} = 0. \tag{29.36}$$

Division by v^8 then gives an approximate general equation for sheet L of the refraction surface for any crystal:

$$\begin{aligned}\lambda_{ijkl}m_i m_j m_k m_l(3\boldsymbol{m}^2 + \boldsymbol{m}^2\lambda_{pqqs}m_p m_s - \lambda_{pqrs}m_p m_q m_r m_s) - \\ - \boldsymbol{m}^4\lambda_{pqqs}m_p m_s - 2\boldsymbol{m}^2\lambda_{ijkl}\lambda_{lspq}m_i m_j m_k m_s m_p m_q = 0.\end{aligned} \tag{29.37}$$

This is an equation of degree eight, which in no way conflicts with the conclusion that the complete exact equation (24.6) for all three surfaces of the refraction surface is only of sixth degree, since (29.37) is only approximate.

---

* Note that the retention of the denominator in (29.17) improves the accuracy, although it also complicates the computation.

The above deals only with the approximate derivation of the velocity and displacement for the quasilongitudinal wave. However, when we know one root of $|v^2 - \Lambda| = 0$, we can easily find the other two roots in the same approximation, whereupon the displacements of the quasitransverse waves may be determined, e.g., from (17.27).

However, there is another and more convenient method for approximate derivation of the properties of quasitransverse waves. This is presented in the next section.

30. Another Form of the Approximate Theory

In the previous section we used (29.3) for the displacement in finding an approximate solution and then, derived from **u'** the velocity of the corresponding wave via (29.4). This involves finding the displacement before we can find the velocity, whereas in many problems we are interested only in the velocity, in which case determination of the displacement is wasted.* However, it is possible to proceed in the reverse order, first finding the velocity and then the corresponding displacement vector.

This is in fact the usual classical approach to the derivation of eigenvalues and eigenvectors for a matrix; in that respect the above approach differs from the usual one. Of course, both are in principle equivalent, but one of them may prove more convenient than the other in a particular case.

Derivation of the velocity amounts to finding the ξ that satisfies (28.13), which we put as

$$(\xi - 1)(\xi^2 - g^2) = h^2 + |\alpha|. \tag{30.1}$$

We now require some relations between the quantities appearing here, which we now consider. Vector $\alpha\mathbf{h}$ resembles any other vector in being resolvable† into three mutually orthogonal vectors **n**, **h**, and [**nh**]. From (27.63)-(27.65) we may put

$$\alpha \boldsymbol{h} = h^2 \boldsymbol{n} + \eta \boldsymbol{h} + \zeta [\boldsymbol{hn}], \tag{30.2}$$

* This occurs, for example, in the derivation of the Debye temperature (Chapter 9).

† Of course, we rule out the case $\mathbf{h} = 0$, when there is a purely longitudinal wave, which allows of an easy exact solution (section 28).

in which

$$\eta = \frac{h\alpha h}{h^2} = \frac{|\alpha|}{h^2}, \qquad \zeta = \frac{[hn]\,\alpha h}{h^2}. \tag{30.3}$$

Here η and ζ are quantities of the first order of smallness in α.

From (28.6) we have that

$$\eta^2 + \zeta^2 = \varkappa^2. \tag{30.4}$$

Then we see that we always have that

$$|\eta| \leqslant \varkappa, \qquad |\zeta| \leqslant \varkappa, \tag{30.5}$$

the first inequality being equivalent to (28.8); $\kappa = 0$ should give $\eta = \zeta = 0$ [see (28.53)-(28.55)]. Condition (28.3) (transversality) now takes the form

$$\zeta = 0, \quad \eta = \pm\varkappa, \tag{30.6}$$

the sign of η being that of $|\alpha|$. A purely longitudinal wave is a particular case of (30.6), because purely transverse waves then also occur, so (30.6) is true also for $\mathbf{h} = 0$, i.e., η remains finite (of course, if $\kappa \neq 0$), in spite of the fact that $|\alpha| = 0$ if $\mathbf{h} = 0$.

From $\mathrm{g}^2 = \kappa^2 + \mathbf{h}^2$ and (30.3) we transform (30.1) to

$$(\xi - 1)(\xi^2 - \varkappa^2) = h^2\xi + |\alpha| = h^2(\xi + \eta). \tag{30.7}$$

Then we see that $\xi = 1$ implies $\mathbf{h}^2(1 + \eta) = 0$; but $1 + \eta$ cannot be zero, because (30.5) shows that we always have $|\eta| \leq \kappa < 1$. Hence $\mathbf{h}^2 = 0$, i.e., $\xi = 1$ only for a purely longitudinal wave. This equality will not be exact for a quasilongitudinal wave, but it will be approximately true, so, putting (30.7) in the form

$$\xi_0 = 1 + \frac{h^2\xi + |\alpha|}{\xi_0^2 - \varkappa^2}, \tag{30.8}$$

we can use this to find ξ_0 approximately by iteration. As our initial value of the unknown we naturally take

$$\xi_0^{(0)} = 1. \tag{30.9}$$

This is the zero approximation; for the first approximation we get

$$\xi_0^{(1)} = 1 + \frac{h^2 + |\alpha|}{1 - \varkappa^2} \tag{30.10}$$

and for the k-th approximation

$$\xi_0^{(k+1)} = 1 + \frac{h^2 \xi_0^{(k)} + |\alpha|}{\xi_0^{(k)2} - \varkappa^2}. \tag{30.11}$$

We discard terms of the fourth order of smallness and above in α in (30.10) to get that

$$\xi_0^{(1)} = 1 + h^2 + |\alpha|. \tag{30.12}$$

Comparison with (29.35) shows that all the remaining terms are regular, so the first approximation given by (30.10) and (30.12) is correct up to the third order inclusive. Similarly it can be shown that the second approximation is correet up to terms of fifth degree in α, and in general the k-th approximation found via the iteration formulas (30.9)-(30.11) contains terms up to degree 2k+1 in α inclusive. This means that the error of the second approximation is of the sixth order in α, so the accuracy of the successive approximations obtained in this way is in agreement with that for the previous section.

From ξ we find the corresponding displacement from (27.66), which from (30.3) may be put as

$$u = n + u' = n + \frac{(\xi + \eta) h + \zeta [hn]}{\xi^2 - \varkappa^2}. \tag{30.13}$$

This formula is exact, so any error in it arises from error in ξ. The ξ in the numerator is multiplied by h, so the degree of the terms retained is one higher than that in $\xi^{(k)}$. Hence, putting in (30.13) that $\xi = \xi_0^{(0)} = 1$, which is true for terms up to the first degree inclusive, we should leave terms up to the second degree in the resulting expression. The result is

$$u_0^{(1)} = n + (1 + \eta) h + \zeta [hn], \tag{30.14}$$

which agrees exactly with the first approximation for $\mathbf{u}_0$ from (29.32) and (30.2). Similarly, it is easily seen that substitution of $\xi_0^{(k)}$ into (30.13) gives us approximation k+1 for the $\mathbf{u}_0$ found from (29.14) or (29.25) and (29.26):

$$u_0^{(k+1)} = n + \frac{\left(\xi_0^{(k)} + \eta\right) h + \zeta\,[hn]}{\xi_0^{(k)2} - \varkappa^2}, \tag{30.15}$$

in which terms of degree 2k+3 and above should be discarded.

The right-hand part of (30.1), $\mathbf{h}^2(1+\eta)$, cannot become negative, because (30.5) shows that $|\eta| \leq \kappa < 1$. Therefore both factors on the left must have the same sign. Of these, $g^2 = (1/2)(\alpha^2)_t$ is a measure of the anisotropy (sections 26 and 27) and is always less than one, so (30.1) implies that either $\xi \geq 1$ or $|\xi| \leq g$. It will be clear from the above that the first case corresponds to the quasi-longitudinal wave, while the second corresponds to quasitransverse waves. The two inequalities become equalities only in the case of a longitudinal normal ($\mathbf{h} = 0$).

Formulas (30.9), (30.11), and (30.15) solve the problem completely for quasilongitudinal waves; for quasitransverse waves we have from (30.7) that

$$\xi - \varkappa = \frac{h^2(\xi + \eta)}{(\xi - 1)(\xi + \varkappa)}, \quad \xi + \varkappa = \frac{h^2(\xi + \eta)}{(\xi - 1)(\xi - \varkappa)}. \tag{30.16}$$

These can be used as iteration formulas. Let ξ in the first be denoted by ξ_1 and that in the second by ξ_2; then we may put

$$\xi_1^{(k+1)} = \varkappa - \frac{h^2\left(\xi_1^{(k)} + \eta\right)}{\left(1 - \xi_1^{(k)}\right)\left(\xi_1^{(k)} + \varkappa\right)}, \quad \xi_1^{(0)} = \varkappa, \tag{30.17}$$

$$\xi_2^{(k+1)} = -\varkappa - \frac{h^2\left(\xi_2^{(k)} + \eta\right)}{\left(1 - \xi_2^{(k)}\right)\left(\xi_2^{(k)} - \varkappa\right)}, \quad \xi_2^{(0)} = -\varkappa. \tag{30.18}$$

We may test the performance of these by reference to the special directions (section 28); here $\zeta = 0$ and $\eta = \pm\kappa$. Let $\eta = \kappa$; then

$$\xi_1^{(k+1)} = \varkappa - \frac{h^2}{1 - \xi_1^{(k)}}. \tag{30.19}$$

In (30.18) we perform the iteration with $\eta = \kappa, \xi_2^{(0)} = -\kappa$ to get

$$\xi_2^{(1)} = \xi_2^{(k)} = \xi_2 = -\varkappa, \tag{30.20}$$

i.e., the zero approximation gives the exact solution for ξ, which corresponds to the result of section 28. If $\eta = -\kappa$, we have

$$\xi_1^{(k)} = \xi_1 = \varkappa, \quad \xi_2^{(k+1)} = -\varkappa - \frac{h^2}{1 - \xi_2^{(k)}}, \tag{30.21}$$

i.e., ξ_1 and ξ_2 exchange parts if η changes sign. Iteration of (30.19) gives

$$\xi_1^{(1)} = \varkappa - \frac{h^2}{1-\varkappa}, \quad \xi_1^{(2)} = \varkappa - \frac{h^2}{1-\varkappa}\left(1 + \frac{h^2}{(1-\varkappa)^2}\right), \tag{30.22}$$

which coincides with (28.30) in the corresponding approximation. Formulas (30.19) and (30.20) also gives the correct result for the acoustic axes. Although the initial values of ξ_1 and ξ_2 are $+\kappa$ and $-\kappa$, respectively, it is easily seen that for the conditions of an acoustic axis [$\eta = \kappa$, $\mathbf{h}^2 = 2\kappa(1+x)$, see section 28] formula (30.19) leads to the result $\xi_1 = -\kappa$ which is obtained directly for ξ_2 from (30.20).

Hence (30.17) and (30.18) give correct results and can be used to find the velocities of both quasitransverse waves. We may deduce from (30.22) that the first approximation is correct up to terms of the third order of smallness in α inclusive, the second up to terms of degree 5, and the k-th up to degree 2k+1 in α. Thus (30.17) and (30.18) give an accuracy for quasitransverse waves comparable with that of (30.11) for quasilongitudinal ones.

In the general case (30.17) and (30.18) give in the first approximation that

$$\xi_1^{(1)} = \varkappa - \frac{h^2}{2\varkappa}(1+\varkappa)(\varkappa+\eta), \quad \xi_2^{(1)} = -\varkappa - \frac{h^2}{2\varkappa}(1-\varkappa)(\varkappa-\eta). \tag{30.23}$$

Here we take only terms up to the third degree in α. Then, by analogy with (29.36) and (29.37), we can derive the equations for both T sheets of the refraction surface; here κ is an irrational function of Λ and $\mathbf{n}$ [see (28.10)], so we have to square (30.23), and, as a result, both equations coincide, since they differ only in the sign of κ.

From ξ_1 and ξ_2 we find expressions for the displacement vectors of the quasitransverse waves. Formula (30.13) is completely general and exact, but it is here inconvenient, since **n** appears separately, though this direction is the least represented in the displacement of a quasitransverse wave. We are interested only in the direction of **u**, so we multiply (30.13) by $(1-\xi)(\xi^2-\kappa^2) = -\mathbf{h}^2(\xi + \eta)$ to transform it to

$$\boldsymbol{u} = C\{(\xi+\eta)((1-\xi)\boldsymbol{h} - h^2\boldsymbol{n}) + \zeta(1-\xi)[\boldsymbol{hn}]\}. \tag{30.24}$$

Correspondingly, (30.7) is altered by replacing κ^2 by $\eta^2 + \zeta^2$ from (30.4) to give

$$(\xi+\eta)[h^2+(1-\xi)(\xi-\eta)] = \zeta^2(1-\xi). \tag{30.25}$$

This is a very convenient form, especially for considering the special directions ($\zeta = 0$). There are three possibilities for $\zeta \to 0$: either the first factor on the left tends to zero, or the second one, or both together. In the first case, ζ^2 is a quantity of the second order of smallness because $\zeta \to 0$, so $\xi + \eta$ will be also of second order, and

$$\frac{\xi+\eta}{\zeta} = 0$$

with

$$\zeta = 0, \quad h^2 - 2\eta(1+\eta) \neq 0. \tag{30.26}$$

Thus here, although $\zeta = 0$ and $\xi + \eta = 0$, the first term in the braces in (30.24) will be of a high order of smallness, so $\mathbf{u} = C[\mathbf{hn}]$, i.e., we have a purely transverse wave. When $\mathbf{h}^2 + (1-\xi)(\xi - \eta) = 0$, the first term persists in (30.24), i.e., $\boldsymbol{u} = C\left(\boldsymbol{n} - \frac{1-\xi}{h^2}\boldsymbol{h}\right)$ in accordance with (28.28). Finally, we have both factors on the left in (30.25) zero together, which corresponds to an acoustic axis; then (30.24) gives **u** is indeterminate. Note that $\xi = 1$ from (30.24) gives $\mathbf{u} = C\mathbf{n}$ directly, i.e., a purely longitudinal wave.

Substitution of the approximate values given for ξ_1 and ξ_2 by (30.17) and (30.18) into (30.24) gives us the displacement vectors of the corresponding quasitransverse waves; (30.24) is an exact relation, so the order of approximation for these vectors will be governed by that for the ξ.

The relationships of this paragraph give a virtually complete solution for elastic waves in crystals, namely determination of the velocities and displacements for all three waves for any direction of the wave normal in any crystal.

Chapter 6

Elastic Waves in Transversely Isotropic Media

31. Covariant Form of the Λ Tensor

Christoffel's equation (15.21) splits up into one linear equation and one quadratic one for any direction of the wave normal only in the case of a hexagonal crystal or a transversely isotropic medium; this is the only case in which it has a comparatively simple general solution.

The tensor for this medium is as in (19.15), but here we shall use another covariant expression having several advantages. A notable feature is that this expression is derived via extremely general arguments, since we use only the following properties of tensor Λ: (1) that it is transversely isotropic, (2) the symmetry, and (3) the homogeneous quadratic dependence on the components of the wave-normal vector **n**.

Transverse isotropy means that there is only one physically distinct direction (the sixfold axis in a hexagonal crystal), which we denote by the unit vector **e**. All directions perpendicular to **e** are completely equivalent in the free medium,* but we envisage one containing a propagating wave whose phase normal is **n**, not a free one. Hence the direction of **n** is distinguished as well as that of **e**. The properties of interest are governed by the second-rank tensor Λ and are dependent on **e** and **n**, apart from possible scalar parameters. Then Λ is a tensor function of **e** and **n**, $\Lambda = \Lambda(\mathbf{e}, \mathbf{n})$, and also of several scalar parameters.

* A medium is considered free if it is subject to no fields or stresses, in particular, if it is free from waves or other processes.

Here considerations of covariance apply, which indicate that the tensor may be expressed via vectors only in a certain way. First we consider the form of the dependence on one vector. By definition, the components of a second-rank tensor behave as do the products of the components of two vectors in response to coordinate transformations; but if Λ is dependent on a sole vector (e.g., **n**), the components of the latter allow us to construct only one combination of tensor type, namely the dyad $\mathbf{n}\cdot\mathbf{n}$. Of course, there is also the tensor $\mathbf{n}^{\times}$ dual to **n** (section 12), whose components are also functions of the n_i, but this tensor by definition is antisymmetric, and this addition drops out for the symmetric tensor Λ. Thus it is clear that the most general dependence of Λ on one vector **n** is of the form

$$\Lambda = a + b\boldsymbol{n}\cdot\boldsymbol{n}, \tag{31.1}$$

in which a and b are any scalars. Scalar a (or, more precisely, the isotropic tensor a, see section 13) may always appear in the expression for Λ, because all the requirements imposed on the dependence of Λ in **n** will still be complied with. Of course, a and b may also be dependent on **n**, but only via an invariant combination of the components n_k, and the sole such combination for one vector is the square $\mathbf{n}^2$. If **n** is unit vector, then $\mathbf{n}^2 = 1$, so a and b are simply numbers independent of the n_k. Expression (31.1) defines Λ for the isotropic medium of (19.10), which is not accidental, since the only distinctive direction is that of **n** for a wave in an isotropic medium. The problem is more complicated for a transversely isotropic medium, but it can still be solved without difficulty. From vectors **n** and **e** we can construct the dyads $\mathbf{e}\cdot\mathbf{e}$, $\mathbf{n}\cdot\mathbf{n}$, $\mathbf{e}\cdot\mathbf{n}$, and $\mathbf{n}\cdot\mathbf{e}$. The symmetry of Λ means that the unsymmetrical dyads $\mathbf{e}\cdot\mathbf{n}$ and $\mathbf{n}\cdot\mathbf{e}$ must appear as the symmetrical combination $\mathbf{e}\cdot\mathbf{n}+\mathbf{n}\cdot\mathbf{e}$, and these dyads then allow us to construct the tensor as

$$\Lambda = a_0 + a_1\boldsymbol{n}\cdot\boldsymbol{n} + a_2\boldsymbol{e}\cdot\boldsymbol{e} + a_3(\boldsymbol{e}\cdot\boldsymbol{n} + \boldsymbol{n}\cdot\boldsymbol{e}), \tag{31.2}$$

in which a_0, a_1, a_2, and a_3 are certain scalars. It is true that here we have not considered all possible combinations of **e** and **n** that are of tensor type; in addition to $\mathbf{e}^{\times}$, $\mathbf{n}^{\times}$, and $[\mathbf{en}]^{\times}$, which are antisymmetric and therefore disregarded, we have the combinations $C(\mathbf{e}\cdot[\mathbf{en}]+[\mathbf{en}]\cdot\mathbf{e})$, $C(\mathbf{n}\cdot[\mathbf{en}]+[\mathbf{en}]\cdot\mathbf{n})$, $[\mathbf{en}]\cdot[\mathbf{en}]$. However, the first two have the property of changing sign on inversion of the coordinate system, because they have odd numbers of vectors, each of which reverses in sign, whereas Λ, being a true even-rank tensor, should not change on inversion (section 8). Of course, we might seek to balance out this change of sign by including in C a scalar* whose sign also reverses, but the only scalar of this type, **p**[**qr**], requires three independent vectors, while we have only two, **e** and **n**. As regards [**en**] · [**en**], we have from (12.43) for $\mathbf{e}^2 = \mathbf{n}^2 = 1$ that

$$[en] \cdot [en] = [en]^2 + en \cdot (e \cdot n + n \cdot e) - (e \cdot e + n \cdot n). \tag{31.3}$$

Thus this merely gives combinations of **e** and **n** that already appear in (31.2).

Then (31.2) (in which a_0, a_1, a_2, and a_3 are certain scalars dependent in the general case on **e** and **n**) is the most general possible form for the dependence of a three-dimensional symmetric second-rank tensor on two vectors **e** and **n**.

So far we have not used property (3), i.e., the homogeneous quadratic dependence of Λ on **n**, which allows us to determine the form of the scalars a_0, a_1, a_2, and a_3, which are invariant and so must be dependent only on invariants constructed from e and **n**, of which there are only three: $\mathbf{e}^2 = 1$, $\mathbf{n}^2 = 1$, and **ne**. Now a_0 is not attached to expressions containing **n**, so a_0 must be quadratically dependent on **n**. This shows that

$$a_0 = b_1 n^2 + b_2 (ne)^2 = b_1 + b_2 (ne)^2. \tag{31.4}$$

Scalar a_1 cannot contain **n** and so is simply a number. For a_2 we have, by analogy with (31.4),

$$a_2 = b_6 + b_4 (ne)^2. \tag{31.5}$$

Scalar a_3 can contain **n** only in linear form, so

$$a_3 = b_3 ne. \tag{31.6}$$

This defines the form of Λ completely for a transversely isotropic medium; (31.4)-(31.6) show that it is dependent on the six

* A pseudoscalar is a quantity that does not alter on rotation of the coordinate system but that changes sign on reflection or inversion.

scalar parameters b_1, b_2, $a_1 = b_3$, b_4, b_5, b_6, whereas these are [see (19.7) and (19.15)] five elastic moduli for a hexagonal crystal; the reason is that (31.2) with (31.4)-(31.6) is somewhat more general than (19.15). To determine how b_1, ..., b_6 are expressed in terms of the elastic moduli we have merely to compare (31.2) with (19.15); here $\mathbf{e} \| x_3$, so $e_1 = e_2 = 0$, $e_3 = 1$, $\mathbf{ne} = n_3$. Equating identical components of (31.2) and (19.15), we get a system of equations:

$$\left.\begin{aligned} a_0 + a_1 n_1^2 &= \lambda_{11} n_1^2 + \lambda_{66} n_2^2 + \lambda_{44} n_3^2, \\ a_0 + a_1 n_2^2 &= \lambda_{66} n_1^2 + \lambda_{11} n_2^2 + \lambda_{44} n_3^2, \\ a_1 n_3 + a_3 &= (\lambda_{13} + \lambda_{44}) n_3, \\ a_0 + a_1 n_3^2 + a_2 + 2a_3 n_3 &= \lambda_{44}(1 - n_3^2) + \lambda_{33} n_3^2, \\ a_1 &= \lambda_{11} - \lambda_{66}, \end{aligned}\right\} \tag{31.7}$$

which is readily solved to give

$$\left.\begin{aligned} a_1 = \lambda_{11} - \lambda_{66}, \quad a_2 = \lambda_{44} - \lambda_{66} + (\lambda_{11} + \lambda_{33} - 2\lambda_{13} - 4\lambda_{44}) n_3^2, \\ a_0 = \lambda_{66} + (\lambda_{44} - \lambda_{66}) n_3^2, \quad a_3 = (\lambda_{13} + \lambda_{44} - \lambda_{11} + \lambda_{66}) n_3. \end{aligned}\right\} \tag{31.8}$$

Comparison with (31.4)-(31.6) shows that (31.2) contain five independent parameters in the case of a hexagonal crystal (as should be so), which are expressed in terms of $\lambda_{\alpha\beta}$ as follows:

$$\left.\begin{aligned} b_1 = \lambda_{66}, \quad b_2 = b_6 = \lambda_{44} - \lambda_{66}, \quad b_3 = \lambda_{11} - \lambda_{66}, \\ b_4 = \lambda_{11} + \lambda_{33} - 2\lambda_{13} - 4\lambda_{44}, \quad b_5 = \lambda_{13} + \lambda_{44} - \lambda_{11} + \lambda_{66}. \end{aligned}\right\} \tag{31.9}$$

Then Λ is expressed as

$$\Lambda = b_1 + b_2 (\boldsymbol{ne})^2 + b_3 \boldsymbol{n} \cdot \boldsymbol{n} + (b_2 + b_4 (\boldsymbol{ne})^2) \boldsymbol{e} \cdot \boldsymbol{e} + b_5 \boldsymbol{ne} (\boldsymbol{e} \cdot \boldsymbol{n} + \boldsymbol{n} \cdot \boldsymbol{e}). \tag{31.10}$$

The advantage of a covariant representation for Λ are obvious; (31.2) is much simpler and more compact than the expanded form of (19.15), and it reveals the structure of Λ, i.e., the general form of the dependence on **e** and **n**. A very important point is that b_1, b_2, etc., are essentially invariants, i.e., they resemble (31.2) as a whole in remaining unchanged under any orthogonal transformation of the coordinates. This cannot be said of the $\lambda_{\alpha\beta}$, which are dependent on the choice of coordinate system; so (31.9) is not of covariant type, since a different choice of basis would cause the parameters b_s to be expressed differently in terms of the $\lambda_{\alpha\beta}$ in the new frame of reference. However, although the equations of

(31.9) are correct only when the x_3 axis is parallel to L^6, the numerical values of the b_s given by them are unaltered, because the b_s are scalars, i.e., invariants.

Although (31.2) is simple, (31.3) allows us to put it in a somewhat different form, which is even more convenient in many calculations. For this purpose we replace the $\mathbf{ne}(\mathbf{e}\cdot\mathbf{n}+\mathbf{n}\cdot\mathbf{e})$ of (31.2) by the quantity $[\mathbf{en}]\cdot[\mathbf{en}]+\mathbf{e}\cdot\mathbf{e}+\mathbf{n}\cdot\mathbf{n}-(1-(\mathbf{en})^2)$, which (31.3) shows to be equal to it. We also introduce the unit vector

$$c=\frac{[en]}{|[en]|}=\frac{[en]}{\sqrt{1-(ne)^2}},\quad c^2=1. \tag{31.11}$$

Combination of expressions containing identical dyads gives us that

$$\Lambda=c_0+c_1 n\cdot n+c_2 e\cdot e+c_3 c\cdot c, \tag{31.12}$$

in which

$$c_0=g_1+g_2 n_3^2,\ c_1=g_3,\ c_2=g_2+g_4 n_3^2,\ c_3=g_5(1-n_3^2) \tag{31.13}$$

and

$$\left.\begin{aligned} g_1&=b_1-b_5=\lambda_{11}-\lambda_{13}-\lambda_{44},\\ g_2&=b_2+b_5=2\lambda_{44}+\lambda_{13}-\lambda_{11},\\ g_3&=b_3+b_5=\lambda_{13}+\lambda_{44},\\ g_4&=b_4=\lambda_{11}+\lambda_{33}-2\lambda_{13}-4\lambda_{44},\\ g_5&=b_5=\lambda_{13}+\lambda_{44}+\lambda_{66}-\lambda_{11}.\end{aligned}\right\} \tag{31.14}$$

From these we may express the $\lambda_{\alpha\beta}$ via the parameters g_s:

$$\left.\begin{aligned} &\lambda_{11}=g_1+g_3,\ \lambda_{13}=g_3-g_1-g_2,\ \lambda_{44}=g_1+g_2,\\ &\lambda_{33}=g_1+2g_2+g_3+g_4,\ \lambda_{66}=g_1+g_5.\end{aligned}\right\} \tag{31.15}$$

Vector $\mathbf{c}$ becomes indeterminate when $\mathbf{n}=\mathbf{e}$, but the term $c_3\mathbf{c}\cdot\mathbf{c}$ then becomes zero, so (31.12) continues to apply to Λ for all cases. This expression is more compact than (31.2); moreover, the fact that $\mathbf{c}$ is a vector perpendicular to $\mathbf{e}$ and $\mathbf{n}$ serves to simplify many calculations.

The principal invariants of (31.12) are

$$\Lambda_t=3c_0+c_1+c_2+c_3, \tag{31.16}$$

$$\boldsymbol{n\Lambda n} = c_0 + c_1 + c_2 + c_3, \tag{31.17}$$

$$(\Lambda^2)_t = (c_0 + c_1 + c_2 + c_3)^2 + 2\left[c_0^2 - c_3(c_1 + c_2) - c_1 c_2 [\boldsymbol{ne}]^2\right]. \tag{31.18}$$

$$\overline{\Lambda}_t = c_0(3c_0 + 2c_3) + (c_1 + c_2)(2c_0 + c_3) + c_1 c_2 [\boldsymbol{ne}]^2, \tag{31.19}$$

$$\boldsymbol{n\Lambda^2 n} = (c_0 + c_1)^2 + c_2(c_2 + 2c_0 + 2c_1)\, n_3^2, \tag{31.20}$$

$$\boldsymbol{n\overline{\Lambda} n} = (c_0 + c_3)(c_0 + c_2 [\boldsymbol{ne}]^2), \tag{31.21}$$

$$|\Lambda| = (c_0 + c_3)[(c_0 + c_1)(c_0 + c_2) - c_1 c_2 n_3^2]. \tag{31.22}$$

These may be calculated readily by the methods of Chapter 2; $|\Lambda|$ will be considered as an example. We use (12.25) and (12.70) to get

$$|\Lambda| = |c_0 + c_3\boldsymbol{c}\cdot\boldsymbol{c}| + [(\overline{c_0 + c_3\boldsymbol{c}\cdot\boldsymbol{c}})(c_1\boldsymbol{n}\cdot\boldsymbol{n} + c_2\boldsymbol{e}\cdot\boldsymbol{e})]_t +$$
$$+ [(c_0 + c_3\boldsymbol{c}\cdot\boldsymbol{c})(\overline{c_1\boldsymbol{n}\cdot\boldsymbol{n} + c_2\boldsymbol{e}\cdot\boldsymbol{e}})]_t;$$

then (12.72) and Table II give

$$|\Lambda| = c_0^2(c_0 + c_3) + c_0[(c_0 + c_3 - c_3\boldsymbol{c}\cdot\boldsymbol{c})(c_1\boldsymbol{n}\cdot\boldsymbol{n} + c_2\boldsymbol{e}\cdot\boldsymbol{e})]_t +$$
$$+ c_1 c_2 [(c_0 + c_3\boldsymbol{c}\cdot\boldsymbol{c})[\boldsymbol{en}]\cdot[\boldsymbol{en}]]_t =$$
$$= c_0^2(c_0 + c_3) + c_0(c_0 + c_3)(c_1 + c_2) + c_1 c_2(1 - n_3^2)(c_0 + c_3),$$

which reduces to (31.22).

The scalar coefficients $c_0, \ldots, c_3$ (or $a_0, \ldots, a_3$) in the expression for Λ may be expressed via independent invariants from the list (31.16)-(31.22), e.g., via Λ_t, $\overline{\Lambda}_t$, $\mathbf{n\Lambda n}$, and $\mathbf{n\overline{\Lambda}n}$.

(31.10) readily yields the explicit form of the fourth-rank λ_{iklm} tensor for a hexagonal crystal. We rewrite (31.10) in terms of subscripts as

$$\Lambda_{ik} = \lambda_{ilkm} n_l n_m = \{(b_1\delta_{lm} + b_2 e_l e_m)\delta_{ik} + b_3\delta_{il}\delta_{km} +$$
$$+ (b_2\delta_{lm} + b_4 e_l e_m) e_i e_k + b_5(e_i e_l \delta_{km} + e_k e_m \delta_{il})\} n_l n_m. \tag{31.23}$$

This implies

$$\lambda_{ilkm} = (b_3 - b_1)\delta_{il}\delta_{km} + b_1(\delta_{ik}\delta_{lm} + \delta_{im}\delta_{lk}) +$$

$$+ (b_5 - b_2)(\delta_{il}e_k e_m + \delta_{km}e_i e_l) +$$

$$+ b_2(\delta_{ik}e_l e_m + \delta_{im}e_k e_l + \delta_{lk}e_i e_m + \delta_{lm}e_i e_k) + b_4 e_i e_l e_k e_m. \tag{31.24}$$

This last expression differs from the expression in braces in (31.23) in being symmetrical in the subscripts, in accordance with the properties of the λ_{iklm} of (6.2), and in such a way as to leave the value of (31.23) unaltered. Expression (31.24) was first given in [31], while the covariant form of (31.12) has been used in [36].

Finally we derive the expression for the tensor $\overline{\Lambda}$ inverse under multiplication with respect to the Λ of (31.12), which we shall need in later work. Starting from (13.34) and (13.37) we get

$$\overline{\Lambda} = \Lambda^2 - \Lambda_t\Lambda + \frac{1}{2}[(\Lambda_t)^2 - (\Lambda^2)_t]. \tag{31.25}$$

However, it is simplest to use (13.33) with $\lambda = -c_0$, $\alpha = c_1\mathbf{n}\cdot\mathbf{n} + c_2\mathbf{e}\cdot\mathbf{e} + c_3\mathbf{c}\cdot\mathbf{c}$, and then to use (12.77). Simple steps give

$$\overline{\Lambda} = (c_0 + c_3)(c_0 + c_1 + c_2 - c_1\boldsymbol{n}\cdot\boldsymbol{n} - c_2\boldsymbol{e}\cdot\boldsymbol{e}) -$$
$$- [g_5(c_0 + c_1 + c_2) - c_1c_2][\boldsymbol{ne}]\cdot[\boldsymbol{ne}]. \tag{31.26}$$

Equation (31.24) shows why the Λ of (31.10) for hexagonal crystals differs from the general transversely isotropic tensor of (31.2) and (31.4)-(31.6) on account of the condition $b_6 = b_2$: the three properties listed in the second paragraph of this chapter, which define (31.2), take no account of the symmetry of the λ_{iklm} tensor, which for transverse isotropy can depend only on δ_{ik} and e_l, which themselves can give rise to only five independent combinations having the appropriate symmetry in the subscripts, these combinations appearing in (31.24). Here the additional condition $b_6 = b_2$ is a consequence of the symmetry with respect to the subscripts in the tensor for the elastic constants.

32. Phase Velocities and Displacements

The basic equation of $\Lambda\mathbf{u} = v^2\mathbf{u}$ is solved without difficulty for the tensor Λ having the form of (31.12) for any direction of the wave normal $\mathbf{n}$. We have $\mathbf{cn} = \mathbf{ce} = 0$, $\mathbf{c}^2 = 1$, so

$$\Lambda\boldsymbol{c} = (c_0 + c_3)\boldsymbol{c}, \tag{32.1}$$

i.e., the displacement and phase velocity of one of the waves corresponding to $\mathbf{n}$ are given by formulas derived from (31.12), (31.13), and (31.14):

$$u_1 = c = \frac{[ne]}{|[ne]|}, \tag{32.2}$$

$$v_1^2 = c_0 + c_3 = (g_1 + g_5) + (g_2 - g_5) n_3^2 = \lambda_{66} + (\lambda_{44} - \lambda_{66}). \tag{32.3}$$

Now $\mathbf{u}_1 \perp \mathbf{n}$, so one of the waves is always purely transverse in a transversely isotropic medium. Knowing one of the solutions to (15.21), it is easy to find the other two, the simplest approach being as follows. The displacement vectors of the three isonormal waves are mutually perpendicular, so the unknown displacements $\mathbf{u}_0$ and $\mathbf{u}_2$ must lie in a plane perpendicular to vector **c**, i.e., in a plane parallel to vectors **n** and **e**. This may be called the meridional plane. Then for $\mathbf{u}_0$ or $\mathbf{u}_2$ we may put

$$u = \eta_1 n + \eta_2 e. \tag{32.4}$$

Substitution into (15.21) gives

$$(c_0 + c_1 n \cdot n + c_2 e \cdot e)(\eta_1 n + \eta_2 e) = \lambda(\eta_1 n + \eta_2 e);$$

comparison of the coefficients to the linearly independent vectors **n** and **e** gives us that

$$\left.\begin{aligned} (c_0 + c_1)\eta_1 + c_1 n_3 \eta_2 = \lambda\eta_1. \\ c_2 n_3 \eta_1 + (c_0 + c_2)\eta_2 = \lambda\eta_2. \end{aligned}\right\} \tag{32.5}$$

The characteristic equation for this system is

$$\begin{vmatrix} c_0 + c_1 - \lambda & c_1 n_3 \\ c_2 n_3 & c_0 + c_2 - \lambda \end{vmatrix} = (\lambda - c_0)^2 - (c_1 + c_2)(\lambda - c_0) + c_1 c_2 [ne]^2 = 0,$$

which implies that

$$\lambda = v^2 = c_0 + \frac{1}{2}\left(c_1 + c_2 \pm \sqrt{(c_1 - c_2)^2 + 4c_1 c_2 n_3^2}\right). \tag{32.6}$$

Then (32.5) gives

$$\frac{\eta_2}{\eta_1} = \frac{\lambda - (c_0 + c_1)}{c_1 n_3} = \frac{c_2 n_3}{\lambda - (c_0 + c_2)}. \tag{32.7}$$

Thus (32.2) and (32.3) are accompanied by the following two

solutions:

$$u_0 = n - \frac{c_1 - c_2 - \sqrt{(c_1 - c_2)^2 + 4c_1c_2n_3^2}}{2c_1n_3} e, \tag{32.8}$$

$$v_0^2 = c_0 + \frac{1}{2}(c_1 + c_2 + \sqrt{(c_1 - c_2)^2 + 4c_1c_2n_3^2}) \tag{32.9}$$

and

$$u_2 = n - \frac{c_1 - c_2 + \sqrt{(c_1 - c_2)^2 + 4c_1c_2n_3^2}}{2c_1n_3} e, \tag{32.10}$$

$$v_2^2 = c_0 + \frac{1}{2}(c_1 + c_2 - \sqrt{(c_1 - c_2)^2 + 4c_1c_2n_3^2}). \tag{32.11}$$

Christoffel's equation is easily solved in this case because any plane parallel to **n** and **e** is a symmetry plane in a transversely isotropic medium. This is obvious when Λ is put in the form of (31.12), because **c** is normal to the (**n**, **e**) plane, and reversal of its direction by reflection does not affect Λ. Hence any direction for **n** lies in a symmetry plane and so is special in the sense of section 17. Thus the general case of propagation of elastic waves for a hexagonal crystal reduces to the particular case considered in sections 17 and 28. From (31.12) we get that

$$\Lambda n = (c_0 + c_1) n + c_2n_3e, \tag{32.12}$$

$$\Lambda^2 n = [(c_0 + c_1)^2 + c_1c_2n_3^2] n + c_2n_3(2c_0 + c_1 + c_2) e. \tag{32.13}$$

Then condition (17.7), $[\mathbf{n}, \Lambda\mathbf{n}]\Lambda^2\mathbf{n} = 0$, is satisfied for any **n**; we find at once from (32.12) that

$$[n, \Lambda n] = c_2n_3[ne], \tag{32.14}$$

and the general formula of (17.16) leads to the velocity of (32.3).

Consider the special cases. First let the wave normal coincide with the sixfold axis: $\mathbf{n} = \mathbf{e}$. Then $n_3 = \mathbf{ne} = 1$ and (31.13), (32.9), and (32.11) give

$$v_0^2 = g_1 + 2g_2 + g_3 + g_4 = \lambda_{33}, \quad v_1^2 = v_2^2 = g_1 + g_2 = \lambda_{44}. \tag{32.15}$$

Then the general theory (section 18) shows that **e** is a longitudinal acoustic axis, with $\mathbf{u}_0 \| \mathbf{n} \| \mathbf{e}$, while (32.2) and (32.10) do not give definite directions for $\mathbf{u}_1$ and $\mathbf{u}_2$, as should be found here.

The second particular case is $\mathbf{n} \perp \mathbf{e}$, $n_3 = \mathbf{ne} = 0$. Then

$$v_0^2 = g_1 + g_3 = \lambda_{11}, \quad v_1^2 = g_1 + g_5 = \lambda_{66}, \quad v_2^2 = g_1 + g_2 = \lambda_{44}, \quad (32.16)$$

and for the displacements

$$u_0 = \boldsymbol{n}, \quad \boldsymbol{u}_1 = [\boldsymbol{ne}], \quad \boldsymbol{u}_2 = \boldsymbol{e}. \quad (32.17)$$

In both of these cases $\mathbf{u}_0$ defines the displacement of the purely longitudinal wave, while $\mathbf{u}_1$ and $\mathbf{u}_2$ do the same for the purely transverse waves. In the general case the $\mathbf{u}_0$ of (32.8) will be the displacement vector of the quasilongitudinal wave, because we always have $c_1 - c_2 = g_3 - g_2 - g_4 n_3^2 > 0$ (see Table III), so the addition to $\mathbf{n}$ in expression (32.8) for $\mathbf{u}_0$ will be less than that for the $\mathbf{u}_2$ of (32.10), i.e., $\mathbf{u}_0$ deviates least from the direction of $\mathbf{n}$. Comparison of (32.9) with (32.11) shows that $v_0^2 > v_2^2$ correspondingly.

Consider now the $\mathbf{n}$ for a hexagonal crystal such that a purely longitudinal wave can propagate along these directions. From (18.2) and (32.14) we have the condition for this as

$$[\boldsymbol{n}, \Lambda\boldsymbol{n}] = c_2 \boldsymbol{ne} \cdot [\boldsymbol{ne}] = 0. \quad (32.18)$$

In addition to the cases $\mathbf{n} = \mathbf{e}$ and $\mathbf{n} \perp \mathbf{e}$ already considered, this gives $c_2 = g_2 + g_4 n_3^2 = 0$, so, from (31.14),

$$n_3^2 = -\frac{g_2}{g_4} = \frac{\lambda_{11} - \lambda_{13} - 2\lambda_{44}}{\lambda_{11} + \lambda_{33} - 2\lambda_{13} - 4\lambda_{44}}. \quad (32.19)$$

But $0 \leq n_3^2 \leq 1$, so this condition can be obeyed only if the elastic constants of the transversely isotropic medium obey the condition

$$-1 < \frac{g_2}{g_4} < 0. \quad (32.20)$$

Table III shows that purely longitudinal waves can propagate in many hexagonal crystals in the directions denoted by ϑ_l, the angle formed by the longitudinal normal with L^6, as well as along the directions $n_3 = 1$ and $n_3 = 0$:

$$n_3 = \cos\vartheta_l = \sqrt{-\frac{g_2}{g_4}}. \quad (32.21)$$

The two quasitransverse waves have the same speed in the

case of an acoustic axis; from (32.3) and (32.11) we have

$$(c_3 - c_1)(c_3 - c_2) = c_1 c_2 n_3^2. \qquad (32.22)$$

It is readily seen from (31.13) that this condition reduces to

$$(1 - n_3^2)[(g_5^2 + g_4 g_5 - g_3 g_4) n_3^2 - (g_5 - g_2)(g_5 - g_3)] = 0. \qquad (32.23)$$

This implies the case $n_3^2 = 1$ ($\mathbf{n} = \mathbf{e}$) considered above and also the condition

$$n_3^2 = \frac{(g_5 - g_2)(g_5 - g_3)}{g_5^2 + g_4(g_5 - g_3)}, \qquad (32.24)$$

which can be satisfied only if

$$0 < \frac{(g_5 - g_2)(g_5 - g_3)}{g_5^2 + g_4(g_5 - g_3)} < 1. \qquad (32.25)$$

Table III shows that ice, magnesium, and quartz are among the crystals that have acoustic axes differing from **e**. This table gives ϑ_a, the angle formed by the acoustic axes with the symmetry axis:

$$\cos \vartheta_a = \sqrt{\frac{(g_5 - g_2)(g_5 - g_3)}{g_4(g_5 - g_3) + g_5^2}}. \qquad (32.26)$$

A hexagonal crystal has rotational symmetry around its six-fold axis as regards the elastic properties, so a circular cone of longitudinal normals or acoustic axes will occur around the **e** axis if (32.20) or (32.25), respectively, is complied with.

This rotational symmetry also means that all the wave surfaces for a hexagonal crystal will be surfaces of rotation, which may be derived simply by finding the section in a symmetry plane passing through the **e** axis, after which rotation around **e** gives the surface. Thus we may use here the results of section 25 for the sections of the wave surfaces by a symmetry plane.

The tensor $\Lambda = \Lambda^n$ of (31.12) takes the form of (25.13) for all wave normals lying in any given meridional plane:

$$\Lambda = \begin{pmatrix} \lambda^n & 0 \\ 0 & n\tau n \end{pmatrix} = \begin{pmatrix} c_0 + c_1 n \cdot n + c_2 e \cdot e & 0 \\ 0 & c_0 + c_3 \end{pmatrix}, \qquad (32.27)$$

Table III. Elastic Constants of Hexagonal Crystals

(The c are in 10^{11} dynes/cm^2; the g are in 10^{11} cm^2/sec^2)

No.	Crystal	c_{11}	c_{33}	c_{44}	c_{66}	c_{13}	g_1	g_2	g_3	g_4	g_5	ϑ_l	ϑ_a	$\|h\|_{max}$
1	Barium titanate													
	$BaTiO_3$	16.6	16.2	4.29	4.47	7.75	0.80	-0.047	2.10	0.024	-0.016			0.0084
2	Beryl													
	$Be_3Al_2(SiO_3)_6$	29.71	26.5	7.54	9.725	7.39	5.56	-2.72	5.61	4.24	-1.90	36°47'		0.051
3	Beryllium Be	29.23	33.64	16.25	13.28	1.4	6.36	2.56	9.70	-2.71	0.93	13°37'	40°23'	0.0617
4	Cadmium Cd	12.1	5.13	1.85	3.65	4.42	0.67	-0.46	0.725	0.11	-0.25			0.30
5	Cadmium sulfide													
	CdS	8.16	8.08	1.43	1.605	4.79	0.402	-0.106	1.29	0.20	-0.070	43°17'	64°39'	0.0023
6	Cancrinite													
	$(Na_2Ca)_4$													
	$(AlSiO_4)_6CaO_3$	5.2	8.26	2.38	2.17	1.24	0.65	0.33	1.48	0.60	0.24			0.20
7	Cobalt Co	30.7	35.81	7.55	7.10	10.27	1.46	-0.60	2.02	1.78	-0.65	57°59'		0.064
8	Ice H_2O	1.333	1.428	0.326	0.365	0.508	0.554	-0.192	0.927	0.490	-0.15	51°15'	72°35'	0.027
9	Magnesium Mg	6.348	6.645	1.842	1.877	2.170	1.313	-0.278	2.255	0.7223	-0.258	51°39'	80°15'	0.018
10	β-Quartz β-SiO_2	11.66	11.04	3.606	4.995	3.28	1.80	-0.441	2.598	0.648	0.083	34°24'	25°41'	0.023
11	Yttrium Y	7.79	7.69	2.431	2.47	2.1	0.73	-0.185	1.01	0.35	-0.18	43°22'	82°51'	0.0048
12	Zinc Zn	17.909	6.880	4.595	7.080	5.537	1.068	-0.4372	1.392	-0.6409	-0.0958			0.31

$$\vartheta_l = \arccos\sqrt{-\frac{g_2}{g_4}}, \qquad \vartheta_a = \arccos\sqrt{\frac{(g_5-g_2)(g_5-g_3)}{g_4(g_5-g_3)+g_5^2}}, \qquad |h|_{\max} = \frac{2g_2+g_4}{g_2+4g_3+\frac{1}{2}g_4-g_5}.$$

if the x_3 axis is set along **c**. To apply the general formulas of section 25 to hexagonal crystals we therefore put

$$\lambda^n = c_0 + c_1 n \cdot n + c_2 e \cdot e =$$

$$= g_1 n^2 + g_2 (ne)^2 + g_3 n \cdot n + (g_2 n^2 + g_4 (ne)^2)\, e \cdot e, \qquad (32.28)$$

$$n\tau n = c_0 + c_3 = (g_1 + g_5)\, n^2 + (g_2 - g_5)\,(ne)^2. \qquad (32.29)$$

Then (25.5) gives the curve formed by sheet T_1 (purely transverse waves) of the refraction surface as

$$m\tau m = (g_1 + g_5)\, m^2 + (g_2 - g_5)\,(me)^2 = 1. \qquad (32.30)$$

Taking $x = x_1 \| \mathbf{e}$, we have

$$me = x, \quad m^2 = x^2 + y^2, \qquad (32.31)$$

so

$$m\tau m = (g_1 + g_2)\, x^2 + (g_1 + g_5)\, y^2 = 1, \qquad (32.32)$$

$$\tau = \begin{pmatrix} g_1 + g_2 & 0 \\ 0 & g_1 + g_5 \end{pmatrix} = (g_1 + g_5) + (g_2 - g_5)\, e \cdot e. \qquad (32.33)$$

Then, from (31.14), we should put in (25.6)-(25.9) that

$$a^2 = g_1 + g_2 = \lambda_{44}, \quad b^2 = g_1 + g_5 = \lambda_{66}. \qquad (32.34)$$

This defines the section of sheet T_1 completely; (25.19) shows that λ_t^m and $|\lambda^m|$ must be found for the other two sheets. From (32.28) we have

$$\lambda_t^m = (2g_1 + g_2 + g_3)\, m^2 + (2g_2 + g_4)\,(me)^2 =$$

$$= (\lambda_{11} + \lambda_{44})\, m^2 + (\lambda_{33} - \lambda_{11})\,(me)^2. \qquad (32.35)$$

We use the two-dimensional formulas (14.24) and (14.28) to get $|\lambda^n|$ as

$$|\lambda^n| = (c_0 + c_1)(c_0 + c_2) - c_1 c_2 (ne)^2 =$$

$$= [(g_1 + g_3)\, n^2 + g_2 (ne)^2][(g_1 + g_2)\, n^2 + (g_2 + g_4)(ne)^2] -$$

$$- g_3 (ne)^2 [g_2 n^2 + g_4 (ne)^2]. \qquad (32.36)$$

Then (31.14) gives

$$|\lambda^m| = Am^4 + Bm^2(me)^2 + C(me)^4, \tag{32.37}$$

in which

$$A = \lambda_{44}\lambda_{66}, \quad B = \lambda_{11}\lambda_{33} - 2\lambda_{44}(\lambda_{11} + \lambda_{33}) - \lambda_{13}^2,$$

$$C = \lambda_{13}^2 + \lambda_{44}(\lambda_{11} + 2\lambda_{13}) + \lambda_{33}(\lambda_{44} - \lambda_{11}). \tag{32.38}$$

From (25.19), (32.31), (32.35), (32.37), and (32.38) we have the equation for the section of sheets L and T_2 of the refraction surface by a meridional plane of a hexagonal crystal:

$$A(x^2 + y^2)^2 + Bx^2(x^2 + y^2) + Cx^4 - Dx^2 - Gy^2 + 1 = 0, \tag{32.39}$$

in which from (32.35)

$$D = \lambda_{33} + \lambda_{44}, \quad G = \lambda_{11} + \lambda_{44}. \tag{32.40}$$

Expressing the λ^n of (32.28) in terms of subscripts, we have

$$\lambda^n_{ab} = \lambda_{acbd} n_c n_d, \tag{32.41}$$

in which

$$\lambda_{acbd} = (g_3 - g_1)\delta_{ac}\delta_{bd} + g_1(\delta_{ab}\delta_{cd} + \delta_{ad}\delta_{bc}) -$$
$$- g_2(e_a e_c \delta_{bd} + \delta_{ac} e_b e_d) + g_2(\delta_{ab} e_c e_d +$$
$$+ \delta_{ad} e_b e_c + \delta_{bc} e_a e_d + \delta_{cd} e_a e_b) + g_4 e_a e_c e_b e_d. \tag{32.42}$$

This last expression is derived via the arguments used for (31.24). It is readily verified that λ_{acbd} has the required symmetry and gives (32.28) when substituted into (32.41). From (25.53) and (32.42) we have

$$\lambda^c_{ad} = \lambda_{abbd} = (2g_1 + g_2 + g_3)\delta_{ad} + (2g_2 + g_4)e_a e_d, \tag{32.43}$$

or, from (31.14), in coordinate-free form,

$$\lambda^c = (\lambda_{11} + \lambda_{44}) + (\lambda_{33} - \lambda_{11})e \cdot e. \tag{32.44}$$

We use (32.41)-(32.44) to get the section of the wave surface for a hexagonal crystal by a plane through the **e** axis, which is assumed to be parallel to the x_1 coordinate axis; hence $e_1 = 1$ and $e_2 = 0$ in this case. Substitution of (32.42) and (32.44) into (25.54) and

(25.51) gives, with $\lambda_{12}^{C} = \lambda_{1112} = \lambda_{2111} = 0$, that

$$\left.\begin{aligned} &A_1 = 2(\lambda_{33}\lambda_{44} + \lambda_{11}^{\mathbf{s}} - \mathbf{s}^2(\lambda_{33} + \lambda_{44})), \quad B_1 = C_2 = 3\lambda_{12}^{\mathbf{s}},\\ &C_1 = B_2 = \lambda_{11}\lambda_{33} + \lambda_{44}^2 - g_3^2 + \lambda_{11}^{\mathbf{s}} + \lambda_{22}^{\mathbf{s}} - \mathbf{s}^2(\lambda_{11} + \lambda_{33} + 2\lambda_{44}),\\ &D_1 = A_2 = \lambda_{12}^{\mathbf{s}}, \quad D_2 = 2(\lambda_{11}\lambda_{44} + \lambda_{22}^{\mathbf{s}} - \mathbf{s}^2(\lambda_{11} + \lambda_{44})). \end{aligned}\right\} \tag{32.45}$$

Here the $\lambda_{\alpha\beta}$ are as in (19.7). From (31.15) and (32.28),

$$\left.\begin{aligned} &\lambda_{11}^{\mathbf{s}} = (g_1 + g_2)\mathbf{s}^2 + (g_2 + g_3 + g_4)s_1^2 = \lambda_{33}s_1^2 + \lambda_{44}s_2^2,\\ &\lambda_{22}^{\mathbf{s}} = g_1\mathbf{s}^2 + g_2 s_1^2 + g_3 s_2^2 = \lambda_{44}s_1^2 + \lambda_{11}s_2^2, \quad \lambda_{12}^{\mathbf{s}} = g_3 s_1 s_2. \end{aligned}\right\} \tag{32.46}$$

The quantities of (25.58) thus take the following values for a hexagonal crystal:

$$\left.\begin{aligned} &(A,B) = 2\alpha\gamma - 3\zeta^2, \quad (C, D) = 2\beta\gamma - 3\zeta^2,\\ &(A, C) = \zeta(6\alpha - \gamma), \quad (B, D) = \zeta(6\beta - \gamma),\\ &(A, D) = 4\alpha\beta - \zeta^2, \quad (B, C) = 9\zeta^2 - \gamma^2. \end{aligned}\right\} \tag{32.47}$$

Here the symbols are

$$\left.\begin{aligned} &\alpha = \lambda_{33}\lambda_{44} - (\lambda_{44}s_1^2 + \lambda_{33}s_2^2), \quad \beta = \lambda_{11}\lambda_{44} - (\lambda_{11}s_1^2 + \lambda_{44}s_2^2),\\ &\gamma = \alpha + \beta - a_0, \quad a_0 = g_3^2 - (\lambda_{11} - \lambda_{44})(\lambda_{33} - \lambda_{44}), \quad \zeta = g_3 s_1 s_2. \end{aligned}\right\} \tag{32.48}$$

Substitution of (32.47) into (25.57) gives us after a few steps that

$$R = (\alpha\beta - \zeta^2)\{[\gamma^2 - 4(\alpha\beta - \zeta^2)]^2 + 36a_0\gamma\zeta^2\} - a_0\zeta^2(\gamma^3 + 27a_0\zeta^2). \tag{32.49}$$

If we take $\mathbf{s}$ (ray-velocity vector) as our radius vector ($s_1 = x$, $s_2 = y$), the above equation becomes the section of the wave surface in a hexagonal crystal by a plane parallel to the $\mathbf{e}$ axis. From (32.48) and (32.49) we see that this is an equation of degree 12 containing only even powers of x and y. A direct calculation shows that the R given by (25.52) will (see section 25) have all the terms of degree above 12 in x and y cancelling out even in the general case, which once again gives us (32.49).

To demonstrate this we apply (32.49) to an isotropic medium, which may be considered as a particular case of a hexagonal crystal. Comparison of (19.6) and (19.7) shows that here we have $\lambda_{33} = \lambda_{11}$, $\lambda_{13} = \lambda_{11} - 2\lambda_{44}$, and $g_3 = \lambda_{13} + \lambda_{44} = \lambda_{11} - \lambda_{44}$, so, from (32.48), $a_0 = 0$

and

$$\left.\begin{aligned}\gamma = \alpha + \beta = 2\lambda_{11}\lambda_{44} - (\lambda_{11} + \lambda_{44})r^2,\\ \alpha\beta - \zeta^2 = \lambda_{11}\lambda_{44}(r^2 - \lambda_{11})(r^2 - \lambda_{44}).\\ \gamma^2 - 4(\alpha\beta - \zeta^2) = (\lambda_{11} - \lambda_{44})^2 r^4,\end{aligned}\right\} \tag{32.50}$$

in which $r^2 = x^2 + y^2 = \mathbf{s}^2$. Then (32.49) becomes

$$(\lambda_{11} - \lambda_{44})^4 r^8 (r^2 - \lambda_{11})(r^2 - \lambda_{44}) = 0,$$

i.e., it splits up as it should (see section 24) into the equations of two circles: $r^2 = \lambda_{11}$ (longitudinal wave) and $r^2 = \lambda_{44}$ (transverse wave).

The transverse isotropy means that (32.49) gives not only the section of the wave surface by a plane parallel to the axis but also the complete equation for the two sheets L and T_2 of this surface. This is readily derived by replacing the $y^2 = s_2^2$ of (32.49) by $y^2 + z^2$, which gives us a surface of rotation obtained by rotating the curve around the symmetry axis $\mathbf{e} \| x_1$.

The above is a discussion of the section of the wave surface for a hexagonal crystal, but the result is actually much more general, being the equation for the section of that surface by any symmetry plane for the cubic, hexagonal, tetragonal, and orthorhombic systems. In fact, from the relations of section 19 for tensor Λ, on the assumption that $n_2 = 0$, we get the following expressions:

Cubic crystal:

$$\Lambda = \begin{pmatrix} \lambda_{11}n_1^2 + \lambda_{44}n_3^2 & 0 & (\lambda_{12} + \lambda_{44})n_1 n_3 \\ 0 & \lambda_{44} & 0 \\ (\lambda_{12} + \lambda_{44})n_1 n_3 & 0 & \lambda_{44}n_1^2 + \lambda_{11}n_3^2 \end{pmatrix}; \tag{32.51}$$

Hexagonal or tetragonal crystal:

$$\Lambda = \begin{pmatrix} \lambda_{11}n_1^2 + \lambda_{44}n_3^2 & 0 & (\lambda_{13} + \lambda_{44})n_1 n_3 \\ 0 & \lambda_{66}n_1^2 + \lambda_{44}n_3^2 & 0 \\ (\lambda_{13} + \lambda_{44})n_1 n_3 & 0 & \lambda_{44}n_1^2 + \lambda_{33}n_3^2 \end{pmatrix}; \tag{32.52}$$

Orthorhombic crystal:

$$\Lambda = \begin{pmatrix} \lambda_{11}n_1^2 + \lambda_{55}n_3^2 & 0 & (\lambda_{13} + \lambda_{55})n_1n_3 \\ 0 & \lambda_{66}n_1^2 + \lambda_{44}n_3^2 & 0 \\ (\lambda_{13} + \lambda_{55})n_1n_3 & 0 & \lambda_{55}n_1^2 + \lambda_{33}n_3^2 \end{pmatrix}. \quad (32.53)$$

The Λ tensors thus coincide precisely for hexagonal and tetragonal crystals if $n_2 = 0$, while a cubic crystal appears as a particular case ($\lambda_{33} = \lambda_{11}$, $\lambda_{66} = \lambda_{44}$) and an orthorhombic crystal differs only in the replacement of λ_{44} by λ_{55}. The curve given for all of these by the wave surface with a symmetry plane is governed by the two-dimensional tensor λ^n of section 25, which is derived by striking out the second row and second column in (32.51)-(32.53), and which has the form

$$\lambda^n = \begin{pmatrix} \lambda_{11}n_1^2 + \lambda_{44}n_3^2 & (\lambda_{13} + \lambda_{44})n_1n_3 \\ (\lambda_{13} + \lambda_{44})n_1n_3 & \lambda_{44}n_1^2 + \lambda_{33}n_3^2 \end{pmatrix}, \quad (32.54)$$

with appropriate replacement of the $\lambda_{\alpha\beta}$ for cubic and orthorhombic crystals. The curves for all these types of crystals are given by (32.49); we merely have to insert the $\lambda_{\alpha\beta}$ appropriate to each.

The two-dimensional tensor τ of (25.2) and (25.12) defines the section of the sheet for the purely transverse waves; it follows from (32.51) that $\mathbf{n}\tau\mathbf{n} = \lambda_{44}$, i.e., $\tau_{ab} = \lambda_{44}\delta_{ab}$, for a cubic crystal, so the section of any wave surface for purely transverse waves by a symmetry plane* is a circle in the case of a cubic crystal [see (25.4), (25.5), and (25.32)].

33. Comparison of a Hexagonal Crystal with an Isotropic Medium

A hexagonal crystal (transversely isotropic medium) is the sole type of anisotropic medium for which the equations of the theory of elastic waves allow of an exact general solution in terms of square roots. All the same, even here it is desirable to use the method of comparison with an isotropic medium, because the exact formulas are sometimes too cumbrous. For this reason, and to illustrate the method of Chapter 5, I consider here some of the

* By this plane we mean one perpendicular to a fourfold axis, or a plane equivalent to this.

relationships for a hexagonal crystal on the basis of comparison with an isotropic medium.

From (27.7), (31.16), and (31.17) we have

$$\left.\begin{aligned} a &= c_0+\frac{1}{2}c_2(1-n_3^2)+\frac{1}{2}c_3=g_1+\frac{1}{2}(g_2+g_5)+\\ &\quad+\frac{1}{2}(g_2-g_5+g_4)n_3^2-\frac{1}{2}g_4n_3^4,\\ b &= c_1+\frac{1}{2}c_2(3n_3^2-1)-\frac{1}{2}c_3=g_3-\frac{1}{2}(g_2+g_5)+\\ &\quad+\frac{1}{2}(3g_2-g_4+g_5)n_3^2+\frac{3}{2}g_4n_3^4,\\ c &= a+b=g_1+g_3+2g_2n_3^2+g_4n_3^4, \end{aligned}\right\} \tag{33.1}$$

whence we get from (26.16) and (26.17) after averaging that

$$\left.\begin{aligned} a_m &= \langle a\rangle = g_1+\frac{2}{3}g_2+\frac{1}{15}g_4+\frac{1}{3}g_5,\\ b_m &= \langle b\rangle = g_3+\frac{2}{15}g_4-\frac{1}{3}g_5. \end{aligned}\right\} \tag{33.2}$$

Then from (26.7)

$$\langle F_m\rangle = \frac{2}{45}\left[7(g_2-g_5)^2+g_4\left(\frac{17}{5}g_4-2g_5\right)+14g_2(g_2+g_4)\right]. \tag{33.3}$$

This defines the mean-square elastic anisotropy of a hexagonal crystal, while $\sqrt{\langle F_m\rangle/\langle\Lambda^2\rangle_t}$ is the corresponding relative anisotropy (section 26). On the other hand, (27.9) gives

$$F=(\Lambda^2)_t-(\Lambda^{0^2})_t=\frac{1}{2}\left[c_3-c_2(1-n_3^2)\right]^2+2c_2^2n_3^2(1-n_3^2)=$$

$$=(1-n_3^2)\left[\frac{1}{2}(1-n_3^2)(g_5-g_2-g_4n_3^2)^2+2n_3^2(g_2+g_4n_3^2)^2\right]. \tag{33.4}$$

Then

$$\langle F\rangle = \frac{4}{15}\left[(g_2-g_5)^2+g_2^2+\frac{1}{3}g_4^2+\frac{2}{7}g_4(4g_2-g_5)\right]. \tag{33.5}$$

Substitution from Table III into (33.3) and (33.5) shows that $\langle F\rangle < \langle F_m\rangle$.

Formula (27.27) gives for a hexagonal crystal that

$$\boldsymbol{h} = \frac{1}{b} c_2 \boldsymbol{en} \cdot [\boldsymbol{n} [\boldsymbol{en}]]. \tag{33.6}$$

Thus vector **h** always lies in the meridional plane, and **h** = 0, which defines the longitudinal normals, reduces to (32.18). From (27.31') and (33.4) we have

$$g^2 = \frac{F}{2b^2} = \frac{[\boldsymbol{ne}]^2}{2b^2} \left\{ \frac{1}{2} [\boldsymbol{ne}]^2 (g_5 - c_2)^2 + 2n_3^2 c_2^2 \right\}. \tag{33.7}$$

Then (27.42) gives

$$\varkappa^2 = g^2 - \boldsymbol{h}^2 = \frac{(1 - n_3^2)^2}{4b^2} (g_5 - c_2)^2, \tag{33.8}$$

so

$$\varkappa = \frac{1}{2} (1 - n_3^2) \left| \frac{g_5 - c_2}{b} \right|. \tag{33.9}$$

Thus, as we noted in section 28, here compliance with the transversality condition causes κ to be determined by a completely rational expression. Tensor Λ' is expressed as follows:

$$\Lambda' = \Lambda - \Lambda^0 = c_0' + c_1' \boldsymbol{n} \cdot \boldsymbol{n} + c_2 \boldsymbol{e} \cdot \boldsymbol{e} + c_3 \boldsymbol{c} \cdot \boldsymbol{c}, \tag{33.10}$$

in which

$$c_0' = -\frac{1}{2} (c_3 + c_2 [\boldsymbol{ne}]^2), \quad c_1' = \frac{1}{2} \{ c_3 - c_2 [3 (\boldsymbol{ne})^2 - 1] \}. \tag{33.11}$$

Thus Λ' differs from Λ only in the replacement of c_0 and c_1 by c_0' and c_1', so (31.16)-(31.22) can be used with the appropriate changes. In particular, we get from (31.22) that

$$|\Lambda'| = \frac{1}{2} (\boldsymbol{ne})^2 [\boldsymbol{ne}]^4 c_2^2 (c_2 - g_5) = b^3 |\alpha|. \tag{33.12}$$

Comparison of (33.12) with (33.6) and (33.9) shows that $|\alpha| = \pm \kappa \mathbf{h}^2$, in accordance with (28.7), the upper (lower) sign corresponding to a positive (negative) value for

$$\frac{c_2 - g_5}{2b} = \frac{g_2 - g_5 + g_4 (\boldsymbol{ne})^2}{2g_3 - g_2 - g_5 + (\boldsymbol{ne})^2 (3g_2 + g_5 - g_4 + 3g_4 (\boldsymbol{ne})^2)}. \tag{33.13}$$

Simple calculations based on (33.6) and (33.9) readily show that the general conditions of (28.34) (for acoustic axes) lead to (32.22), which was derived by direct solution of (32.3) and (32.11). Similarly, we may verify for a hexagonal crystal all the other general relations of section 28 for the special directions of any crystal.

The approximate methods of sections 29 and 30 retain their value for hexagonal crystals although an exact solution is available here, because the approximate formulas give rational formulas for **u** and v^2 that are of high accuracy. Moreover, all directions in a hexagonal crystal are special, which leads to simplification of the general formulas of sections 29 and 30.

The first approximation for the velocity of the quasilongitudinal wave is given by (28.13), (30.12), (33.6), and (33.12) as

$$v_0^2 = a + b\xi_0 = a + b(1 + h^2 + |\alpha|) =$$

$$= a + b + \frac{(\boldsymbol{en})^2 [\boldsymbol{en}]^2 c_2^2}{2b^2}(2b + (c_2 - g_5)[\boldsymbol{en}]^2) =$$

$$= c_0 + c_1 + c_2 n_3^2 + 4\,\frac{n_3^2(1-n_3^2)\,c_2^2\,(c_1 + c_2 n_3^2 - c_3)}{[2c_1 + c_2(3n_3^2 - 1) - c_3]^2}. \tag{33.14}$$

The displacement is given by (30.14) with $\zeta = 0$ and

$$\eta = \frac{|\alpha|}{h^2} = \frac{[\boldsymbol{ne}]^2 (c_2 - g_5)}{2b}; \tag{33.15}$$

the result being

$$\boldsymbol{u}_0 = \boldsymbol{n} + \frac{c_2 \boldsymbol{en}}{2b^2}\{2b + (c_2 - g_5)[\boldsymbol{ne}]^2\}\,[\boldsymbol{n}\,[\boldsymbol{en}]\,]. \tag{33.16}$$

The conclusions of section 30 show that we make an error not exceeding the fourth power of $|\mathbf{h}|$ in taking only the first approximation; (33.6) implies that $\mathbf{h} = 0$ for $\mathbf{n} \perp \mathbf{e}$ and $\mathbf{n} \| \mathbf{e}$, so we may assume that for $\vartheta = \arccos(\mathbf{ne}) = \pi/4$ ($\mathbf{ne} = 1/\sqrt{2}$) we have $|\mathbf{h}|$ close to its maximal value, i.e.,

$$|\boldsymbol{h}|_{\max} \sim \frac{2g_2 + g_4}{4g_3 + g_2 + \frac{1}{2}g_4 - g_5} = \frac{2(\lambda_{33} - \lambda_{11})}{6(\lambda_{13} + \lambda_{44}) - 2\lambda_{66} + \lambda_{11} + \lambda_{33}}. \tag{33 17}$$

Table III gives values of this for some hexagonal crystals, which show that the error of the first approximation is only $0.31^4 \approx 0.01$ even for the highly anisotropic crystal of zinc, while this error is less than 10^{-4} for cobalt (medium anisotropy), i.e., is negligibly small. Thus the formulas of (33.14) and (33.16) (first approximation) may be considered as exact for most hexagonal crystals.

The velocity and displacement for the quasitransverse waves may be found via (30.23) and (30.24), but it is simpler to use (33.14) and (33.16) with the exact solutions of (32.2) and (32.3) for transverse waves. We know v_0^2 and v_1^2, so v_2^2 is found from

$$v_2^2 = \Lambda_t - v_0^2 - v_1^2 = c_0 + c_2\,[\boldsymbol{ne}]^2\left(1 - \frac{4c_2 n_3^2 (c_1 - c_3 + c_2 n_3^2)}{\left[2c_1 + c_2\,(3n_3^2 - 1) - c_3\right]^2}\right). \quad (33.18)$$

The direction of the corresponding displacement is found as the vector product of the displacement of (32.2) and (33.16):

$$\boldsymbol{u}_2 = [\boldsymbol{u}_0\,[\boldsymbol{en}]\,] = [\boldsymbol{n}\,[\boldsymbol{en}]\,] - \frac{c_2 \boldsymbol{en}\cdot[\boldsymbol{en}]^2}{2b^2}\,[2b + (c_2 - g_5)\,[\boldsymbol{en}]^2]\,\boldsymbol{n}. \quad (33.19)$$

Equations (33.14) and (33.18) may be considered as the polar equations of the quasilongitudinal and quasitransverse sheets of the phase-velocity surface in a hexagonal crystal ($n_3 = \cos\vartheta$).

34. Mean Transverse Anisotropy

The mean elastic anisotropy of section 26 was derived by reference to the isotropic medium most similar in elastic properties. The covariant form of (31.12) is a simple general representation of the Λ tensor for a transversely isotropic medium (hexagonal crystal), and Christoffel's equation has exact solutions, so this approach may be extended to this case, which means that the analogy with the method of section 26 allows us to choose the parameters of the tensor of (31.12) in such a way that the latter differs on average as little as possible from tensor Λ. This defines the fictitious transversely isotropic medium most similar on average to the given crystal as regards elastic properties. Let the tensor Λ^t for this medium be put in the form of (31.12):

$$\Lambda_t = (g_1 + g_2 n_3^2) + g_3 \boldsymbol{n}\cdot\boldsymbol{n} + (g_2 + g_4 n_3^2)\,\boldsymbol{e}\cdot\boldsymbol{e} + g_5\,[\boldsymbol{ne}]^2\,\boldsymbol{c}\cdot\boldsymbol{c}. \quad (34.1)$$

By analogy with (26.6), parameters $g_1, \ldots, g_5$ are sought from the minimum of

$$\langle F^t\rangle = \langle(\Lambda^t - \Lambda)^2\rangle_t = \langle \Lambda'^{t2}\rangle_t. \quad (34.2)$$

Here the direction of **e** must first be chosen. The choice is obvious for cubic, tetragonal, and trigonal crystals; **e** lies along a higher-order symmetry axis. A monoclinic crystal has only one distinct direction, namely the L^2 axis or the normal to the symmetry plane, and this is naturally taken as the direction of **e**. The choice is also unambiguous for orthorhombic and triclinic crystals: for orthorhombic ones, the direction of **e** is that of a twofold axis (normal to a symmetry plane), while for triclinic ones it is that of the longitudinal normals (section 18). In both cases the choice is not unique, so a selection must be made from among the possibilities on the basis of the least difference between the velocities of the transverse waves. The squares of the latter are v_1^2 and v_2^2 for this direction in an orthorhombic or triclinic crystal, so the direction of **e** here is that for which $|v_1^2 - v_2^2|/(v_1^2 + v_2^2)$ is minimal. Condition (34.2) gives rise to the following equations ($s = 1, 2, 3, 4, 5$):

$$\frac{\partial \langle F^{t} \rangle}{\partial g_s} = 2 \left\langle (\Lambda^{t} - \Lambda) \frac{\partial \Lambda^{t}}{\partial g_s} \right\rangle_{t} = 0. \tag{34.3}$$

Tensor Λ^t is given by (34.1) as a homogeneous linear function of the g_s, so Euler's theorem gives

$$\sum_{s=1}^{5} g_s \frac{\partial \Lambda^{t}}{\partial g_s} = \Lambda^{t}. \tag{34.4}$$

Multiplication of (34.3) by g_s and summation over s gives us a relation analogous to (26.20):

$$\langle (\Lambda^{t} - \Lambda) \Lambda^{t} \rangle_{t} = \langle \Lambda^{t\,2} \rangle_{t} - \langle \Lambda \Lambda^{t} \rangle_{t} = 0. \tag{34.5}$$

(34.1) and (34.3) together imply five equations:

$$\left. \begin{gathered} \langle \Lambda^{t} - \Lambda \rangle_{t} = 0, \ \langle n_3^2 (\Lambda^{t} - \Lambda) \rangle_{t} + \langle \boldsymbol{e} (\Lambda^{t} - \Lambda) \boldsymbol{e} \rangle = 0, \\ \langle \boldsymbol{n} (\Lambda^{t} - \Lambda) \boldsymbol{n} \rangle = 0, \\ \langle n_3^2 \boldsymbol{e} (\Lambda^{t} - \Lambda) \boldsymbol{e} \rangle = 0, \ \langle [\boldsymbol{ne}]^2 \boldsymbol{c} (\Lambda^{t} - \Lambda) \boldsymbol{c} \rangle = 0. \end{gathered} \right\} \tag{34.6}$$

We use (34.1) with $\langle n_3^2 \rangle = 1/3$, $\langle n_3^4 \rangle = 1/5$ [see (26.16) and (26.17)] to get

$$\left.\begin{aligned}
\langle\Lambda^{t}\rangle_{t} &= \frac{1}{3}(9g_1+6g_2+3g_3+g_4+2g_5),\\
\langle n_3^2\Lambda^{t}\rangle_{t} &= \frac{1}{15}(15g_1+14g_2+5g_3+3g_4+2g_5),\\
\langle e\Lambda^{t}e\rangle &= \frac{1}{3}(3g_1+4g_2+g_3+g_4),\\
\langle n\Lambda^{t}n\rangle &= \frac{1}{15}(15g_1+10g_2+15g_3+3g_4),\\
\langle n_3^2 e\Lambda^{t}e\rangle &= \frac{1}{15}(5g_1+8g_2+3g_3+3g_4),\\
\langle [ne]^2 c\Lambda^{t}c\rangle &= \frac{2}{15}(5g_1+g_2+4g_5).
\end{aligned}\right\} \tag{34.7}$$

The notation to be used is

$$\left.\begin{aligned}
\langle\Lambda\rangle_{t} = \frac{1}{3}d_1,\quad \langle n_3^2\Lambda\rangle_{t}+\langle e\Lambda e\rangle = \frac{2}{15}d_2,\quad \langle n\Lambda n\rangle = \frac{1}{15}d_3,\\
\langle n_3^2 e\Lambda e\rangle = \frac{1}{15}d_4,\quad \langle [ne]^2 c\Lambda c\rangle = \frac{2}{15}d_5.
\end{aligned}\right\} \tag{34.8}$$

Then system (34.6) becomes

$$\left.\begin{aligned}
9g_1+6g_2+3g_3+g_4+2g_5 &= d_1,\\
15g_1+17g_2+5g_3+4g_4+g_5 &= d_2,\\
15g_1+10g_2+15g_3+3g_4 &= d_3,\\
5g_1+8g_2+3g_3+3g_4 &= d_4,\\
5g_1+g_2+4g_5 &= d_5.
\end{aligned}\right\} \tag{34.9}$$

The solutions are

$$\left.\begin{aligned}
g_1 &= \frac{1}{112}(213d_1-90d_2-28d_3+77d_4-84d_5),\\
g_2 &= \frac{1}{112}(-225d_1+114d_2+28d_3-105d_4+84d_5),\\
g_3 &= \frac{1}{4}(-5d_1+2d_2+d_3-2d_4+2d_5),\\
g_4 &= \frac{1}{16}(55d_1-30d_2-8d_3+35d_4-20d_5),\\
g_5 &= \frac{1}{8}(-15d_1+6d_2+2d_3-5d_4+8d_5).
\end{aligned}\right\} \tag{34.10}$$

Formulas (34.8) and (34.10) define completely the elastic moduli of the transversely isotropic medium most similar on average to the given crystal as regards elastic properties. The tensor of (34.1) for that medium allows us, by analogy with section 26, to discuss the mean-square transverse elastic anisotropy of the crystal. As a measure of this we may [see (26.27), (34.2), and (34.5)] use the expression

$$\sqrt{\langle F^t\rangle} = \sqrt{\langle \Lambda^2\rangle_t - \langle \Lambda^{t2}\rangle_t} = \sqrt{\langle \Lambda'^{t2}\rangle_t}. \tag{34.11}$$

The relative mean-square transverse elastic anisotropy [see (26.28)] is similarly defined by its measure

$$\Delta^t = \sqrt{\frac{\langle \Lambda^2\rangle_t - \langle \Lambda^{t2}\rangle_t}{\langle \Lambda^2\rangle_t}}. \tag{34.12}$$

A transversely isotropic medium has five elastic constants $g_1, \ldots, g_5$, so the Λ^t closest to Λ may be chosen on the basis of five parameters, instead of the mere two available for an isotropic medium. Then we would expect that Λ^t of (34.1) and (34.10) will differ less from Λ than does the Λ_m of (26.2) and (26.13), and also that the mean transverse anisotropy of (34.11) or (34.12) will be less than the mean anisotropy of (26.27) or (26.28). This will be demonstrated in Chapter 8 by reference to actual crystals.

It is clear that the actual Λ may be satisfactorily replaced by the corresponding Λ^t if the relative transverse anisotropy of (34.12) is small enough; many topics may be discussed approximately on this basis.

35. Comparison with a Transversely Isotropic Medium*

In Chapter 5 we compared the elastic properties of a given crystal with those of the most similar isotropic medium in two ways: by averaging over all directions of **n** (section 26) and for a specified direction of **n** (section 27). Similarly, there are two ways of comparing a crystal with a transversely isotropic medium. The above section corresponds to the first (average) approach, while in this section we deal with the approach for a fixed wave normal **n**. Here the averaging operation is omitted, and Λ^0 is sought (see sections 27 and 34) from the minimum in

$$F = \langle (\Lambda^0 - \Lambda)^2\rangle_t. \tag{35.1}$$

* This section constitutes the results of [36].

Here we take Λ^0 in the form of (32.12):

$$\Lambda^0 = a_0 + a_1 n \cdot n + a_2 e \cdot e + a_3 c \cdot c, \tag{35.2}$$

in which a_0, a_1, a_2, and a_3 are certain functions of $\mathbf{n}$, $\mathbf{c} = [\mathbf{en}]/|[\mathbf{en}]|$, and unit vector $\mathbf{e}$ is chosen as in the previous section. Here instead of the five constant coefficients g_s (s = 1, ..., 5) of section 34 we have to find four functions a_s (s = 0, 1, 2, 3) of $\mathbf{n}$. The subsequent steps are very much as in sections 27 and 34. From (35.1) we have that

$$\left((\Lambda^0 - \Lambda)\frac{\partial \Lambda^0}{\partial a_s}\right)_t = 0 \tag{35.3}$$

and

$$\sum_{s=0}^{3} a_s \left((\Lambda^0 - \Lambda)\frac{\partial \Lambda^0}{\partial a_s}\right)_t = (\Lambda^{0^2})_t - (\Lambda\Lambda^0)_t = 0, \tag{35.4}$$

so

$$F = (\Lambda'^2)_t = (\Lambda\Lambda')_t = (\Lambda^2)_t - (\Lambda^{0^2})_t, \tag{35.5}$$

in which

$$\Lambda' = \Lambda - \Lambda^0. \tag{35.6}$$

The system of (35.3) has the form

$$\Lambda'_t = n\Lambda' n = e\Lambda' e = c\Lambda' c = 0; \tag{35.7}$$

which is expanded to give

$$\left.\begin{aligned} 3a_0 + a_1 + a_2 + a_3 = \Lambda_t, \quad a_0 + a_1 n_3^2 + a_2 = e\Lambda e, \\ a_0 + a_1 + a_2 n_3^2 = n\Lambda n, \quad a_0 + a_3 = c\Lambda c. \end{aligned}\right\} \tag{35.8}$$

The solution is

$$\left.\begin{aligned} a_0 &= \frac{(1 + n_3^2)[ec]\,\Lambda\,[ec] - n\Lambda n + n_3^2 e\Lambda e}{2n_3^2}, \\ a_1 &= \frac{n\Lambda n - n_3^2 e\Lambda e - [ne]^2 \cdot [ec]\,\Lambda\,[ec]}{2n_3^2\,[ne]^2}, \end{aligned}\right\} \tag{35.9}$$

$$\left.\begin{aligned} a_2 &= \frac{(1-2n_3^2)\,\boldsymbol{n}\Lambda\boldsymbol{n} + n_3^2\boldsymbol{e}\Lambda\boldsymbol{e} - [\boldsymbol{ne}]^2\,[\boldsymbol{ec}]\,\Lambda\,[\boldsymbol{ec}]}{2n_3^2\,[\boldsymbol{ne}]^2}, \\ a_3 &= \frac{3n_3^2\boldsymbol{c}\Lambda\boldsymbol{c} - n_3^2\Lambda_t + \boldsymbol{n}\Lambda\boldsymbol{n} - [\boldsymbol{ec}]\,\Lambda\,[\boldsymbol{ec}]}{2n_3^2}. \end{aligned}\right\} \tag{35.9}$$

Here we have used $\Lambda_t = \{\Lambda(\mathbf{e}\cdot\mathbf{e}+\mathbf{c}\cdot\mathbf{c}+[\mathbf{ec}]\cdot[\mathbf{ec}])\}_t = \mathbf{e}\Lambda\mathbf{e}+\mathbf{c}\Lambda\mathbf{c}+[\mathbf{ec}]\,\Lambda\,[\mathbf{ec}]$. The solution of (35.9) becomes an indeterminacy of type 0/0 for all a_s if $n_3 = 0$, because for $\mathbf{ne} = 0$ we have [see (12.47)] for any tensor A that $A_t = \mathbf{n}A\mathbf{n}+\mathbf{e}\Lambda\mathbf{e}+\mathbf{c}\Lambda\mathbf{c}$, so any three of the equations of (35.7) imply the fourth, and system (35.8) contains only three independent equations for four variables. For $\mathbf{ne} = 0$ we have from (31.3) that $\mathbf{c}\cdot\mathbf{c} = 1-\mathbf{e}\cdot\mathbf{e}-\mathbf{n}\cdot\mathbf{n}$, i.e., the last term in (35.2) is expressed via the first three, so we may put

$$\Lambda^0 = a_0 + a_1\boldsymbol{n}\cdot\boldsymbol{n} + a_2\boldsymbol{e}\cdot\boldsymbol{e}.$$

Performing the calculations of (35.3)-(35.9) with this expression for Λ^0, we get

$$a_0 = \boldsymbol{c}\Lambda\boldsymbol{c},\quad a_1 = \boldsymbol{n}\Lambda\boldsymbol{n} - \boldsymbol{c}\Lambda\boldsymbol{c},\quad a_2 = \boldsymbol{e}\Lambda\boldsymbol{e} - \boldsymbol{c}\Lambda\boldsymbol{c}. \tag{35.10}$$

The same result is obtained directly from (35.8) by putting $n_3 = a_3 = 0$.

We can transform (35.9) to the following more compact and convenient form by using (31.3):

$$\left.\begin{aligned} a_0 &= [\boldsymbol{ec}]\,\Lambda\,[\boldsymbol{ec}] + \frac{1}{n_3}\,[\boldsymbol{ne}]\,[\boldsymbol{e},\,\Lambda\boldsymbol{e}], \\ a_1 &= \frac{[\boldsymbol{en}]\,[\boldsymbol{e},\,\Lambda\boldsymbol{e}]}{n_3(1-n_3^2)},\quad a_2 = \frac{[\boldsymbol{ne}]\,[\boldsymbol{n},\Lambda\boldsymbol{n}]}{n_3(1-n_3^2)}, \\ a_3 &= 2\boldsymbol{c}\Lambda\boldsymbol{c} - \Lambda_t + \frac{1}{n_3}\,\boldsymbol{e}\Lambda\boldsymbol{n}. \end{aligned}\right\} \tag{35.11}$$

These relations do not follow from (31.3) and (35.7) with $\mathbf{ne} = 0$, but (35.2) and (35.6) give

$$\mathbf{e}\Lambda'\mathbf{n} = \mathbf{e}\Lambda\mathbf{n} - n_3(a_0 + a_1 + a_2).$$

Substitution into (35.11) readily gives the result that (35.12) is obeyed for $\mathbf{ne} = 0$ also (see below).

The definition of (31.11) for **c** becomes meaningless for **n** = **e** and **ne** = $n_3 = 1$, and expression (35.2) for Λ^0 becomes simply $a + b\mathbf{n} \cdot \mathbf{n}$, so the above formulas are unsuitable in their present form for directions of **n** lying close to **e**.

Returning to the general case, we see that (31.3) and (35.7) together give (for **ne** ≠ 0) that

$$e\Lambda' n = n\Lambda' e = 0. \tag{35.12}$$

From (35.7) and (35.12) we have that $\mathbf{n} \perp \Lambda'\mathbf{n} \perp \mathbf{e}$, $\mathbf{e} \perp \Lambda'\mathbf{e} \perp \mathbf{n}$, so

$$\Lambda' e \,\|\, \Lambda' n \,\|\, [en] \,\|\, c. \tag{35.13}$$

We now put

$$\Lambda' c = w, \tag{35.14}$$

whereupon (35.7) implies that vector **w** must be perpendicular to **c**, i.e., must lie in the meridional plane:

$$wc = c\Lambda' c = 0. \tag{35.15}$$

From (35.13), $\Lambda'\mathbf{e} = k_1\mathbf{c}$, $\Lambda'\mathbf{n} = k_2\mathbf{c}$; scalar multiplication of these by **c** gives $k_1 = \mathbf{we}$, $k_2 = \mathbf{wn}$, so we may put the result of multiplying Λ' by the three independent vectors **e**, **n**, and **c** as

$$\Lambda' e = we \cdot c, \qquad \Lambda' n = wn \cdot c, \qquad \Lambda' c = w. \tag{35.16}$$

We have shown in section 12 that these results define Λ' uniquely. Use of (12.58) gives

$$\Lambda' = \frac{1}{[en]\,c} \left(w \cdot [en] + we \cdot c \cdot [nc] + wn \cdot c \cdot [ce]\right). \tag{35.17}$$

Use of the definition of (31.11) for **c** and property (35.15) gives

$$we \cdot [nc] + wn \cdot [ce] = [c,\, wn \cdot e - we \cdot n] = [c\,[w\,[en]]] = w \cdot |[en]|,$$

so (35.17) becomes*

$$\Lambda' = w \cdot c + c \cdot w. \tag{35.18}$$

---

*Note that (35.18) is correct for [**ne**] ≠ 0; the case [**ne**] = 0 requires a special discussion.

Tensor Λ' is defined by (35.6) as the difference between tensor Λ for the given crystal and the Λ^0 of (35.2) for the transversely isotropic medium most similar to it; (35.18) shows that this difference is completely determined by one vector **w**, with $\mathbf{w} \perp \mathbf{c}$ from (35.15), so Λ' is dependent on two parameters only, if **c** is taken as given. It is readily shown that the Λ' defined by (35.18) has the properties of (35.7) and (35.12), and also [see (12.70)] has the property

$$|\Lambda'| = 0. \tag{35.19}$$

From (35.18) we have also that

$$F = (\Lambda'^2)_t = 2w^2. \tag{35.20}$$

Thus the length of **w** directly defines the transverse anisotropy of the crystal (section 34); (35.2), (35.6), and (35.18) show that the Λ for any crystal may be put as

$$\Lambda = \Lambda^0 + \Lambda' = a_0 + a_1 \boldsymbol{n} \cdot \boldsymbol{n} + a_2 \boldsymbol{e} \cdot \boldsymbol{e} + a_3 \boldsymbol{c} \cdot \boldsymbol{c} + \boldsymbol{w} \cdot \boldsymbol{c} + \boldsymbol{c} \cdot \boldsymbol{w}, \tag{35.21}$$

with the as given by (35.11) and $\mathbf{wc} = 0$. For comparison we note that (28.60) with $\kappa = 0$ implies that the tensor may be put as $\Lambda = a + b(\mathbf{n} \cdot \mathbf{n} + \mathbf{h} \cdot \mathbf{n} + \mathbf{n} \cdot \mathbf{h})$; but the condition $\kappa = 0$ is a very severe restriction, whereas (35.21) is always applicable, i.e., for all crystals with a general direction for **n**. This shows that comparison with a transversely isotropic medium in the general case allows, as we would expect, a closer approximation to tensor Λ than does comparison with an isotropic medium. Chapter 7 demonstrates this by reference to particular crystals.

Tensor Λ' has the properties of the Λ' of section 27, and also the property $|\Lambda'| = 0$, so we may put, by analogy with (27.33) and (27.38), that

$$\overline{\Lambda'_t} = -\frac{1}{2}(\Lambda'^2)_t = -w^2, \tag{35.22}$$

$$\Lambda'^3 + \overline{\Lambda'_t}\Lambda' = \Lambda'(\Lambda'^2 - w^2) = 0. \tag{35.23}$$

The latter equation shows that the eigenvalues of Λ' are 0 and $\pm|\mathbf{w}|$. From (12.72) and (35.18) we also have that

$$\overline{\Lambda'} = -[cw] \cdot [cw], \quad n\overline{\Lambda'}n = -([wn]\,c)^2 = -[wn]^2. \qquad (35.24)$$

By analogy with section 27, we can transform the basic equation $\Lambda\mathbf{u} = \lambda\mathbf{u}$ by putting

$$u = c + u', \quad u'c = 0, \quad \lambda = a_0 + a_3 + \lambda'. \qquad (35.25)$$

Substituting of (35.21) and (35.25) into Christoffel's equation and multiplication by **c** gives us a relation analogous to (27.20):

$$\lambda' = wu'. \qquad (35.26)$$

Then $\Lambda\mathbf{u} = \lambda\mathbf{u}$ becomes

$$(a_3 - a_1 n \cdot n - a_2 e \cdot e)\,u' = w - wu' \cdot u'. \qquad (35.27)$$

All the vectors **n, e, w,** and **u'** appearing here lie in the meridional plane, because they are all perpendicular to **c**, so (35.27) is only a two-dimensional vector equation, and we can apply the relationships of section 14. We put

$$A = a_3 - a_1 n \cdot n - a_2 e \cdot e, \qquad \gamma = A^{-1}. \qquad (35.28)$$

Then from (14.16) and (14.19)

$$|A| = \frac{1}{2}\,[(A_t)^2 - (A^2)_t] = a_3 b_0 + a_1 a_2 (1 - n_3^2). \qquad (35.29)$$

$$\overline{A} = A_t - A = b_0 + a_1 n \cdot n + a_2 e \cdot e, \qquad (35.30)$$

$$\gamma = A^{-1} = \frac{\overline{A}}{|A|} = \frac{1}{\Delta}(b_0 + a_1 n \cdot n + a_2 e \cdot e), \qquad (35.31)$$

in which

$$\left.\begin{aligned} b_0 &= a_3 - a_1 - a_2, \\ \Delta = |A| &= (a_3 - a_1)(a_3 - a_2) - a_1 a_2 n_3^2. \end{aligned}\right\} \qquad (35.32)$$

Thus (35.27) is transformed to a form entirely analogous to (29.3):

$$u' = \gamma w - wu' \cdot \gamma u'. \qquad (35.33)$$

Matrix γ of (35.31) differs from matrix β of (29.2) in being symmetric. We have $\mathbf{c}\gamma = (b_0/\Delta)\mathbf{c}$, so

$$c\gamma w = c\gamma u' = 0, \tag{35.34}$$

because all the vectors of (35.33), as before, lie in the (**e**, **n**) meridional plane.

If vector γ**w** is small, as is usually the case, then (35.33) resembles (29.3) allows of approximate solution by iteration. Putting

$$u'_{(1)} = \gamma w = \frac{1}{\Delta}(b_0 w + a_1 wn \cdot n + a_2 we \cdot e), \tag{35.35}$$

we get the successive approximation via the recurrence formula

$$u'_{(k+1)} = \gamma w - wu'_{(k)} \cdot \gamma u'_{(k)}. \tag{35.36}$$

The k-th approximation gives the correction to the eigenvalue of (35.26) for the quasitransverse wave as

$$\lambda'_{(k)} = wu'_{(k)}. \tag{35.37}$$

As in section 29, it is easily verified that the k-th approximation gives the terms correctly up to quantities of order $|w|^{2k+1}$ inclusive. We have from (14.18) that

$$\gamma^2 = \gamma_t \gamma - |\gamma|,$$

while (35.31) and (35.32) imply that

$$\gamma_t = (2b_0 + a_1 + a_2)/\Delta = (b_0 + a_3)/\Delta,$$

$$|\gamma| = 1/|A| = 1/\Delta,$$

so

$$\gamma^2 w = \frac{(b + a_3)\gamma w - w}{\Delta}. \tag{35.38}$$

By analogy with (29.25) we conclude that we have for any approximation that

$$u'_{(k)} = \eta_{(k)}\gamma w + \zeta_{(k)} w, \tag{35.39}$$

while (35.36) and (35.38) imply the following recurrence formulas:

$$\left.\begin{aligned}\eta_{(k+1)} &= 1 - \frac{b+a_3}{\Delta}\eta_{(k)}(\eta_{(k)}\boldsymbol{w}\gamma\boldsymbol{w} + \zeta_{(k)}\boldsymbol{w}^2),\\ \zeta_{(k+1)} &= (\eta_{(k)}\boldsymbol{w}\gamma\boldsymbol{w} + \zeta_{(k)}\boldsymbol{w}^2)\left(\zeta_{(k)} - \frac{\eta_{(k)}}{\Delta}\right).\end{aligned}\right\} \tag{35.40}$$

We have for the zero approximation that $\mathbf{u}_0' = 0$, $\eta_{(0)} = \zeta_{(0)} = 0$; while (35.40) gives for the first approximation that $\eta_{(1)} = 1$, $\zeta_{(1)} = 0$, in accordance with (35.35). In the second approximation

$$\boldsymbol{u}'_{(2)} = \gamma\boldsymbol{w} - \boldsymbol{w}\gamma\boldsymbol{w}\cdot\gamma^2\boldsymbol{w} = \left(1 - \frac{b+a_3}{\Delta}\boldsymbol{w}\gamma\boldsymbol{w}\right)\gamma\boldsymbol{w} - \frac{\boldsymbol{w}\gamma\boldsymbol{w}}{\Delta}\boldsymbol{w}. \tag{35.41}$$

From (35.37), the correction $\lambda'_{(1)}$ to the eigenvalue in the first approximation is

$$\lambda'_{(1)} = \boldsymbol{w}\gamma\boldsymbol{w} = \frac{1}{\Delta}(b_0\boldsymbol{w}^2 + a_1(\boldsymbol{wn})^2 + a_2(\boldsymbol{we})^2). \tag{35.42}$$

For the second approximation

$$\lambda'_{(2)} = \boldsymbol{w}\boldsymbol{u}'_{(2)} = \boldsymbol{w}\gamma\boldsymbol{w}\left(1 - \frac{\boldsymbol{w}^2 + (b_0 + a_3)\,\boldsymbol{w}\gamma\boldsymbol{w}}{\Delta}\right). \tag{35.43}$$

As in section 29, this approximate method requires the previous determination of the displacement vector, which is then used to determine the phase velocity. Moreover, (35.36) and (35.37) allow us to find the displacement and velocity of only one of the quasitransverse waves, namely that for which the displacement vector deviates least from **c**. This is clear from the fact that the iteration process will converge to precisely that wave, since it has the least $|\gamma\mathbf{w}|$. The displacement and velocity for each of the other isonormal waves may be found to the same accuracy once they are known for one wave; but the approach of section 30 (for an isotropic reference medium) is also convenient, this method having the distinction that we first find the velocity and then find the corresponding displacement. Moreover, the method is suitable for direct determination of the parameters of any of the three waves.

Consider now the explicite form for the characteristic equation $|\lambda - \Lambda| = 0$, in which Λ is defined by (35.21):

$$|\lambda - \Lambda| = |\lambda - \Lambda^0 - \Lambda'| =$$

$$|\lambda - \Lambda^0| - ((\overline{\lambda - \Lambda^0})\,\Lambda')_t + ((\lambda - \Lambda^0)\,\overline{\Lambda}')_t = 0. \tag{35.44}$$

Here we have used (12.25) and (35.19). We introduce the symbol

$$x = \lambda - a_0; \tag{35.45}$$

to get from (31.22) and (35.27) that

$$|\lambda - \Lambda^0| = (x - a_3)[(x - a_1)(x - a_2) - a_1 a_2 n_3^2], \tag{35.46}$$

$$((\lambda - \Lambda^0)\overline{\Lambda}')_t = -[\boldsymbol{c}\boldsymbol{w}](x - a_1 \boldsymbol{n}\cdot\boldsymbol{n} - a_2 \boldsymbol{e}\cdot\boldsymbol{e})[\boldsymbol{c}\boldsymbol{w}] =$$

$$= -x\boldsymbol{w}^2 + a_1([\boldsymbol{w}\boldsymbol{n}]\boldsymbol{c})^2 + a_2([\boldsymbol{w}\boldsymbol{e}]\boldsymbol{c})^2. \tag{35.47}$$

But **c** is perpendicular to **e**, **n**, and **w**, so $[\mathbf{wn}]\|\mathbf{c}$ and $[\mathbf{we}]\|\mathbf{c}$, which means that

$$([\boldsymbol{w}\boldsymbol{n}]\boldsymbol{c})^2 = [\boldsymbol{w}\boldsymbol{n}]^2, \quad ([\boldsymbol{w}\boldsymbol{e}]\boldsymbol{c})^2 = [\boldsymbol{w}\boldsymbol{e}]^2. \tag{35.48}$$

Further, from (13.34)

$$\overline{\lambda - \Lambda^0} = (\overline{\lambda - \Lambda^0})_t + (\lambda - \Lambda^0)^2 - (\lambda - \Lambda^0)_t(\lambda - \Lambda^0)$$

and, since $(\lambda - \Lambda^0)\mathbf{c}$ is parallel to **c**,

$$((\overline{\lambda - \Lambda^0})\Lambda')_t = (\overline{\Lambda}^0\Lambda')_t = 0. \tag{35.49}$$

Then (35.44) becomes

$$(x - a_3)[(x - a_1)(x - a_2) - a_1 a_2 n_3^2] = x\boldsymbol{w}^2 - a_1[\boldsymbol{w}\boldsymbol{n}]^2 - a_2[\boldsymbol{w}\boldsymbol{e}]^2. \tag{35.50}$$

Here $\mathbf{w} = 0$ gives the equation for a hexagonal crystal:

$$(x - a_3)[(x - a_1)(x - a_2) - a_1 a_2 n_3^2] = 0, \tag{35.51}$$

which has the roots

$$x_1 = a_3, \quad x_\pm = \frac{1}{2}\left(a_1 + a_2 \pm \sqrt{(a_1 - a_2)^2 + 4a_1 a_2 n_3^2}\right), \tag{35.52}$$

which correspond to (32.3), (32.9), and (32.11). If $|\mathbf{w}|$ is small but not zero, we can use the iteration formulas

$$x_{(k+1)} = a_3 + \frac{\boldsymbol{w}^2 x_{(k)} - a_1[\boldsymbol{w}\boldsymbol{n}]^2 - a_2[\boldsymbol{w}\boldsymbol{e}]^2}{(x_{(k)} - a_1)(x_{(k)} - a_2) - a_1 a_2 n_3^2}, \quad x_{(0)} = a_3, \tag{35.53}$$

$$x_{(k+1)} = x_{\pm} + \frac{w^2 x_{(k)} - a_1[wn]^2 - a_2[we]^2}{\left(x_{(k)} - a_3\right)\left(x_{(k)} - x_{\mp}\right)}, \quad x_{(0)} = x_{\pm}. \tag{35.54}$$

The first of these corresponds to the quasitransverse wave whose displacement vector is closest in direction to **c**. The first approximation from (35.53) gives the same result as (35.42).

To find the displacements of the waves we note that (35.25), (35.26), and (35.45) give

$$wu' = x - a_3, \tag{35.55}$$

so (35.27) becomes

$$(x - a_1 n \cdot n - a_2 e \cdot e)\, u' = w, \tag{35.56}$$

whereupon (35.28), (35.31), and (35.32) imply that

$$u' = \frac{(x - a_1 - a_2)\, w + a_1 wn \cdot n + a_2 we \cdot e}{x\left(x - a_1 - a_2\right) + a_1 a_2\left(1 - n_3^2\right)}. \tag{35.57}$$

This exact formula defines the displacement of any of the waves from a known solution x to (35.50). Substitution of the approximate values $x_{(k)}$ given by (35.53) and (35.54) gives us $\mathbf{u}'_{(k)}$ to the same accuracy. Here, as in section 30, the k-th approximation for x is correct to terms of order $|w|^{2k+1}$ inclusive, while the k-th approximation for **u'** is correct to ones of order $|w|^{2k+2}$ inclusive.

To conclude we note that (35.15) and (35.21) give

$$w = [c\,[\Lambda c,\ c]\,]. \tag{35.58}$$

This defines **w** directly via the initial tensor Λ and the vector **c**; it is analogous to (27.27).

Chapter 7

Elastic Waves in Crystals of the Higher Systems

In this chapter the methods developed in previous sections are applied to crystals of the cubic, tetragonal, and trigonal systems.

36. Cubic Crystals

Section 19 gives the form of tensor Λ for cubic crystals; from (19.12)-(19.14)

$$\Lambda = c_1 + c_2 \boldsymbol{n} \cdot \boldsymbol{n} + c_3 \nu, \tag{36.1}$$

$$c_1 = \lambda_{44}, \quad c_2 = \lambda_{12} + \lambda_{44}, \quad c_3 = \lambda_{11} - \lambda_{12} - 2\lambda_{44}, \tag{36.2}$$

$$\nu = \begin{pmatrix} n_1^2 & 0 & 0 \\ 0 & n_2^2 & 0 \\ 0 & 0 & n_3^2 \end{pmatrix}. \tag{36.3}$$

Here the coordinate axes should coincide with L^2 or L^4 axes.

From (13.34), (13.35), and (13.37) we see that matrix ν satisfies the following conditions:

$$\nu_t = \boldsymbol{n}^2 = 1, \; (\nu^2)_t = \boldsymbol{n}\nu\boldsymbol{n} = n_1^4 + n_2^4 + n_3^4 = N, \tag{36.4}$$

$$(\nu^3)_t = \boldsymbol{n}\nu^2\boldsymbol{n} = n_1^6 + n_2^6 + n_3^6, \quad |\nu| = n_1^2 n_2^2 n_3^2, \tag{36.5}$$

$$\bar{\nu}_t = \frac{1}{2}\left[(\nu_t)^2 - (\nu^2)_t\right] = \frac{1}{2}(1 - N) = n_1^2 n_2^2 + n_2^2 n_3^2 + n_3^2 n_1^2, \tag{36.6}$$

$$\boldsymbol{n}\bar{\nu}\boldsymbol{n} = 3|\nu| = \bar{\nu}_t + \boldsymbol{n}\nu^2\boldsymbol{n} - \boldsymbol{n}\nu\boldsymbol{n} = (\nu^3)_t - \frac{1}{2}(3N - 1). \tag{36.7}$$

$$3N-1=(n_1^2-n_2^2)^2+(n_2^2-n_3^2)^2+(n_3^2-n_1^2)^2. \tag{36.8}$$

These equations, with those of section 13, give us the principal invariants of tensor Λ, which are

$$\Lambda_t=3c_1+c_2+c_3, \tag{36.9}$$

$$n\Lambda n=c_1+c_2+c_3N, \tag{36.10}$$

$$(\Lambda^2)_t=(c_1+c_2)^2+2c_1(c_1+c_3)+c_3(2c_2+c_3)N, \tag{36.11}$$

$$\bar{\Lambda}_t=3c_1^2+2c_1(c_2+c_3)+\frac{c_3}{2}(2c_2+c_3)(1-N), \tag{36.12}$$

$$n\Lambda^2 n=(c_1+c_2)^2+2c_3(c_1+c_2)N+c_3^2(\nu^3)_t=(n\Lambda n)^2+c_3^2[n,\nu n]^2, \tag{36.13}$$

$$n\bar{\Lambda}n=c_1^2-c_3^2+\frac{c_3^2}{2}(2c_1+c_3)(1-N)+c_3^2 n\nu^2 n, \tag{36.14}$$

$$|\Lambda|=c_1^2(c_1+c_2+c_3)+\frac{1}{2}c_1c_3(2c_2+c_3)(1-N)+c_3^2(3c_2+c_3)|\nu|. \tag{36.15}$$

Then from (26.16) and (26.17) we get the mean elastic anisotropy as

$$\langle n_k^2\rangle=\frac{1}{3},\quad \langle n_k^4\rangle=\frac{1}{5},\quad \langle N\rangle=0.6. \tag{36.16}$$

Thus for a cubic crystal

$$\langle\Lambda_t\rangle=\Lambda_t=3c_1+c_2+c_3,\quad \langle n\Lambda n\rangle=c_1+c_2+0.6c_3, \tag{36.17}$$

and from (26.13)

$$\left.\begin{aligned} a_m&=\frac{1}{2}\langle\Lambda_t-n\Lambda n\rangle=c_1+0.2c_3,\\ b_m&=\frac{1}{2}\langle 3n\Lambda n-\Lambda_t\rangle=c_2+0.4c_3,\end{aligned}\right\} \tag{36.18}$$

so the Λ_m of (26.2) for the isotropic medium most similar to a cubic crystal on average is

$$\Lambda_m=c_1+0.2c_3+(c_2+0.4c_3)\,n\cdot n. \tag{36.19}$$

The difference between Λ and Λ_m is

$$\Lambda'_m = \Lambda - \Lambda_m = c_3\left(\nu - \frac{1+2n\cdot n}{5}\right). \tag{36.20}$$

This is proportional to c_3. For F_m we have from (26.6) that

$$F_m = \langle \Lambda'^2_m \rangle_t = \frac{c_3^2}{25}(1+5N). \tag{36.21}$$

The form of (36.1) shows directly that the difference from an isotropic medium arises from the $c_3\nu$ term and that this difference vanishes for $c_3 = 0$. We therefore call c_3 the anisotropy parameter of a cubic crystal; but section 26 shows, from (26.6) and (26.7), that a measure of the mean anisotropy of a cubic crystal is

$$\sqrt{\langle F_m \rangle} = \sqrt{\langle \Lambda'^2_m \rangle_t} = 0.4c_3. \tag{36.22}$$

On the other hand, if we assume that the anisotropy is characterized by the $c_3\nu$ term, a measure of this is

$$\sqrt{\langle (c_3\nu)^2 \rangle_t} = \sqrt{0.6}\, c_3 = 0.775c_3, \tag{36.23}$$

which is nearly twice (36.22). Thus the actual anisotropy, as defined from section 26, is much less than it might appear at first sight.

The F_m of (36.21) is dependent on the direction of **n**, so the extreme values are of interest as well as the mean; these correspond to maxima and minima in N. We seek a turning point from the function $f = N - 2\lambda(\mathbf{n}^2 - 1)$, after which we obtain (with no summation over k) the equations

$$n_k(n_k^2 - \lambda) = 0. \tag{36.24}$$

Multiplication of each by its n_k and summation over k gives $\lambda = N$. From (36.24) we have three possibilities:

I. $$n_{k_1} = n_{k_2} = 0, \quad n_{k_3}^2 = N = 1, \tag{36.25}$$

II. $$n_{k_1} = 0, \quad n_{k_2}^2 = n_{k_3}^2 = N = \frac{1}{2}, \tag{36.26}$$

Table IV. Elastic Constants of Cubic Crystals

(The c are in 10^{11} dynes/cm^2)

No.	Crystal	Chemical formula	c_{11}	c_{12}	c_{44}	c_1	c_2	c_3	Δ_m	$\sqrt{\langle h^2 \rangle}$	$\lvert h \rvert_{max}$	κ
1	Aluminum	Al	12.30	7.08	3.09	1.13	3.72	-0.351	0.029	0.0191	0.0278	1°34'
2		K-Al-SO_4	2.465	1.025	0.865	0.493	1.078	-0.165	0.0041	0.0319	0.0467	2°37'
3		Rb-Al-SO_4	2.535	1.033	0.844	0.448	0.9963	-0.0987	0.026	0.0201	0.0293	1°39'
4		Cs-Al-SO_4	3.115	1.539	0.839	0.420	1.190	-0.0510	0.012	0.0085	0.0123	0°42'
5		Tl-Al-SO_4	2.540	1.130	0.814	0.351	0.8372	-0.0939	0.031	0.0229	0.0333	1°53'
6		NH_4-Al-SO_4	2.520	1.090	0.811	0.494	1.158	-0.117	0.027	0.0205	0.0229	1°42'
7		CH_3NH_3-Al-SO_4	2.971	1.732	0.584	0.368	1.458	0.0447	0.0093	0.0059	0.0084	0°27'
8		K-Ga-SO_4	2.356	0.994	0.849	0.447	0.9710	-0.177	0.049	0.0384	0.0564	3°91'
9	Alums	Rb-Ga-SO_4	2.450	0.996	0.853	0.421	0.9131	-0.124	0.036	0.0281	0.0411	2°19'
10		Cs-Ga-SO_4	3.069	1.533	0.816	0.384	1.104	-0.0451	0.012	0.00810	0.0117	0°40'
11		NH_4-Ga-SO_4	2.395	1.029	0.805	0.451	1.028	-0.137	0.036	0.0274	0.0400	2°15'
12		CH_3NH_3-Ga-SO_4	2.898	1.686	0.562	0.327	0.1309	0.0513	0.012	0.00752	0.0107	0°37'
13		Rb-In-SO_4	2.366	0.954	0.826	0.392	0.8448	-0.114	0.036	0.0278	0.0406	2°17'
14		Cs-In-SO_4	2.957	1.407	0.816	0.369	1.005	-0.0371	0.010	0.0073	0.0105	0°36'
15		Cs-Fe-SO_4	3.038	1.484	0.841	0.407	1.126	-0.0620	0.016	0.0110	0.0159	0°54'
16		K-Al-SeO_4	2.330	0.970	0.775	0.390	0.8787	-0.0957	0.028	0.0222	0.0323	1°50'
17		Rb-Al-SeO_4	2.426	0.996	0.778	0.368	0.8396	-0.0596	0.018	0.0143	0.0207	1°10'
18		Cs-Al-SeO_4	2.608	1.178	0.742	0.334	0.8633	-0.0243	0.0075	0.00555	0.00799	0°27'
19		NH_4-Al-SeO_4	2.384	1.040	0.752	0.398	0.9492	-0.0847	0.023	0.0181	0.0262	1°29'
20		CH_3NH_3-Al-SeO_4	2.736	1.008	0.543	0.297	0.8489	0.351	0.096	0.0925	0.0954	5°32'
21		Cs-Ga-SeO_4	2.530	1.130	0.756	0.323	0.8053	-0.0478	0.016	0.0119	0.0172	0°58'
22	Ammonium bromide	NH_4Br	3.38	0.91	0.685	0.281	0.65	0.45	0.14	0.11	0.14	8°15'
23	Ammonium chloride	NH_4Cl	3.79	0.97	0.83	0.54	1.18	0.760	0.13	0.0999	0.135	7°51'
24	Barium fluoride	BaF	9.01	4.03	2.49	0.516	1.35	0	0	0	0	0
25	Barium nitrate	$Ba(NO_3)_2$	6.04	1.86	1.22	0.377	0.951	0.537	0.12	0.0899	0.122	7°17'
26	Calcium fluoride	CaF_2	16.44	5.02	3.47	1.09	2.67	1.41	0.11	0.0850	0.116	6°45'

27	Cesium bromide	CsBr	3.10	0.84	0.75	0.17	0.357	0.17	0.10	0.078	0.11	6°14'
28	Cesium chloride	CsCl	3.64	0.92	0.80	0.20	0.431	0.281	0.13	0.101	0.136	7°55'
29	Cesium iodide	CsI	2.45	0.71	0.62	0.14	0.294	0.11	0.081	0.064	0.088	5°7'
30	Chromite	$FeCr_2O_4$	32.25	14.37	11.67	2.5	5.8	-1.2	0.058	0.045	0.066	3°41'
31	Cobalt	Co	30.37	15.43	7.47	0.855	2.620	0	0	0	0	0
32	Copper	Cu	17.620	12.494	8.177	0.9067	2.292	-1.245	0.19	0.135	0.212	10°50'
33	Diamond	C	94.9	15.1	51.2	14.6	18.9	-6.44	0.073	0.0770	0.116	6°19'
34	Galena	PbS	10.2	3.8	2.5	0.33	0.84	0.19	0.054	0.040	0.056	3°12'
35	Gallium antimonide	GaSb	8.85	4.04	4.32	0.769	1.49	-0.682	0.13	0.109	0.168	8°52'
36	Gallium arsenide	GaAs	11.88	5.38	5.94	1.12	2.13	-1.01	0.14	0.114	0.177	9°16'
37	Germanium	Ge	12.89	4.83	6.71	1.3	2.16	-1.00	0.13	0.111	0.172	9°10'
38	Hexamethylene tetraamine	$C_6H_{12}N_6$	1.643	0.433	0.515	0.385	0.708	0.13	0.041	0.034	0.048	2°48'
39	Gold	Au	20.163	16.967	4.544	0.2332	1.104	-0.3023	0.10	0.0600	0.0894	4°56'
40	Indium antimonide	InSb	6.72	3.67	3.02	0.522	1.16	-0.517	0.14	0.106	0.163	8°37'
41	Iron	Fe	23.7	14.1	11.60	1.473	3.26	-1.73	0.16	0.131	0.204	10°30'
42	Lead	Pb	4.75	4.03	1.44	0.127	0.482	-0.190	0.15	0.0798	0.139	7°28'
43	Lithium	Li	1.574	1.333	1.158	2.117	4.553	-3.793	0.30	0.244	0.412	17°35'
44	Lithium bromide	LiBr	3.94	1.88	1.91	0.550	1.09	-0.507	0.11	0.111	0.171	9°1'
45	Lithium chloride	LiCl	4.94	2.26	2.49	1.20	2.30	-1.11	0.14	0.117	0.181	9°28'
46	Lithium fluoride	LiF	12.46	4.24	6.49	2.45	4.06	-1.80	0.12	0.105	0.161	8°32'
47	Lithium iodide	LiI	2.85	1.40	1.35	0.332	0.677	-0.308	0.13	0.108	0.167	8°47'
48	Magnesium oxide	MgO	28.9	8.8	15.5	4.33	6.79	-3.04	0.11	0.106	0.164	8°39'
49	Magnetite	Fe_3O_4	27.3	10.6	9.71	1.89	3.94	-0.528	0.035	0.0276	0.0403	2°16'
50	Molybdenum	Mo	45.5	17.6	11.0	1.08	2.80	0.58	0.051	0.037	0.052	3°1'
51	Nickel	Ni	24.36	14.94	11.96	1.3	3.0	-1.6	0.17	0.13	0.21	10°60'
52	Palladium	Pd	22.213	17.71	7.137	0.600	2.09	-0.821	0.14	0.091	0.138	7°26'
53	Potassium	K	0.414	0.330	0.264	0.306	0.689	-0.512	0.27	0.208	0.344	15°43'
54	Potassium bromide	KBr	4.18	0.56	0.52	0.18	0.383	0.915	0.28	0.238	0.298	16°18'
55	Potassium chloride	KCl	4.032	0.66	0.628	0.317	0.649	1.07	0.23	0.193	0.248	13°58'

TABLE IV (continued)

No.	Crystal	Chemical formula	c_{11}	c_{12}	c_{44}	c_1	c_2	c_3	Δ_m	$\sqrt{<h^2>}$	$\lvert h \rvert_{max}$	κ
56	Potassium cyanide	KCN	1.94	1.18	0.15	0.97	0.856	0.30	0.048	0.059	0.082	4°46'
57	Potassium fluoride	KF	6.58	1.49	1.28	0.507	1.10	1.00	0.16	0.131	0.173	10°1'
58	Potassium iodide	KI	2.71	0.45	0.364	0.116	0.26	0.489	0.26	0.21	0.27	14°50'
59	Pyrite	FeS_2	36.2	-4.64	10.52	2.10	1.18	3.96	0.22	0.280	0.343	18°4'
60	Rubidium bromide	RbBr	3.185	0.48	0.385	0.115	0.26	0.577	0.29	0.23	0.29	15°55'
61	Rubidium chloride	RbCl	3.645	0.61	0.475	0.170	0.388	0.745	0.26	0.212	0.269	14°59'
62	Rubidium fluoride	RbF	5.7	1.25	0.91	0.32	0.75	0.91	0.20	0.16	0.21	11°57'
63	Rubidium iodide	RbI	2.585	0.375	0.281	0.0791	0.185	0.4637	0.30	0.244	0.305	16°35'
64	Silicon	Si	16.57	6.39	7.96	3.42	6.16	-2.47	0.11	0.0929	0.142	7°35'
65	Silver	Ag	13.149	9.733	5.109	0.4804	1.396	0.6396	0.16	0.109	0.169	9°45'
66	Silver bromide	AgBr	5.62	3.28	0.728	0.113	0.619	0.137	0.065	0.0395	0.0554	5°12'
67	Silver chloride	AgCl	6.01	3.62	0.625	0.112	0.764	0.205	0.080	0.0473	0.0660	3°50'
68	Sodium	Na	0.615	0.496	0.592	0.610	1.12	-1.10	0.34	0.314	0.568	20°20'
69	Sodium bromate	$NaBrO_3$	5.45	1.91	1.50	0.449	1.02	0.162	0.038	0.0290	0.0410	2°22'
70	Sodium bromide	NaBr	3.87	0.97	0.97	0.30	0.606	0.300	0.099	0.0806	0.110	6°25'
71	Sodium chlorate	$NaClO_3$	4.99	1.41	1.17	0.470	1.04	0.498	0.10	0.0786	0.108	6°16'
72	Sodium chloride	NaCl	5.750	0.986	1.327	0.5922	1.032	0.9415	0.15	0.130	0.173	10°0'
73	Sodium fluoride	NaF	9.70	2.56	2.80	1.00	1.92	0.552	0.061	0.0503	0.0700	4°4'
74	Sodium iodide	NaI	2.93	0.73	0.737	0.201	0.401	0.20	0.099	0.081	0.11	6°25'
75	Spinel	$MgAl_2O_4$	30.05	15.37	15.86	4.4	8.7	-4.7	0.16	0.14	0.21	10°53'
76	Strontium nitrate	$Sr(NO_3)_2$	4.73	2.18	1.46	0.489	1.22	-0.124	0.028	0.0207	0.0301	1°42'
77	Thallium bromide	TlBr	3.78	1.48	0.756	0.100	0.296	0.104	0.084	0.0602	0.0834	4°51'
78	Thallium chloride	TlCl	4.01	1.53	0.760	0.109	0.327	0.14	0.10	0.070	0.096	5°36'
79	Thorium	Th	5.72	4.89	4.78	0.427	0.863	-0.779	0.33	0.276	0.477	20°6'
80	Tungsten	W	51.5	20.7	15.1	0.782	1.85	0.03	0.0043	0.0033	0.0047	0°16'
81	Vanadium	V	22.79	11.87	4.25	0.706	2.68	0.402	0.042	0.0276	0.0390	2°15'
82	Zinc blende	ZnS	10.79	7.22	4.12	1.01	2.775	-1.14	0.14	0.0961	0.147	7°50'

III. $$n_1^2 = n_2^2 = n_3^2 = N = \frac{1}{3}, \tag{36.27}$$

in which k_1, k_2, and k_3 are all different and take the values 1, 2, and 3. Then $N_{min} = 1/3$ (case III) and $N_{max} = 1$ (case I), so (36.21) gives

$$\frac{8}{75} c_3^2 \leqslant F_m \leqslant \frac{6}{25} c_3^2. \tag{36.28}$$

The extremal N and F_m from (36.25)-(36.27) correspond to symmetry axes of orders 4, 2, and 3, respectively.

From (36.11) we have

$$\langle \Lambda^2 \rangle_t = (c_1 + c_2)^2 + 2c_1(c_1 + c_3) + 0.6c_3(2c_2 + c_3).$$

Substitution of this with (36.22) into (26.28), with the numerical parameters of [34], gives us $\Delta_m = \sqrt{\langle F_m \rangle / \langle \Lambda^2 \rangle_t}$; Table IV lists this for some cubic crystals. The range of Δ_m given in this table covers nearly the entire range for cubic crystals with known elastic constants. It is clear that aluminum and the alums have low elastic anisotropy, whereas brass and lithium are relatively highly anisotropic.

We may also compare a cubic crystal with an isotropic medium for a fixed direction of the wave normal (section 27); then we get from (27.1)-(27.3), (27.7), (27.9), (36.9), and (36.10) that

$$a = c_1 + \frac{1}{2} c_3 (1 - N), \quad b = c_2 + \frac{1}{2} c_3 (3N - 1), \tag{36.29}$$

$$\Lambda' = c_3 \left[\nu - \frac{1}{2} (1 - N) - \frac{1}{2} (3N - 1) \boldsymbol{n} \cdot \boldsymbol{n} \right], \tag{36.30}$$

$$F = \langle \Lambda'^2 \rangle_t = \frac{1}{2} c_3^2 (1 - N)(3N - 1). \tag{36.31}$$

We see that $F = 0$, which implies $\Lambda' = 0$, for $N = 1$ ($\mathbf{n} \| L^4$) and for $N = 1/3$ ($\mathbf{n} \| L^3$), so $\Lambda = \Lambda^0$ for waves along the fourfold and threefold axes, and a cubic crystal in these directions does not differ from an isotropic medium. This agrees with the general deductions of section 18 on high-order symmetry axes in any crystal. From (36.28), we see that F_m is not zero for any **n**. The various methods of relating a cubic crystal to an isotropic medium may be compared by finding the mean of the F of (36.31). We shall see (section 49) that

$$\langle n_k^8 \rangle = \frac{1}{9}, \quad \langle n_{k_1}^4 n_{k_2}^4 \rangle = \frac{1}{105}, \tag{36.32}$$

so

$$\langle N^2 \rangle = \frac{41}{105} \tag{36.33}$$

and

$$\langle F \rangle = \frac{4}{35} c_3^2. \tag{36.34}$$

Comparison with (36.22) and (36.23) shows that $\Lambda^0 = a + \mathrm{b}\mathbf{n}\cdot\mathbf{n}$ [in which a and b are defined by (36.29)] differs less from Λ than does the Λ_m of (36.19) or the tensor $\mathbf{c}_1 + \mathbf{c}_2\mathbf{n}\cdot\mathbf{n}$. The definition of (27.2) shows that F cannot be negative, so the zero values of F found for N=1 and N=1/3 represent the absolute minimum of F. The maximum F is given by dF/dN=0, this being $c_3^2/6$ for N=2/3. Hence

$$0 \leqslant F \leqslant \frac{c_3^2}{6}, \tag{36.35}$$

and this maximum F is less than the maximum F_m of (36.28), which again demonstrates the advantages of the unaveraged tensor Λ^0.

Consider now the special directions. Here (17.7) gives

$$[\boldsymbol{n}, \Lambda\boldsymbol{n}]\,\Lambda^2\boldsymbol{n} = [\boldsymbol{n}, \nu\boldsymbol{n}]\,\nu^2\boldsymbol{n} = \begin{vmatrix} n_1 & n_2 & n_3 \\ n_1^2 & n_2^2 & n_3^2 \\ n_1^3 & n_2^3 & n_3^3 \end{vmatrix} =$$

$$= n_1 n_2 n_3 (n_1^2 - n_2^2)(n_2^2 - n_3^2)(n_3^2 - n_1^2) = 0. \tag{36.36}$$

Then a cubic crystal has purely transverse waves for $n_k = 0$ or $n_{k_1} = n_{k_2}$, which conditions define symmetry planes; this agrees with the general conclusions of section 17, but (36.36) also implies that the special direction of a cubic crystal lie only in symmetry planes. Thus the cone of (17.10) splits up into 9 symmetry planes for a cubic crystal.

The quantities of section 27 are expressed as follows for a

cubic crystal [see (36.29)-(36.31) and (36.4)-(36.7)]:

$$h = \alpha n = \frac{1}{b}\Lambda' n = \frac{c_3}{b}(\nu n - Nn), \tag{36.37}$$

$$h^2 = \frac{c_3^2}{b^2}(n\nu^2 n - N^2), \tag{36.38}$$

$$g^2 = \frac{1}{2}(\alpha^2)_t = \frac{1}{2b^2}(\Lambda'^2)_t = \frac{1}{2b^2}F = \frac{c_3^2}{4b^2}(1 - N)(3N - 1), \tag{36.39}$$

$$\varkappa^2 = g^2 - h^2 = \frac{c_3^2}{4b^2}(N^2 + 4N - 4n\nu^2 n - 1), \tag{36.40}$$

$$|\alpha| = \frac{1}{b^3}|\Lambda'| = \frac{c_3^3}{2b^3}\left[(5 - 9N)|\nu| + \frac{1}{2}(1 - N)^2(2N - 1)\right]. \tag{36.41}$$

Longitudinal normals have (section 28) $\mathbf{h} = 0$, which (36.37) shows is equivalent to (36.24); only symmetry axes (or normals to symmetry planes) can be longitudinal axes in a cubic crystal, which agrees with the general conclusions of section 18. Equations (36.24) are equivalent to $\mathbf{h} = 0$ and are conditions for turning points in N, and hence in the $\mathbf{n}\Lambda\mathbf{n}$ of (36.10) also, which is also in agreement with a general property demonstrated in section 18: $\mathbf{n}\Lambda\mathbf{n}$ is extremal for longitudinal normals.

The acoustic axes are (section 18) special directions that all satisfy (36.36), so they also can lie only in symmetry planes for a cubic crystal. One of the components of the wave-normal vector is zero for a symmetry plane coincident with a coordinate plane, e.g., $n_3 = 0$. In that case

$$n_1^2 + n_2^2 = 1, \quad N = n_1^4 + n_2^4 = 1 - 2n_1^2 n_2^2, \tag{36.42}$$

$$n\nu^2 n = n_1^6 + n_2^6 = 1 - 3n_1^2 n_2^2 = \frac{3N - 1}{2}, \quad |\nu| = 0; \tag{36.43}$$

in which (36.38), (36.40), and (36.41) give that $\mathbf{h}^2$, κ, and $|\alpha|$ take the following values:

$$\left.\begin{aligned} h^2 = \frac{c_3^2}{4b^2}(1 - N)(2N - 1), \quad \varkappa = \left|\frac{c_3}{b}\right|\frac{1 - N}{2}, \\ |\alpha| = \frac{c_3^3}{4b^3}(1 - N)^2(2N - 1). \end{aligned}\right\} \tag{36.44}$$

Two components of **n** must coincide, e.g., $n_2 = n_3$, for a diagonal symmetry plane. Then

$$n_1^2 + 2n_2^2 = 1, \quad N = n_1^4 + 2n_2^4, \quad \boldsymbol{n}\nu^2\boldsymbol{n} = n_1^6 + 2n_2^6, \quad |\nu| = n_1^2 n_2^4 \qquad (36.45)$$

and

$$\left.\begin{aligned} h^2 = 2\frac{c_3^2}{b^2} n_1^2 n_2^2 (n_1^2 - n_2^2)^2, \quad \varkappa = \left|\frac{c_3}{b}\right| n_2^2 \,|\, n_1^2 - n_2^2 \,|, \\ |\alpha| = 2\frac{c_3^3}{b^3} |\nu| (n_1^2 - n_2^2)^3. \end{aligned}\right\} \qquad (36.46)$$

Then, in accordance with section 28, κ has a rational expression and complies with the condition $|\alpha| = \pm \kappa h^2$.

Now it is seen that condition (28.34), $\mathbf{h}^2 - 2\kappa^2 = 2\kappa$, which gives the acoustic axes, for the case of (36.42)-(36.44) becomes

$$(1 - N)\left|\frac{c_3}{b}\right|\left[\left|\frac{c_3}{b}\right|(3N - 2) - 2\right] = 0. \qquad (36.47)$$

The expression in square brackets cannot become zero, because (36.42) gives $1/2 \leq N \leq 1$, and even the most highly anisotropic crystals (Table IV) have $|c_3/b| < 3/2$. Therefore we must have $N = 1$, which means $n_1 = 0$ or $n_2 = 0$ from (36.42), i.e., a fourfold axis. Similarly for the case of (36.45) and (36.46) we have

$$\left|\frac{c_3}{b}\right| n_2^2 \,|\, n_1^2 - n_2^2 \,| \left[\left|\frac{c_3}{b}\right| (n_1^2 - n_2^2)^2 - 1\right] = 0. \qquad (36.48)$$

Here again the expression in square brackets cannot become zero, so the only possibilities are $n_2 = 0 = n_3$ (L^4 axis) or $n_1^2 = n_2^2 = n_3^2$ (L^3 axis). Thus the acoustic axes of a cubic crystal cannot be other than symmetry axes of order above two, so a cubic crystal has 7 acoustic axes. In both cases the condition for an acoustic axis coincides with the condition $\kappa = 0$; hence only longitudinal acoustic axes can occur in a cubic crystal, for which $\mathbf{h} = \kappa = \alpha = 0$ (section 28). These axes are directions along which intersect three (L^3) or four (L^4) different sheets (planes) of the cone of special directions of (36.36).

The relationships of section 28 readily give the speed and displacements of waves whose normals lie in symmetry planes; these are obtained by substituting the values of (36.44) or (36.46), respectively, in the general formulas (28.27)-(28.30). However, we now turn to the use of the general approximate methods of sections 29 and 30 with cubic crystals.

37. Approximate Theory for Cubic Crystals

First we consider comparison with an isotropic medium. The results of sections 29 and 30 show that it is particularly important as regards accuracy in the approximate relations that $|\mathbf{h}|$, κ, and $|\alpha|$ should be small. An especially large part is played by $\mathbf{h}$, which appears in most of the formulas, the accuracy of the methods of sections 29 and 30 improving as $|\mathbf{h}|$ decreases. Formula (36.38) gives $\mathbf{h}^2$, while (36.29) shows that b varies little with $\mathbf{n}$, because (Table IV) $|c_3|$ is usually much less than $|c_2|$. Therefore we usually need consider only the factor $\mathbf{n}\nu^2\mathbf{n}-N^2$ in order to find $\langle\mathbf{h}^2\rangle$ or $\mathbf{h}^2_{max}$. Simple calculations give (see section 49) that

$$\langle \boldsymbol{n}\nu^2\boldsymbol{n} - N^2\rangle = \langle [\boldsymbol{n}, \nu\boldsymbol{n}]^2\rangle = \frac{4}{105} = 0.0381, \quad [\boldsymbol{n}, \nu\boldsymbol{n}]^2_{max} = 0.079, \tag{37.1}$$

the maximum value being attained for an $\mathbf{n}$ whose components n_i, n_k, and n_l are defined by

$$n_i^2 = n_k^2 = 0.093, \quad n_l^2 = 0.814. \tag{37.2}$$

Then the maximal $|\mathbf{h}|$, and thus the maximal deviation of the displacement of the quasilongitudinal wave from $\mathbf{n}$, occurs when $\mathbf{n}$ lies in a symmetry plane perpendicular to a twofold axis ($|n_i| = |n_k|$) and forms an angle 25°35' with L^4. These results are universal in the sense that this maximal deviation will occur for any cubic crystal of not too high anisotropy near the directions of (37.2), of which there are 12 (four per L^4 axis). From (37.2), we have $N = 0.68$ and $b = b_1 = c_2 + 0.52c_3$, whereas $\langle b\rangle = c_2 + 0.4c_3$. The result is

$$\sqrt{\langle h\rangle^2} = 0.195\frac{|c_3|}{\langle b\rangle}, \quad \sqrt{(h^2)_{max}} = 0.28\frac{|c_3|}{b_1}. \tag{37.3}$$

Table IV lists the characteristic quantities for all cubic crystals whose parameters are known. Column 10 gives $\Delta_m = 2|c_3|/5\sqrt{\langle\Lambda^2\rangle t}$, the values of which show that for nearly all these crystals we have

$$\sqrt{\langle h^2\rangle} < \Delta_m < \sqrt{(h^2)_{max}}. \tag{37.4}$$

The exceptions are the 14 crystals Al, RbBr, V, AgBr, Pd, Au, AgCl, Pb, TlBr, TlCl, and alums containing CH_3NH_3.

From (29.11) we see that $|\beta \mathbf{h}|$ is the maximal value of $\mathbf{u}'$ in the first approximation. The corresponding $\mathbf{n}$ of (37.2) lies in a symmetry plane, so (28.3) and (28.7) are obeyed:

$$[\boldsymbol{nh}]\,\alpha \boldsymbol{h} = 0, \quad |\alpha| = \pm \boldsymbol{h}^2 \varkappa. \tag{37.5}$$

This, with (29.11), (29.12), and (29.24), allows us to simplify the expression for the length of $\mathbf{u}'$ in the first and second approximations and to reduce these to the following form [see (29.15), (29.20), and (29.22)]:

$$|\boldsymbol{u}_1'| = |\beta \boldsymbol{h}| = \frac{|\boldsymbol{h}|}{1 \mp \varkappa}, \quad |\boldsymbol{u}_2'| = |\boldsymbol{u}_1'| - |\boldsymbol{u}_1'|^3. \tag{37.6}$$

The upper (lower) sign corresponds to $|\alpha| > 0$ ($|\alpha| < 0$). From (36.41) we conclude that the sign of $|\alpha|$ coincides with that of c_3 if (37.5) is obeyed, the latter being given in column 9 of Table IV. We have $|\mathbf{n}| = 1$, so $|\mathbf{u}'|$ equals the tangent of the angle χ formed by $\mathbf{u}$ with $\mathbf{n}$. Column 13 gives $\chi_{max} = \tan^{-1} |\mathbf{u}_2'|_{max}$.

From section 29, we have that terms in $|\mathbf{h}^3|$ and above may be discarded in the first approximation, while those of $|\mathbf{h}|^5$ and above may be discarded in the second; in general, the error is of order $|\mathbf{h}|^{2k+1}$ in the k-th approximation. The relative error in the experimental determination of elastic constants is usually at least 10^{-3}, whereas Table IV shows that $<|\mathbf{h}|>^3 < 0.014$ even for the most anisotropic crystal (lithium) in the first approximation, while the second approximation gives an error of 10^{-3}, which is of the order of the accuracy of the elastic constants. This means that there is no need to go beyond the second approximation for any cubic crystal; Table IV shows that the first approximation suffices for many, because the relative error does not exceed 10^{-3}.

Crystals may be classified as of low, medium, or high anisotropy. The low type corresponds to satisfactory results *on average* from the zero approximation for the velocity; medium, to the requirement for the first approximation; and high, to requirement of the second or higher approximation.

Then for a cubic crystal of low anisotropy we have

$$\langle \mathbf{h}^2 \rangle \leqslant 10^{-3}, \quad \sqrt{\langle \mathbf{h}^2 \rangle} \leqslant 0.0316. \tag{37.7}$$

Column 11 of Table I shows that 25 crystals out of the 82 (30%) satisfy this condition: Al, BaF_2, W, V, Co, Fe_3O_4, $NaBrO_3$, $Sr(NO_3)_2$, and the large group of alums. The least anisotropic is tungsten ($\Delta_m = 0.0043$), while the most anisotropic in this group is vanadium ($\Delta_m = 0.04$).

For a crystal of medium anisotropy

$$\langle \mathbf{h}^2 \rangle^2 \leqslant 10^{-3}, \quad 0.0316 < \sqrt{\langle \mathbf{h}^2 \rangle} \leqslant 0.178. \tag{37.8}$$

This is the largest group (57% of the total); it includes ZnS, NaI, Cu, Ag, and Mo. The Δ_m range from 0.041 (K-Al-SO_4) to 0.20 (RbF).

For a crystal of high anisotropy

$$\langle \mathbf{h}^2 \rangle^2 > 10^{-3}, \quad \sqrt{\langle \mathbf{h}^2 \rangle} > 0.178. \tag{37.9}$$

There are only 11 of these: RbCl, RbBr, RbI, Na, Li, FeS_2, K, KCl, KBr, KI, and Th. The Δ_m range from 0.22 (FeS_2) to 0.34 (Na).

In the cases of BaF_2 and Co the c_3, and hence all quantities including c_3, such as Δ_m, are zero. This is due to inaccuracy in the presently available elastic constants.

The data give rise to some conclusions on the elements and groups in these crystals [52]. For instance, all alkali metals and alkali halides have marked elastic anisotropy; most of them belong to the high group (the metals and the compounds of rubidium, potassium, and sodium), while the rest belong to the more strongly anisotropic subgroup of the medium group. All compounds of a given element are roughly equal in anisotropy, but fluorine always causes a reduction. The greater the anisotropy of the alkali metal itself, the less that of the halides. For instance, the Δ_m for compounds of sodium ($\Delta_m = 0.34$) with F, Cl, Br, and I are, respectively, 0.061, 0.15, 0.099, and 0.099 while for those of potassium ($\Delta_m = 0.27$) the values are 0.16, 0.23, 0.28, and 0.26.

All the alums are similar in anisotropy, nearly all belonging to the low group, except for K-Al-SO_4 and K-Ca-SO_4, whose anisotropy is raised somewhat by the presence of K.

The anisotropy is related to the position of the element in the periodic system; for instance, all elements in the first group (the alkali metals, Cu, Ag, and Au) have marked anisotropy, which falls fairly smoothly along the group. The medium-anisotropy type includes all the cubic elements in group four, but here the anisotropy increases along the group. No similar statements can be made about other groups, since these contain only one or two cubic elements.

A much simpler form can be given to the approximate equation for the L sheet of the refraction surface of (29.36) and (29.37) in the case of a cubic crystal. We divide (29.36) by v^8 and put it as

$$3m^2 \cdot \boldsymbol{m}\Lambda^m\boldsymbol{m} - m^4\Lambda_t^m + m^2\Lambda_t^m\boldsymbol{m}\Lambda^m\boldsymbol{m} - (\boldsymbol{m}\Lambda^m\boldsymbol{m})^2 - 2m^2\boldsymbol{m}(\Lambda^m)^2\boldsymbol{m} = 0 \quad (37.10)$$

while (36.1), (36.3), (36.10), and (36.13) give for a cubic crystal that

$$\Lambda_t^m = (3c_1 + c_2 + c_3)\, m^2, \quad \boldsymbol{m}\Lambda^m\boldsymbol{m} = (c_1 + c_2)\, m^4 + c_3 (m_1^4 + m_2^4 + m_3^4),$$

$$\boldsymbol{m}(\Lambda^m)^2\boldsymbol{m} = (c_1 + c_2)^2 m^6 + 2c_3 (c_1 + c_2)\, m^2 \times$$

$$\times (m_1^4 + m_2^4 + m_3^4) + c_3^2 (m_1^6 + m_2^6 + m_3^6),$$

so with $\mathbf{m} = \mathbf{r}$, $m_1 = x$, $m_2 = y$, $m_3 = z$, we get the following equation for sheet L of surface M in a cubic crystal

$$c_3 (x^4 + y^4 + z^4)\,[3r^2 - c_3 (x^4 + y^4 + z^4) + (c_3 - 3c_1 - 5c_2)\, r^4] -$$

$$- 2c_3^2 r^2 (x^6 + y^6 + z^6) + r^6\,[2c_2 - c_3 + (c_1 + c_2)(c_3 - 2c_2)\, r^2] = 0. \quad (37.11)$$

Consider now the parameters of the hexagonal crystal on average closest to the given cubic crystal (section 34). The terms $c_1 + c_2\mathbf{n}\cdot\mathbf{n}$ of the Λ of (36.1) already coincide with the analogous terms in the expression for the Λ^t of (34.1), so we need merely to find the Λ^t closest on average to the ν of (36.3). As our **e** to represent the L^6 axis of the hexagonal crystal we take $x_3 \| L^4$. Then in this case

$$\boldsymbol{e} = (0, 0, 1), \quad \boldsymbol{n} = (n_1, n_2, n_3), \quad [\boldsymbol{en}] = (-n_2, n_1, 0) \quad (37.12)$$

and

$$\boldsymbol{e}\nu\boldsymbol{e} = n_3^2, \quad [\boldsymbol{en}]\,\nu\,[\boldsymbol{en}] = 2n_1^2 n_2^2. \quad (37.13)$$

Then from (34.8) we have

$$d_1 = 3, \quad d_2 = 5, \quad d_3 = 9, \quad d_4 = 3, \quad d_5 = 1, \tag{37.14}$$

after which (34.10) gives

$$g_1 = -g_2 = \frac{3}{4}, \quad g_3 = 0, \quad g_4 = \frac{7}{4}, \quad g_5 = -\frac{1}{2}. \tag{37.15}$$

Then the tensor $\Lambda^{0,t}$ closest on average to ν is

$$\Lambda^{0,t} = \frac{3}{4}\left([\boldsymbol{ne}]^2 - \left(1 - \frac{7}{3}n_3^2\right)\boldsymbol{e}\cdot\boldsymbol{e} - \frac{2}{3}[\boldsymbol{ne}]\cdot[\boldsymbol{ne}]\right), \tag{37.16}$$

while the Λ^t closest to the Λ of (36.1) is

$$\Lambda^t = c_1 + c_2\boldsymbol{n}\cdot\boldsymbol{n} + c_3\Lambda^{0,t}. \tag{37.17}$$

Then simple steps give

$$\langle F^t \rangle = \langle (\Lambda - \Lambda^t)^2 \rangle_t = c_3^2 \langle (\nu - \Lambda^{0,t}) \rangle_t = c_3^2/15. \tag{37.18}$$

Comparison with (36.22) and (36.34) shows that the Λ^t of (37.17) differs less from Λ than does the Λ_m of (36.19) and the $\Lambda^0 = a + b\mathbf{n}\cdot\mathbf{n}$ of (36.29). Column 10 of Table IV gives values of $\Delta^t = \sqrt{\langle F^t \rangle / \langle \Lambda^2 \rangle_t} = 0.31\,|c_3|/\sqrt{\langle \Lambda^2 \rangle_t}$ for some cubic crystals.

Section 35 deals with the comparison with a transversely isotropic medium; here we have only to find the Λ^0 closest to the ν of (36.3) for any fixed $\mathbf{n}$. As before, we take $\mathbf{e}\|x_3$, and use (37.12) and (37.13) together with

$$[\boldsymbol{ec}] = \frac{1}{\sqrt{1-n_3^2}}(-n_1, -n_2, 0), \quad [\boldsymbol{ec}]\,\nu\,[\boldsymbol{ec}] = \frac{n_1^4 + n_2^4}{1 - n_3^2}, \tag{37.19}$$

to get from (35.58) that

$$a_0 = [\boldsymbol{ec}]\,\nu\,[\boldsymbol{ec}], \quad a_1 = 0, \quad a_2 = n_3^2 - a_0, \quad a_3 = 1 - n_3^2 - 2a_0. \tag{37.20}$$

Then, from (35.2), the tensor for the transversely isotropic medium most similar to the ν of (36.3) for a given $\mathbf{n}$ is

$$\Lambda_{\nu}^{0}=a_0+(n_3^2-a_0)\,\boldsymbol{e}\cdot\boldsymbol{e}+(1-n_3^2-2a_0)\,\boldsymbol{c}\cdot\boldsymbol{c}, \tag{37.21}$$

while the Λ^0 most similar to Λ is

$$\Lambda^0=c_1+c_2\boldsymbol{n}\cdot\boldsymbol{n}+c_3\Lambda_{\nu}^{0}. \tag{37.22}$$

The difference between these is

$$\Lambda'=\Lambda-\Lambda^0=c_3(\nu-\Lambda_{\nu}^{0}). \tag{37.23}$$

Here

$$\Lambda'\boldsymbol{e}=c_3(\nu\boldsymbol{e}-n_3^2\boldsymbol{e}), \tag{37.24}$$

so from (35.14)

$$\boldsymbol{c}\Lambda'\boldsymbol{e}=\boldsymbol{w}\boldsymbol{e}=0. \tag{37.25}$$

Also, **wc** = 0 from (35.15), so

$$\boldsymbol{w}=C\,[\boldsymbol{e}\boldsymbol{c}]. \tag{37.26}$$

Thus **w** always lies in a plane passing through the axis $\mathbf{e}\|L^4$ and the wave normal **n**. Multiplication by [**ec**] gives

$$C=\boldsymbol{w}\,[\boldsymbol{e}\boldsymbol{c}]=\boldsymbol{c}\Lambda'\,[\boldsymbol{e}\boldsymbol{c}]=c_3\boldsymbol{c}\nu\,[\boldsymbol{e}\boldsymbol{c}]=c_3n_1n_2\frac{n_1^2-n_2^2}{n_1^2+n_2^2}=|\boldsymbol{w}|. \tag{37.27}$$

Then we have for Λ' the representation of (35.18)

$$\Lambda'=C([\boldsymbol{e}\boldsymbol{c}]\cdot\boldsymbol{c}+\boldsymbol{c}\cdot[\boldsymbol{e}\boldsymbol{c}]), \tag{37.28}$$

while the following expression [see (35.21)] is correct for the entire Λ of (36.1) for a cubic crystal:

$$\Lambda=c_1+a_0c_3+c_2\boldsymbol{n}\cdot\boldsymbol{n}+c_3(n_3^2-a_0)\,\boldsymbol{e}\cdot\boldsymbol{e}+$$
$$+c_3^2(1-n_3^2-2a_0)\,\boldsymbol{c}\cdot\boldsymbol{c}+C(\boldsymbol{c}\cdot[\boldsymbol{e}\boldsymbol{c}]+[\boldsymbol{e}\boldsymbol{c}]\cdot\boldsymbol{c}). \tag{37.29}$$

The transverse anisotropy is now given by (35.20) as

$$F=(\Lambda'^2)_t=2\boldsymbol{w}^2=2C^2=\frac{c_3^2}{8}\sin^4\vartheta\sin^2 4\varphi, \tag{37.30}$$

in which we have the usual spherical coordinates

$$n_1 = \sin\vartheta\cos\varphi, \quad n_2 = \sin\vartheta\sin\varphi, \quad n_3 = \cos\vartheta. \tag{37.31}$$

From (37.30) it is clear that F, and hence Λ', must become zero for $n_1 = n_2$ or $n_1 = 0$ or $n_2 = 0$, i.e., for all directions lying in the four symmetry planes passing through L^4. Then in these planes Λ^0 coincides exactly with Λ, and the properties of the cubic crystal do not differ from those of a hexagonal one. The largest F occurs for directions of **n** lying in a plane perpendicular to L^4 and bisecting the angles between the above meridional planes, i.e., forming angles of $\pm\pi/8$ with the x_1 or x_2 axis. For these directions

$$F = F_{\max} = \frac{c_3^2}{8}. \tag{37.32}$$

For the mean F of (37.30) we have

$$\langle F\rangle = \langle\Lambda'^2\rangle_t = \frac{c_3^2}{30}. \tag{37.33}$$

Comparison with (36.22), (36.34), and (37.18) shows that the tensor of (37.22) is the best approximation to Λ.

38. Tetragonal Crystals

The Λ tensors of (19.15) and (19.18) are for hexagonal and tetragonal crystals, respectively; the only difference between them lies in the form of Λ_{12}, namely in which

$$\Lambda_{12}^{\text{tet}} = \Lambda_{12}^{\text{hex}} - cn_1n_2, \tag{38.1}$$

in which

$$c = \lambda_{11} - \lambda_{12} - 2\lambda_{66}. \tag{38.2}$$

Thus we may put the Λ for a tetragonal crystal in the form

$$\Lambda = \Lambda_h - cn_1n_2\theta, \quad \Lambda_h = c_0 + c_1\boldsymbol{n}\cdot\boldsymbol{n} + c_2\boldsymbol{e}\cdot\boldsymbol{e} + c_3\boldsymbol{c}\cdot\boldsymbol{c}, \tag{38.3}$$

in which c_0, c_1, c_2, and c_3 are as in (31.12)-(31.14) for a hexagonal crystal, $e\|x_3\|L^4$, and θ is the following very simple matrix:

$$\theta = \begin{pmatrix} 0 & 1 & 0 \\ 1 & 0 & 0 \\ 0 & 0 & 0 \end{pmatrix}. \tag{38.4}$$

whose properties are

$$\theta_t = 0, \quad \theta^2 = 1 - e \cdot e, \quad \bar{\theta} = \theta^2 - 1 = -e \cdot e, \quad \theta^3 = \theta. \tag{38.5}$$

It is readily seen that

$$\left.\begin{array}{c} \theta e = 0, \quad n\theta n = -[ne]\,\theta\,[ne] = 2n_1n_2, \quad [ne]\,\theta n = n_2^2 - n_1^2, \\ c\theta^2 c = 1, \quad n\theta^2[ne] = 0, \quad n\theta^2 n = [ne]^2 = 1 - n_3^2. \end{array}\right\} \tag{38.6}$$

The c of (38.2) differs from the c_3 of (36.2) for a cubic crystal only in that λ_{44} is replaced by λ_{66}.

Consider the special directions possible in a tetragonal crystal. Here the general condition of (17.7) becomes

$$[n,\ \Lambda n]\,\Lambda^2 n = [n,\ \Lambda_h n - cn_1n_2\theta n] \times \left[\Lambda_h^2 - cn_1n_2(\theta\Lambda_h + \Lambda_h\theta) + c^2n_1^2n_2^2\theta^2\right] n = 0. \tag{38.7}$$

Λ_h is the tensor as for a hexagonal crystal, so $[\mathbf{n},\ \Lambda_h\mathbf{n}]\,\Lambda_h^2\mathbf{n} = 0$ from (32.12) and (32.13), and thus (38.7) becomes

$$cn_1n_2\left\{[n,\ \Lambda_h n](cn_1n_2\theta^2 - \theta\Lambda_h - \Lambda_h\theta)\,n - [n,\ \theta n]\,(\Lambda_h - cn_1n_2\theta)^2\, n\right\} = 0. \tag{38.8}$$

From (38.3)-(38.6) we readily get

$$\left.\begin{array}{l} \Lambda_h n = (c_0 + c_1)\,n + c_2n_3e, \quad [n,\ \Lambda_h n] = c_2n_3[ne], \\ \Lambda_h\theta n = c_0\theta n + c_1 n\theta n \cdot n + g_5[ne]\,\theta n \cdot [ne], \\ \theta\Lambda_h n = (c_0 + c_1)\,\theta n, \quad [n,\ \theta n]\,\theta^2 n = n_3(n_2^2 - n_1^2), \\ [n,\ \theta n]\,\Lambda_h^2 n = -c_2n_3(2c_0 + c_1 + c_2)\,[ne]\,\theta n. \end{array}\right\} \tag{38.9}$$

These show that (38.7) becomes

$$cn_1n_2n_3(n_1^2 - n_2^2)\left[c_2(c_3 - c_2) + c\,(2g_5 + c)\,n_1^2n_2^2\right] = 0. \tag{38.10}$$

Then a tetragonal crystal has purely transverse waves in the three coordinate planes $n_1 = 0$, $n_2 = 0$, and $n_3 = 0$, as well as in the two planes $n_1 = \pm n_2$, passing through $L^4 \| x_3$. Moreover, such waves can propagate along directions lying in the surface of a cone of fourth degree whose equation is obtained by equating the expression in square brackets in (38.10) to zero. From (31.13) this becomes

$$An_3^4 + Bn_3^2 - Cn_1^2n_2^2 + D = 0, \tag{38.11}$$

in which

$$\left.\begin{aligned} A &= g_4(g_4 + g_5), & B &= 2g_2g_4 + g_5(g_2 - g_4), \\ C &= c(2g_5 + c), & D &= g_2(g_2 - g_5). \end{aligned}\right\} \tag{38.12}$$

Thus the cone of special directions of (17.10) for a tetragonal crystal splits up into 5 planes and the cone of fourth degree of (38.11), whose axis is $\mathbf{e}\|L^4$.

The condition for longitudinal normals in a tetragonal crystal is

$$[\boldsymbol{n}, \Lambda\boldsymbol{n}] = c_2n_3[\boldsymbol{ne}] - cn_1n_2[\boldsymbol{n}, \theta\boldsymbol{n}] = 0. \tag{38.13}$$

Both terms must be zero individually if $[\mathbf{ne}]$ is not parallel to $[\mathbf{n}, \theta\,\mathbf{n}]$. Consider first the case of these two vectors parallel, which occurs if

$$[[\boldsymbol{ne}][\boldsymbol{n}, \theta\boldsymbol{n}]] = [\boldsymbol{ne}]\theta\boldsymbol{n}\cdot\boldsymbol{n} = (n_2^2 - n_1^2)\,\boldsymbol{n} = 0. \tag{38.14}$$

Here

$$[\boldsymbol{n}, \theta\boldsymbol{n}] = K[\boldsymbol{ne}], \quad K = \frac{[\boldsymbol{ne}][\boldsymbol{n}, \theta\boldsymbol{n}]}{[\boldsymbol{ne}]^2} = -\frac{2n_1n_2n_3}{1 - n_3^2}.$$

We have from (38.14) that $1 - n_3^2 = 2n_1^2$, so the condition for the planes $n_1^2 - n_2^2 = 0$ becomes

$$(c_2 + cn_1^2)\,n_3[\boldsymbol{ne}] = 0. \tag{38.15}$$

Neglecting the obvious case $\mathbf{n}\|\mathbf{e}$, this is satisfied for $n_3 = 0$ and

$$c_2 + cn_1^2 = g_2 + g_4n_3^2 + \frac{1}{2}c\,(1 - n_3^2) = 0, \tag{38.16}$$

i.e., when

$$n_3^2 = \frac{c + 2g_2}{c - 2g_4}, \tag{38.17}$$

if the parameters obey

$$0 < \frac{c+2g_2}{c-2g_4} < 1. \tag{38.18}$$

Both vectors of (38.13) must be zero if $n_1^2 \neq n_2^2$; since

$$[n,\ \theta n]^2 = (n_1^2 - n_2^2)^2 + n_3^2(1 - n_3^2),$$

the equation $[\mathbf{n},\ \theta\mathbf{n}] = 0$ leads to the case of (38.14) already considered. We are thus left with the two conditions

$$c_2 n_3 = 0, \quad c n_1 n_2 = 0, \tag{38.19}$$

which imply that longitudinal normals lie in the planes $n_1 = 0$ and $n_2 = 0$ at the points where these meet the equatorial plane ($n_3 = 0$) and may also lie at an angle ϑ to the symmetry axis ($\cos\vartheta = n_3$), this being defined by

$$c_2 = g_2 + g_4 n_3^2 = 0. \tag{38.20}$$

This condition coincides with the analogous condition of (32.19) and (32.20) for a hexagonal crystal.

Conditions (38.16) correspond to intersection of the planes $n_1^2 = n_2^2$ with the cone of (38.11), while (38.20) corresponds to intersection of the planes $n_1 = 0$ and $n_2 = 0$ with the same cone. In fact, if

$$n_1^2 = n_2^2 = \frac{1 - n_3^2}{2},$$

(38.11) [see (31.13)] becomes

$$(c_2 + c n_1^2)(2g_5 n_1^2 + c n_1^2 - c_2) = 0,$$

i.e., is satisfied subject to condition (38.16). This is obvious for the planes $n_1 = 0$ and $n_2 = 0$ with (38.20) [see (38.10)], so these longitudinal normals correspond to mutual intersection of the separate sheets of the cone of special directions. The four planes $n_1 = 0$, $n_2 = 0$, $n_1 = \pm n_2$ along the $L^4 \| \mathbf{e}$ axis and meet the plane $n_3 = 0$ along four equally separated directions. These five longitudinal normals occur in any tetragonal crystal; $\mathbf{e}$ is a longitudinal acoustic axis. There can also be four further longitudinal normals, which correspond to intersection of the cone of (38.10) with the planes $n_1^2 - n_2^2 = 0$ [see (38.16)–(38.18)] or the planes $n_1 = 0$, $n_2 = 0$ [see (38.20)].

Thus a tetragonal crystal can have 5, 9, or 13 longitudinal normals.

The number and location of the acoustic axes may be found from (18.19), if the tensor inverse to the Λ of (38.3) is found. From (12.64), since $\theta_t = 0$, we have

$$\overline{\Lambda} = \overline{\Lambda_h - cn_1n_2\theta} =$$

$$\overline{\Lambda}_h + c^2n_1^2n_2^2\overline{\theta} - cn_1n_2[\Lambda_h\theta + \theta\Lambda_h - (\Lambda_h)_t\theta - (\theta\Lambda_h)_t], \tag{38.21}$$

or, from (31.26), (38.5), and (38.6),

$$\begin{aligned}\overline{\Lambda} = (c_0 + c_3)(c_0 + c_1 + c_2 - c_1\boldsymbol{n}\cdot\boldsymbol{n} - c_2\boldsymbol{e}\cdot\boldsymbol{e}) - \\ - [\boldsymbol{en}]^2[g_5(c_0 + c_1 + c_2) - c_1c_2]\,\boldsymbol{c}\cdot\boldsymbol{c} - \\ - c^2n_1^2n_2^2\boldsymbol{e}\cdot\boldsymbol{e} - cn_1n_2\big[c_1(\boldsymbol{n}\cdot\boldsymbol{n}\theta + \theta\boldsymbol{n}\cdot\boldsymbol{n}) + c_3(\boldsymbol{c}\cdot\boldsymbol{c}\theta + \theta\boldsymbol{c}\cdot\boldsymbol{c}) - \\ - (c_0 + c_1 + c_2 + c_3)\theta + 2n_1n_2(g_5 - c_1)\big].\end{aligned} \tag{38.22}$$

To write (18.19) for the acoustic axes it is sufficient in (38.22) to replace c_0 by $c_0' = c_0 - v_1^2$ and equate the result to zero. Let $\overline{\Lambda}_1$ be the tensor derived from (38.22) with c_0 replaced by c_0'. Then

$$\overline{\Lambda}_1 = 0. \tag{38.23}$$

But $\overline{\Lambda}_1$ is a symmetric tensor, so this splits up into six equations corresponding to zero values for the components of $\overline{\Lambda}_1$, though not all of these equations are independent. For a real symmetric matrix A to be zero it is necessary and sufficient for all the eigenvalues to be zero [see (13.20)]; but the eigenvalues are the roots of $\lambda^3 - A_t\lambda^2 + \overline{A}_t\lambda - |A| = 0$, from (13.4), and can all equal zero only if all of the following three invariants are zero:

$$A_t = \overline{A}_t = |A| = 0. \tag{38.24}$$

To have $A = 0$ for $A = \widetilde{A} = A^*$, it is necessary and sufficient for the three equations of (38.24) to be obeyed. If, on the other hand, the matrix is inverse with respect to some other matrix ($A = \overline{B}$), then only two conditions are necessary (not three) in order to give $A = 0$; in fact, from (12.18) and (12.20)

$$|A| = |B|^2, \quad \overline{A} = \overline{\overline{B}} = |B|B. \tag{38.25}$$

Then if $A = \widetilde{A} = A^* = \overline{B}$, we have from $|A| = |B|^2 = 0$ automatically that $\overline{A}_t = |B|B_t = 0$, so here it is sufficient to have $A_t = |A| = 0$. This is precisely the case we have here, so condition (38.23) is equivalent to the following two:

$$\overline{\Lambda}_{1t}=0, \quad |\overline{\Lambda}_1| = |\Lambda_1|^2 = 0. \tag{38.26}$$

The equation $|\Lambda_1|=|\Lambda - v_1^2|=0$ is simply the condition for v_1^2 to be an eigenvalue of Λ. These two invariant conditions are necessary and sufficient for the presence of an acoustic axis. In section 28 we arrive at the two invariant conditions of (28.34) for acoustic axes by a different route. We expand (38.26) via (12.25), (38.21), and (38.22) to get

$$\overline{\Lambda}_{1t}=c_0'(c_0'+c_3)(2c_0'+c_3)(c_0'+c_1+c_2)+c_1c_2[\boldsymbol{ne}]^2+ \\ + cn_1^2n_2^2(2c_1-c-2g_5)=0, \tag{38.27}$$

$$|\Lambda_1|=(c_0'+c_3)[(c_0'+c_1)(c_0'+c_2)-c_1c_2n_3^2]+ \\ +2cn_1^2n_2^2[(c_0'+c_2)(c_1-c_3)-c_1c_2n_3^2]-c^2n_1^2n_2^2(c_0'+c_2+c_1n_3^2)=0. \tag{38.28}$$

However, analysis of the necessary and sufficient conditions of (38.27) and (38.28) is greatly facilitated if to this we add other equations implied by (38.23), which of course are also necessary conditions for acoustic axes. To derive these equations we may multiply $\overline{\Lambda}_1$ from left and right by any vectors and equate the result to zero, in accordance with (38.23). The simplest relationships are obtained if we multiply $\overline{\Lambda}_1$ by the pairs (**e**, [**ne**]), ([**ne**], [**e**[**ne**]]), and (**e**, [**e**[**ne**]]), for many terms then cancel out, and we get

$$\boldsymbol{e}\overline{\Lambda}_1[\boldsymbol{ne}]=\boldsymbol{e}\overline{\Lambda}[\boldsymbol{ne}]=-cc_1n_1n_2n_3(n_1^2-n_2^2)=0, \tag{38.29}$$

$$[\boldsymbol{ne}]\overline{\Lambda}_1[\boldsymbol{e}[\boldsymbol{ne}]]=cn_1n_2(n_1^2-n_2^2)(c_0'+c_2+c_1n_3^2)=0, \tag{38.30}$$

$$\boldsymbol{e}\overline{\Lambda}_1[\boldsymbol{e}[\boldsymbol{ne}]]=-c_1n_3[(c_0+c_3)[\boldsymbol{ne}]^2+2cn_1^2n_2^2]=0. \tag{38.31}$$

To these we may add

$$\boldsymbol{e}\overline{\Lambda}_1\boldsymbol{e}=(c_0'+c_3)(c_0'+c_1[\boldsymbol{ne}]^2)+c^2n_1^2n_2^2(2c_1-c-2g_5)=0, \tag{38.32}$$

$$\boldsymbol{c}\overline{\Lambda}_1\boldsymbol{c}=c_0'(c_0'+c_1+c_2)+c_1c_2[\boldsymbol{ne}]^2-\frac{2cn_1^2n_2^2}{1-n_3^2}(c_0'+c_2+c_1n_3^2)=0. \tag{38.33}$$

Each of the conditions (38.29)-(38.33) is necessary, so from (38.29) we conclude directly that the acoustic axes in a tetragonal crystal cannot lie outside the planes $n_1 = 0$, $n_2 = 0$, $n_3 = 0$, or $n_1 = \pm n_2$; for $n_1 = 0$ or $n_2 = 0$ the Λ of (38.3) reduces to (31.12), the tensor for a hexagonal crystal, for which (32.24) and (32.25) serve to determine the question of acoustic axes distinct from **e**. Thus there are four acoustic axes lying in the coordinate planes $n_1 = 0$ and $n_2 = 0$ if the parameters of the tetragonal crystal satisfy (32.25).

Let $n_3 = 0$ (but $n_1 \neq 0$, $n_2 \neq 0$, $n_1^2 \neq n_2^2$); then (38.30) gives

$$c_0' + c_2 = 0. \tag{38.34}$$

Under these conditions $|\Lambda_1| = 0$, while (38.27), gives ($n_1 = \cos\varphi$, $n_2 = \sin\varphi$)

$$n_1^2 n_2^2 = \frac{1}{4}\sin^2 2\varphi = \frac{(g_2 - g_3)(g_2 - g_5)}{c\,(2g_5 - 2g_3 + c)} = \frac{(\lambda_{44} - \lambda_{11})(\lambda_{66} - \lambda_{44})}{c\,(\lambda_{11} + \lambda_{12})}. \tag{38.35}$$

The same result is obtained directly from (38.32), so if

$$0 \leqslant \frac{(g_2 - g_3)(g_2 - g_5)}{c\,(c + 2g_5 - 2g_3)} \leqslant \frac{1}{4}, \tag{38.36}$$

the crystal has four acoustic axes in planes perpendicular to the L^4 axis, whose azimuths φ are given by (38.35).

Finally we have the third possibility

$$n_1^2 = n_2^2, \quad 1 - n_3^2 = 2n_1^2, \tag{38.37}$$

which gives us from (38.31) that

$$c_0' + n_1^2(c + 2g_5) = 0. \tag{38.38}$$

Here $|\Lambda_1| = 0$, while (38.27) or (38.33) gives

$$n_1^2 = \frac{(c + g_5)(g_2 + g_3 + g_4) - g_3(g_2 + g_4)}{c\,(c + g_3 + 2g_4 + 3g_5) + 2g_5(g_4 + g_5) - 2g_3 g_4}. \tag{38.39}$$

Thus four acoustic axes lie in the planes $n_1 = \pm n_2$ if the right side of this last equation lies between 0 and 1. It has been shown [18] that condition (38.36) cannot be compatible with conditions (32.25) and (38.39) simultaneously, so the following are the only possibilities for a tetragonal crystal: 1) none of conditions (32.25), (38.36), and (38.39) is obeyed, and there is only one acoustic axis, along L^4; 2) one of these three conditions is obeyed, and the crystal has five acoustic axes; and 3) (32.25) and (38.39) are obeyed, or (38.36) with either (32.25) or (38.39), when there are nine acoustic axes.

The most natural reference medium for a tetragonal crystal is a hexagonal one (transversely isotropic medium). From (38.3) we see that Λ_h coincides exactly in structure with tensor (31.12) for a transversely isotropic medium, so it is simplest to consider the addition $-cn_1n_2\theta$ as the tensor characterizing the difference between tetragonal and hexagonal crystals. Hence

$$\Lambda'^{t} = -cn_1n_2\theta, \quad (\Lambda'^{t2})_t = 2c^2n_1^2n_2^2$$

and the transverse anisotropy of a tetragonal crystal is

$$\langle F^t \rangle = \langle \Lambda'^{t2} \rangle_t = \frac{2}{15} c^2 = 0.133c^2. \tag{38.40}$$

However, the method of section 34 allows us to isolate from the tensor $n_1n_2\theta$ a similar part transversely isotropic on average. We put $\Lambda = n_1n_2\theta$ in (34.8) and apply (38.6),

$$d_1 = d_2 = d_4 = 0, \quad d_3 = -2d_5 = 2,$$

whereupon we get from (34.10) that

$$g_1 = -g_2 = g_4 = -g_5/2 = 1/4, \quad g_3 = 0.$$

Hence the tensor of (34.1) most similar on average to $cn_1n_2\theta$ is

$$\Lambda^t = c(1 - n_3^2)(1 - \mathbf{e}\cdot\mathbf{e} - 2\mathbf{e}\cdot\mathbf{c})/4, \tag{38.41}$$

whence we have

$$\langle F^t \rangle = \langle (cn_1n_2\theta - \Lambda^t)^2 \rangle = c^2/15, \tag{38.42}$$

which represents half the quantity of (38.40).

An even more precise approximation is obtained by seeking the tensor most similar to θ by the method of section 35. We have from (38.6) that

$$c\theta c = -[ec]\,\theta\,[ec] = -\frac{2n_1n_2}{1-n_3^2},$$

so (35.11) gives*

$$a_0' = -a_2' = -\frac{1}{2}a_3' = \frac{2n_1n_2}{1-n_3^2}, \quad a_1' = 0, \tag{38.43}$$

which implies that

$$\Lambda' = \Lambda - \Lambda^0 = -cn_1n_2\left[\theta - \frac{2n_1n_2}{1-n_3^2}(1 - e\cdot e - 2c\cdot c)\right]. \tag{38.44}$$

We have $\Lambda'\mathbf{e} = 0$, so $\mathbf{we} = \mathbf{c}\Lambda'\mathbf{e} = 0$ [see (35.14)]; thus, as for (37.26) for cubic crystals, we have

$$w = \Lambda' c = C\,[ec], \tag{38.45}$$

in which

$$C = w\,[ec] = c\Lambda'\,[ec] = -cn_1n_2c\theta\,[ec] = c\,\frac{n_1n_2\left(n_1^2 - n_2^2\right)}{n_1^2+n_2^2} = |w|. \tag{38.46}$$

This value of C differs from (36.27) for a cubic crystal only in having c_3 replaced by c; with allowance for this change, for the conclusions of the previous section from (37.27) onwards are correct for tetragonal crystals as well as cubic ones, except that (37.29) for Λ is replaced by

$$\Lambda = a_0 + a_1 n\cdot n + a_2 e\cdot e + a_3 c\cdot c + C\,(c\cdot[ec] + [ec]\cdot c), \tag{38.47}$$

in which

$$a_0 = c_0 - cn_1n_2a_0', \quad a_1 = c_1, \quad a_2 = c_2 + cn_1n_2a_0',$$
$$a_3 = c_3 + 2cn_1n_2a_0'. \tag{38.48}$$

In particular, (37.33), with $|c_3|$ replaced by $|c|$, applies to a tetragonal crystal, so for $|c| = |c_3|$ a cubic crystal and a tetragonal one differ identically from the transversely isotropic medium most similar to them for a given **n**.

The $\mathbf{u}_1'$ (first approximation) given by (35.35) for the displacement vector of the transverse wave **c** is

---

* Primes denote this a_s for $\Lambda = \theta$.

Table V. Elastic Constants of Tetragonal Crystals

(The c are in 10^{11} dynes/cm^2; the g are in 10^{11} cm^2/sec^2)

No.	Crystal	c_{11}	c_{33}	c_{44}	c_{66}	c_{12}	c_{13}	g_1	g_2	g_3	g_4	g_5	c	$\lvert\gamma W\rvert_{max}$
1	Ammonium dihydrogen phosphate $(NH_4)H_2PO_4$	6.85	3.24	0.861	0.602	-0.18	1.78	2.33	-1.86	1.47	1.71	-2.00	3.23	0.44
2	Barium titanate $BaTiO_3$	27.5	16.5	5.43	11.3	17.9	15.1	1.15	-0.255	3.40	-1.31	0.717	-2.15	0.14
3	Indium In	4.45	4.44	0.655	1.22	3.944	4.051	-0.0351	0.125	0.645	-0.251	0.202	-0.265	0.12
4	Pentaerythritol $C(CH_2OH)_4$	0.6667	0.80	0.35	0.4032	-0.3068	0.050	0.336	0.105	0.50	-0.0420	0.172	0.211	0.23
5	Nickel sulfate hexahydrate $NiSO_4 \cdot 6H_2O$	3.21	2.93	1.16	1.78	2.31	0.21	0.889	-0.329	0.662	0.522	-0.0290	-1.29	0.39
6	Potassium dihydrogen arsenate KH_2AsO_4	5.31	3.7	1.2	0.7	-0.6	-0.2	1.50	-1.08	0.349	1.61	-1.26	1.57	0.96
7	Potassium dihydrogen phosphate KH_2PO_4	7.23	6.18	1.28	0.618	0.439	1.98	1.70	-1.15	1.39	1.85	-1.43	2.38	0.36
8	Rutile TiO_2	27.3	48.4	12.5	19.4	17.6	14.9	-0.026	3.28	7.14	-1.1	5.08	-7.58	0.32
9	White tin Sn	8.391	9.665	1.754	0.7407	4.870	2.810	0.5257	-0.2848	0.6269	0.7445	-0.4239	0.2802	0.076
10	Zircon $ZrSiO_4$	7.35	4.600	1.380	1.600	0.900	1.36	1.01	-0.708	0.601	0.814	-0.660	0.7125	0.20

$$u'_1 = \gamma w = \frac{1}{\Delta}(b_0 w + a_1 wn \cdot n + a_2 we \cdot e), \tag{38.49}$$

where from (35.32) and (38.48)

$$\left.\begin{aligned} b_0 &= a_3 - a_1 - a_2 = c_3 - c_1 - c_2 + \frac{2cn_1^2 n_2^2}{1 - n_3^2}, \\ \Delta &= (a_3 - a_1)(a_3 - a_2) - a_1 a_2 n_3^2 = a_3 b_0 + a_1 a_2 (1 - n_3^2). \end{aligned}\right\} \tag{38.50}$$

From (38.45) we get

$$u' = \frac{C}{\Delta}(b_0 [ec] + a_1 |[ne]| n). \tag{38.51}$$

The sum of the squares of the components of Λ', i.e., the quantity of (37.30), serves as a measure of the difference between Λ and Λ^0 as a function of **n**:

$$F = (\Lambda'^2)_t = 2w^2 = 2c^2 n_1^2 n_2^2 \frac{(n_1^2 - n_2^2)^2}{(n_1^2 + n_2^2)^2} = \frac{c^2}{8} \sin^4 \vartheta \sin^2 4\varphi, \tag{38.52}$$

the maximal value occurring for $\vartheta = \pi/2$ (i.e., in the equatorial plane of the crystal) and for $\varphi = (2k+1)(\pi/8)$. The value is

$$F_{max} = \frac{c^2}{8}. \tag{38.53}$$

Then from (38.45) and (38.46)

$$w = C[ec] = -\frac{c}{4} n \tag{38.54}$$

and from (31.13), (31.14), (38.43), (38.48), (38.50), and (38.51)

$$u'_1 = \gamma w = \frac{cn}{4(a_1 - a_3)} = \frac{cn}{4(\lambda_{11} - \lambda_{66}) - 2c}. \tag{38.55}$$

Then $|\gamma w|^{2k+1}$, in which

$$|\gamma w|_{max} = \frac{|c|}{2(2\lambda_{11} - 2\lambda_{66} - c)} = \frac{|c|}{2(\lambda_{11} + \lambda_{12})}, \tag{38.56}$$

gives the order of the error from the k-th approximation (section 35). Table V gives data for some tetragonal crystals on the basis of the parameters of [34]. If a relative error of 10^{-3} will suffice, it is clear that the first approximation is satisfactory for tin, indium, and barium titanate, since that approximation has an error of the order of $|\gamma \mathbf{w}|^3$.

39. Comparison with a Hexagonal Crystal

Section 35 presents the method of comparing a given crystal with the transversely isotropic medium (hexagonal crystal) most similar to it for a given **n**. We have seen in section 35 that this approach provides an approximate method of deriving the velocities and displacements of the transverse waves.

However, the method of section 35 is analogous to that of sections 27 and 28 (comparison with an isotropic medium), and it can be used in an exact analysis as well as in approximate treatments. The essential point is that the tensor Λ for any crystal (for $n_3 \neq 1$) may be put in the universal form of (35.21):

$$\Lambda = \Lambda^0 + \Lambda', \quad \Lambda' = \boldsymbol{w} \cdot \boldsymbol{c} + \boldsymbol{c} \cdot \boldsymbol{w}, \quad \boldsymbol{w} = \Lambda' \boldsymbol{c}, \tag{39.1}$$

$$\Lambda^0 = a_0 + a_1 \boldsymbol{n} \cdot \boldsymbol{n} + a_2 \boldsymbol{e} \cdot \boldsymbol{e} + a_3 \boldsymbol{c} \cdot \boldsymbol{c}, \tag{39.2}$$

in which the a_s (s = 0, 1, 2, 3) are defined by (35.9) and (35.11). The representation of (39.1) and (39.2) is completely general and applies to any crystal; moreover, any direction may be taken as that of vector **e**. Although in section 35 we derived the natural conditions* for the choice of **e**, all the general relationships of that section apply for any **e**. The convenience of this representation arises because both parts of the general Λ (Λ^0 and Λ') are expressed in fairly compact form and have fairly simple properties.

As regards the special directions, we may say that they are defined for any crystal by

$$[\boldsymbol{n}, \Lambda \boldsymbol{n}] \Lambda^2 \boldsymbol{n} = [\boldsymbol{n}, (\Lambda^0 + \Lambda') \boldsymbol{n}] (\Lambda^0 + \Lambda')^2 \boldsymbol{n} = 0, \tag{39.3}$$

with

$$[\boldsymbol{n}, \Lambda^0 \boldsymbol{n}] \Lambda^{0^2} \boldsymbol{n} = 0 \tag{39.4}$$

---

* The choice of **e** proposed in section 35 has the general feature that **e** is a longitudinal normal for any crystal.

as for a hexagonal crystal (section 32), because Λ^0 has the same structure. Then

$$[\boldsymbol{n},\ \Lambda^0\boldsymbol{n}](\Lambda^0\Lambda' + \Lambda'\Lambda^0 + \Lambda'^2)\boldsymbol{n} + $$
$$+ [\boldsymbol{n},\ \Lambda'\boldsymbol{n}](\Lambda^{0^2} + \Lambda^0\Lambda' + \Lambda'\Lambda^0 + \Lambda'^2)\boldsymbol{n} = 0. \quad (39.5)$$

From (39.1), (39.2), (35.15), and (35.16) we have

$$[\boldsymbol{n},\ \Lambda^0\boldsymbol{n}] = a_2 n_3 [\boldsymbol{ne}], \quad (39.6)$$

$$\Lambda'^2 = \boldsymbol{w}\cdot\boldsymbol{w} + w^2\boldsymbol{c}\cdot\boldsymbol{c}. \quad (39.7)$$

Simple steps give

$$[\boldsymbol{n},\ \Lambda\boldsymbol{n}]\,\Lambda^2\boldsymbol{n} = a_2 n_3 |[\boldsymbol{ne}]| [(a_3 - a_2)\,\boldsymbol{wn} + a_2 n_3 \boldsymbol{we}] + (\boldsymbol{wn})^2\cdot\boldsymbol{w}\,[\boldsymbol{nc}]. \quad (39.8)$$

Since

$$\boldsymbol{n}\Lambda^0\boldsymbol{c} = \boldsymbol{e}\Lambda^0\boldsymbol{c} = \boldsymbol{c}\Lambda^0\,[\boldsymbol{nc}] = 0, \quad (39.9)$$

then

$$\left.\begin{aligned} \boldsymbol{wn} &= \boldsymbol{c}\Lambda'\boldsymbol{n} = \boldsymbol{c}(\Lambda - \Lambda^0)\,\boldsymbol{n} = \boldsymbol{c}\Lambda\boldsymbol{n}, \\ \boldsymbol{we} &= \boldsymbol{c}\Lambda\boldsymbol{e}, \qquad \boldsymbol{w}\,[\boldsymbol{nc}] = \boldsymbol{c}\Lambda\,[\boldsymbol{nc}]. \end{aligned}\right\} \quad (39.10)$$

The general transversality condition then becomes

$$\boldsymbol{n}\Lambda\boldsymbol{c}\,[a_2(a_3 - a_2)\,n_3 |[\boldsymbol{ne}]| + \boldsymbol{n}\Lambda\boldsymbol{c}\cdot\boldsymbol{c}\Lambda\,[\boldsymbol{nc}]\,] + a_2^2 n_3^2 \boldsymbol{e}\Lambda\,[\boldsymbol{ne}] = 0. \quad (39.11)$$

Substitution of the parameters for tetragonal crystals [see (38.43)],

$$a_3 = c_3 + \frac{4cn_1^2n_2^2}{1 - n_3^2}, \qquad a_2 = c_2 + \frac{2cn_1^2n_2^2}{1 - n_3^2}, \quad (39.12)$$

gives (38.10) directly.

From (39.6) and (39.7) we also have the general relation

$$[\boldsymbol{n},\ \Lambda\boldsymbol{n}] = a_2 n_3\,[\boldsymbol{ne}] + \boldsymbol{wn}\cdot[\boldsymbol{nc}] = a_2 n_3\,[\boldsymbol{ne}] + \boldsymbol{n}\Lambda\boldsymbol{c}\cdot[\boldsymbol{nc}]. \quad (39.13)$$

Therefore for a longitudinal normal we must have

$$a_2^2 n_3^2 (1 - n_3^2) + (\boldsymbol{n}\Lambda\boldsymbol{c})^2 = [\boldsymbol{n},\ \Lambda\boldsymbol{n}]^2 = 0. \quad (39.14)$$

Both terms on the left are positive, so we obtain two conditions:

$$a_2 n_3 (1 - n_3^2) = 0, \quad n\Lambda c = 0. \tag{39.15}$$

Therefore in the general case we get from (35.11) the conditions for longitudinal normals in any crystal as

$$n\Lambda [ne] = 0, \tag{39.16}$$

$$[\mathbf{ne}][\mathbf{n}, \Lambda\mathbf{n}] = 0. \tag{39.17}$$

For tetragonal crystals these give

$$n_1 n_2 (n_1^2 - n_2^2) = 0, \tag{39.18}$$

$$c_2 (1 - n_3^2) + 2c n_1^2 n_2^2 = 0. \tag{39.19}$$

From (38.20) we have for $n_1 = 0$ (or $n_2 = 0$) that $a_2 = c_2 = 0$; if $n_1^2 = n_2^2 = (1 - n_3^2)/2$, we get (38.17) from (39.18).

The representation of (39.1) and (39.2) may be used also to find the acoustic axes; the condition for these is

$$\overline{\Lambda}_1 = \overline{\Lambda_1^0 + \Lambda'} = \overline{\Lambda_1^0 + w \cdot c + c \cdot w} = 0, \tag{39.20}$$

in which

$$\Lambda_1^0 = \Lambda^0 - v_1^2 = a_0' + a_1 n \cdot n + a_2 e \cdot e + a_3 c \cdot c, \quad a_0' = a_0 - v_1^2, \tag{39.21}$$

and v_1^2 is the squared velocity of the purely transverse wave. From (12.64) or (13.34) and (13.37),

$$\overline{\alpha} = \alpha^2 - \alpha_t \alpha + \frac{1}{2}((\alpha_t)^2 - (\alpha^2)_t), \tag{39.22}$$

we have

$$\overline{\Lambda_1^0 + \Lambda'} = \overline{\Lambda}_1^0 + \Lambda' \Lambda_1^0 + \Lambda_1^0 \Lambda' + \Lambda'^2 - \Lambda_{1t}^0 \Lambda' - w^2 = 0. \tag{39.23}$$

Section 38 shows that condition (39.20) is equivalent to

$$\overline{\Lambda}_{1t} = \overline{\Lambda}_{1t}^0 - w^2 = 0, \tag{39.24}$$

$$|\Lambda_1| = |\Lambda_1^0 + w \cdot c + c \cdot w| = |\Lambda_1^0| - [cw]\,\Lambda_1^0\,[cw] = 0. \tag{39.25}$$

However, these latter equations are best associated with ones obtained by equating to zero expressions of the form $\mathbf{s}_1\bar{\Lambda}_1\mathbf{s}_2$, in which $\mathbf{s}_1$ and $\mathbf{s}_2$ are arbitrary vectors; in this way, in particular, we get from (39.23) after multiplication from both sides by $\mathbf{c}$ [see (35.7) and (35.15)] that

$$c\bar{\Lambda}_1 c = c\bar{\Lambda}_1^0 c = 0. \tag{39.26}$$

Now $\Lambda_1^0 \mathbf{c} = (a_0' + a_3)\mathbf{c}$, so

$$|\Lambda_1^0| = c\,|\Lambda_1^0|\,c = c\Lambda_1^0\bar{\Lambda}_1^0 c = (a_0' + a_3)\,c\bar{\Lambda}_1^0 c, \tag{39.27}$$

so (39.26) implies that

$$|\Lambda_1^0| = |\Lambda^0 - v_1^2| = 0. \tag{39.28}$$

This is an interesting result. Condition (39.25) implies that v_1^2 is an eigenvalue of tensor Λ, while (39.28) implies that the same v_1^2 is an eigenvalue of tensor Λ^0, which characterizes the transversely isotropic medium most similar (for a given **n**) to the given crystal. In other words, the velocity of the transverse waves along an **n** acoustic axis is as for the transversely isotropic medium most similar to it for that **n** in the sense of section 35. Hence comparison with that medium (section 35), although somewhat formal, enables one to represent any crystal via a medium of simpler type that reproduces some of its physical properties.

Use of another pair of vectors $\mathbf{s}_1$ and $\mathbf{s}_2$ gives us from (39.20) and (39.23) that

$$e\bar{\Lambda}_1 c = a_1 n_3 nw - (a_0' + a_1)\,ew = 0, \tag{39.29}$$

$$n\bar{\Lambda}_1 c = a_2 n_3 ew - (a_0' + a_2)\,nw = 0, \tag{39.30}$$

$$e\bar{\Lambda}_1\,[e\,[ne]\,] = ew \cdot (nw - n_3 ew) - a_1 (a_0' + a_3)\,n_3\,[ne]^2 = 0. \tag{39.31}$$

From (39.25) and (39.28),

$$[cw]\,\Lambda_1^0\,[cw] = a_0' w^2 + a_1 (\,[nw]\,c)^2 + a_2 (\,[ew]\,c)^2 = 0.$$

Now $\mathbf{c}\|[\mathbf{nw}]\|[\mathbf{ew}]$ $(\mathbf{nc}=\mathbf{ec}=\mathbf{wc}=0)$, so instead of this we may put

$$a_0'w^2+a_1[nw]^2+a_2[ew]^2=0 \tag{39.32}$$

or

$$v_1^2=[cw_1]\Lambda[cw_1]=a_0+a_1[nw_1]^2+a_2[ew_1]^2,\quad w_1=w/|w|. \tag{39.32'}$$

All of conditions (39.26) and (39.28)-(39.32) are necessary, i.e., they must be complied with for an acoustic axis; they allow us to examine the various possibilities, which for an acoustic axis must lead to the necessary and sufficient conditions of (39.24) and (39.25).

The above relationships are quite general, being applicable to any crystal and any direction $\mathbf{e}\neq\mathbf{n}$. The particular case $\mathbf{n}\|\mathbf{e}$ may be examined without difficulty directly from the detailed form of the tensor Λ (section 19).

For a tetragonal crystal we have $\mathbf{we}=0$ from (38.45) and $\mathbf{nw}=\mathbf{n}\Lambda\mathbf{c}=[cn_1n_2(n_1^2-n_2^2)]/(1-n_3^2)$, so (39.29) reduces to (38.29). It is readily shown that we can derive from (39.24)-(39.26) all the relationships of the previous section that define the orientation of the acoustic axes of tetragonal crystals. However, a tetragonal crystal is of relatively high symmetry, so the relationships of sections 17 and 18 can be applied directly, the expressions remaining fairly simple. However, even for trigonal crystals the operations of sections 17 and 18 become cumbrous and complicated, and this becomes even more pronounced for the crystals of the lower systems. In all such cases the relationships of the present section substantially simplify the work; this is illustrated in the next section by reference to trigonal crystals.

40. Trigonal Crystals

Tensor Λ for a trigonal crystal is given by (19.19) as of the form

$$\Lambda=\Lambda_h+c0,\quad c=2\lambda_{14}, \tag{40.1}$$

in which

$$\Lambda_h=c_0+c_1n\cdot n+c_2e\cdot e+c_3c\cdot c \tag{40.2}$$

is the Λ tensor of (31.12)-(31.14) for a transversely isotropic medium, while θ is the following tensor in the components of the wave-normal vector:

$$\theta = \tilde{\theta} = \begin{pmatrix} n_2 n_3 & n_1 n_3 & n_1 n_2 \\ n_1 n_3 & -n_2 n_3 & \frac{n_1^2 - n_2^2}{2} \\ n_1 n_2 & \frac{n_1^2 - n_2^2}{2} & 0 \end{pmatrix}; \tag{40.3}$$

here vector **e** is directed along the L^3 axis and $x_3 \| \mathbf{e} \| L^3$. Tensor θ has the following properties:

$$\left.\begin{aligned} &\theta_t = \boldsymbol{e}\theta\boldsymbol{e} = 0, \ \ \boldsymbol{n}\theta\boldsymbol{n} = -2\,[\boldsymbol{ne}]\,\theta\,[\boldsymbol{ne}] = 4n_3\boldsymbol{n}\theta\boldsymbol{e} = 2n_2n_3(3n_1^2 - n_2^2), \\ &\boldsymbol{n}\theta\,[\boldsymbol{ne}] = 3n_3\boldsymbol{e}\theta\,[\boldsymbol{ne}] = \frac{3}{2}\,n_1n_3(3n_2^2 - n_1^2), \\ &\boldsymbol{n}\theta^2\,[\boldsymbol{ne}] = \frac{n_1n_2}{4}(3n_1^2 - n_2^2)(3n_2^2 - n_1^2), \\ &(\theta^2)_t = \frac{1}{2}(1 - n_3^2)(1 + 3n_3^2) = -2\bar{\theta}_t, \\ &[\boldsymbol{n},\ \theta\boldsymbol{n}]\,\theta^2\boldsymbol{n} = \frac{n_1}{8}(n_1^2 - 3n_2^2)\left[9n_3^2(1 - n_3^2)(1 - 3n_3^2) - n_2^2(3n_1^2 - n_2^2)^2\right]. \end{aligned}\right\} \tag{40.4}$$

We may find the Λ tensor for the transversely isotropic medium most similar to a trigonal crystal simply by considering θ; (34.8), (34.10), and (40.4) show that all the d_s and g_s become zero, so the Λ_h of (40.2) is the closest on average to Λ, and

$$\Lambda'^{t} = c\theta, \quad (\Lambda'^{t2})_t = c^2(\theta^2)_t = \frac{1}{2}\,c^2(1 - n_3^2)(1 + 3n_3^2), \tag{40.5}$$

while the transverse anisotropy is

$$\langle F^t \rangle = \langle \Lambda'^{t2} \rangle_t = 2c^2/15. \tag{40.6}$$

Comparison of θ with such a medium for a fixed direction of the wave normal gives us on the basis of (35.11) and (40.4) that

$$a_1' = \frac{n_2(3n_1^2 - n_2^2)}{2n_3(1 - n_3^2)},$$

$$a_0' = -(1 - 3n_3^2)\,a_1', \quad a_2' = (1 - 4n_3^2)\,a_1', \quad a_3' = (1 - 5n_3^2)\,a_1'. \tag{40.7}$$

Then

$$\Lambda' = c\{\theta - a_1'[(3n_3^2 - 1) + n\cdot n + $$
$$+ (1 - 4n_3^2)e\cdot e + (1 - 5n_3^2)c\cdot c]\}, \tag{40.8}$$

$$w = \Lambda' c = c(\theta c + 2n_3^2 a_1' c). \tag{40.9}$$

Coefficients a_0, a_1, a_2, and a_3 are, respectively,

$$\left.\begin{aligned} a_0 &= c_0 - c(1 - 3n_3^2)a_1', \quad a_1 = c_1 + ca_1', \\ a_2 &= c_2 + c(1 - 4n_3^2)a_1', \quad a_3 = c_3 + c(1 - 5n_3^2)a_1'. \end{aligned}\right\} \tag{40.10}$$

The special directions are considered via (39.11). It is easily seen that

$$\left.\begin{aligned} wn &= n\Lambda c = cn\theta c = \frac{3}{2}cn_1n_3\frac{3n_2^2 - n_1^2}{|[ne]|}, \\ we &= e\Lambda c = ce\theta c = \frac{1}{2}cn_1\frac{3n_2^2 - n_1^2}{|[ne]|}, \\ w[nc] &= c\Lambda[nc] = cc\theta[nc] = \frac{c}{2}\frac{n_1(3n_3^2 - 1)(3n_2^2 - 1)}{1 - n_3^2}. \end{aligned}\right\} \tag{40.11}$$

Substitution of (40.7), (40.10), and (40.11) into (39.11) gives us the condition for the special directions of a trigonal crystal as

$$n_1(3n_2^2 - n_1^2)\Big\{n_3^2c_2(3c_3 - 2c_2) + \frac{1}{2}n_2n_3(3n_1^2 - n_2^2)c[3g_5(1 - 4n_3^2) - c_2] + $$
$$+ \frac{c^2}{4}[n_2^2(3n_1^2 - n_2^2)^2 - 9n_3^2(1 - n_3^2)(1 - 3n_3^2)]\Big\} = 0. \tag{40.12}$$

The case $n_3 = 0$ requires special consideration, because the expressions for the a_s' of (40.7) becomes infinitely large; but in the case they should be taken as zero, as may be seen by putting $n_3 = 0$ in (40.3), whereupon θ becomes

$$\theta = \theta^0 = \begin{vmatrix} 0 & 0 & n_1n_2 \\ 0 & 0 & \frac{n_1^2 - n_2^2}{2} \\ n_1n_2 & \frac{n_1^2 - n_2^2}{2} & 0 \end{vmatrix}. \tag{40.13}$$

It is readily seen that $\theta_t^0 = \mathbf{n}\theta^0\mathbf{n} = \mathbf{e}\theta^0\mathbf{e} = \mathbf{c}\theta^0\mathbf{c} = 0$, so equations (35.8) show that the tensor of an isotropic medium most similar to θ^0 is zero. In other words, $\mathbf{n} \perp \mathbf{e}$ in (40.1) implies that the Λ_h of (40.2) will be Λ^0, so $\Lambda' = c\theta^0$. It should not be though that, since the general expressions of (40.7) increase without limit for $n_3 \to 0$, there is no continuous transition from $n_3 \neq 0$ to $n_3 = 0$, for the second term in the braces in (40.8) for $n_3 \to 0$ becomes $a_1'(-1 + \mathbf{n} \cdot \mathbf{n} + \mathbf{e} \cdot \mathbf{e} + \mathbf{c} \cdot \mathbf{c})$, and, although $a_1' \to \infty$, the tensor $(\mathbf{n} \cdot \mathbf{n} + \mathbf{e} \cdot \mathbf{e} + \mathbf{c} \cdot \mathbf{c} - 1)$ tends to zero for $\mathbf{ne} = 0$. The result, as above, is $\Lambda' = c\theta^0$. In addition, it is simple to consider the transition to $n_3 = 0$ in (40.7)-(40.9) by reference to (40.1), (40.2), and (40.13). The transversality condition, with the condition that here

$$\Lambda_h \boldsymbol{n} = (c_0 + c_1)\boldsymbol{n}, \quad \theta^0 \boldsymbol{n} = \frac{n_2}{2}(3n_1^2 - n_2^2)\boldsymbol{e}, \tag{40.14}$$

becomes

$$\frac{1}{8} c^3 n_1 n_2^2 (3n_2^2 - n_1^2)(3n_1^2 - n_2^2)^2 = 0, \tag{40.15}$$

i.e., coincides with (40.12) for $n_3 = 0$, so (40.12) is the entirely general transversality condition. This implies that purely transverse waves propagate in a trigonal crystal in the coordinate plane $n_1 = 0$ and in the two planes for which $3n_2^2 - n_1^2 = 0$, i.e.,

$$\frac{n_2}{n_1} = \tan\varphi = \pm\frac{1}{\sqrt{3}}. \tag{40.16}$$

These planes form angles of $\pm\pi/3$ with the plane $n_1 = 0$; the three planes together pass through $L^3 \| \mathbf{e} \| x_3$ and lie mutually at 120°, in accordance with the symmetry of the crystal.

Section 9 shows that tensor Λ for trigonal crystals (classes 18-20 of Table I) takes the form of (19.19) or (40.1)-(40.3), respectively, when a coordinate axis lies along the L^2 axis (or along the normal to a symmetry plane). It is therefore completely natural (see section 16) to find the transversality condition satisfied when $n_1 = 0$, i.e., when the wave normal lies in a symmetry plane. Trigonal crystals of classes 16 and 17 of Table I give a Λ tensor and an elastic-modulus matrix exactly as for the more symmetrical trigonal classes when the x_1 axis is chosen in accordance with the second condition of (9.20). This matrix entirely defines the elastic properties, so all properties of all trigonal crystals are identical along the appropriate directions. From this we may conclude that

Table VI. Elastic Constants of Trigonal Crystals

(The c are in 10^{11} dynes/cm^2; the g are in 10^{11} cm^2/sec^2)

No.	Crystal	c_{11}	c_{33}	c_{44}	c_{66}	c_{13}	c_{14}	g_1	g_2	g_3	g_4	g_5	c
1	Aluminum phosphate	10.5	13.4	2.31	3.785	6.93	-1.27	0.491	0.409	3.60	0.312	0.984	-0.990
2	Antimony	7.916	4.498	2.852	2.717	2.615	1.060	0.3664	0.06029	0.8179	-0.6320	0.04010	0.3172
3	Bismuth	8.497	5.943	1.459	1.765	2.860	-0.573	0.4263	-0.2774	0.4407	0.2943	-0.2462	-0.1169
4	Calcite	13.74	8.01	3.42	4.67	4.50	-2.03	2.15	-0.885	2.92	-0.343	-0.424	-1.50
5	Corundum	46.50	56.30	23.30	17.05	11.70	10.10	2.904	2.980	8.838	-3.485	1.402	5.101
6	Dextrose-NaBr	2.06	2.40	0.634	0.765	0.79	-0.03	0.376	-0.00118	0.843	0.204	0.0763	-0.0355
7	Dextrose-NaCl	2.20	1.77	0.771	0.555	0.75	-0.03	0.434	0.0588	0.973	-0.393	-0.0793	-0.0384
8	Dextrose-NaI	2.58	2.06	0.771	0.53	0.49	-0.03	0.708	-0.294	0.677	0.309	-0.423	-0.0322
9	Hematite	24.2	22.8	8.50	9.35	1.60	-1.3	2.69	-1.07	1.93	1.87	-0.906	-0.496
10	Mercury	3.600	5.051	1.290	0.356	3.030	0.470	-0.5315	0.1484	0.3189	-0.1897	0.07943	0.06939
11	α-Quartz	8.76	10.68	5.72	4.08	1.33	1.73	0.737	1.73	3.04	-2.63	1.02	1.49
12	Tourmaline	27.2	16.5	6.5	11.6	3.5	-0.68	5.55	-3.45	3.23	3.45	-1.81	-0.439
13	Sodium nitrate	8.67	3.74	2.13	3.52	1.60	0.82	2.19	-1.25	1.65	0.306	-0.629	0.727

purely longitudinal waves will propagate along the x_1 axis in classes 16 and 17, since this occurs in classes 18-20. Moreover, this property (longitudinality in a plane perpendicular to L^3) may serve in the low-symmetry classes to define the direction of the x_1 axis [12]. These arguments apply also to tetragonal crystals of the lower symmetry classes.

Then the general condition of (40.12) for the special directions may be formulated as follows for a trigonal crystal: the cone of special directions in every class splits up into three planes that pass through the L^3 axis and that are perpendicular to longitudinal normals lying in the plane $n_3 = 0$, together with a cone of sixth degree (with L^3 as axis):

$$n_3^2 c_2 (3c_3 - 2c_2) + \frac{1}{2} n_2 n_3 (3n_1^2 - n_2^2) c \left[3g_5 (1 - 4n_3^2) - c_2\right] + \\ + \frac{1}{4} c^2 \left[n_2^2 (3n_1^2 - n_2^2)^2 - 9n_3^2 (1 - n_3^2)(1 - 3n_3^2)\right] = 0. \tag{40.17}$$

The general conditions (39.16) and (39.17) for longitudinal normals give [see (40.7), (40.10), and (40.11)] for a trigonal crystal that

$$n_1 n_3 (3n_2^2 - n_1^2) = 0, \tag{40.18}$$

$$2c_2 n_3 (1 - n_3^2) + c n_2 (3n_1^2 - n_2^2)(1 - 4n_3^2) = 0. \tag{40.19}$$

For $n_1 = 0$, $n_2^2 = 1 - n_3^2$ the last equation becomes

$$2c_2 n_3 - c n_2 (1 - 4n_3^2) = 0.$$

If $n_1^2 = 3n_2^2$, $4n_2^2 = 1 - n_3^2$, then (40.18) implies that

$$2c_2 n_3 + 2c n_2 (1 - 4n_3^2) = 0.$$

Now $2n_2 = \sqrt{1 - n_3^2}$, so in both cases we get the same equation

$$4 (g_2 + g_4 n_3^2)^2 n_3^2 - c^2 (1 - n_3^2)(1 - 4n_3^2)^2 = 0. \tag{40.20}$$

This is what would be expected from the presence of the L^3 axis. Equation (40.20) is cubic in n_3^2 and has one or three real roots, which correspond to three or nine longitudinal normals. Moreover, the case $n_3 = 0$ requires separate consideration; from (40.14)

$$[\boldsymbol{n}, \Lambda\boldsymbol{n}] = c\,[\boldsymbol{n}, \theta^0\boldsymbol{n}] = \frac{c}{2}\,n_2(3n_1^2 - n_2^2)\,[\boldsymbol{ne}], \tag{40.21}$$

so three further longitudinal normals lie in the equatorial plane:

$$n_2 = 0, \quad n_2 = \pm\sqrt{3}\,n_1, \tag{40.22}$$

the directions of these not being given by (40.18) and (40.19) for $n_3 = 0$. It is readily seen that (40.18) and (40.19) correspond to the intersection of the planes $n_1 = 0$, $n_1^2 = 3n_2^2$ with the cone of (40.17); the three longitudinal normals in the plane $\mathbf{ne} = 0$ of (40.22) also lie in the cone of (40.17), but they are not lines of intersection with other sheets of the general cone of special directions of (40.12). Hence not all the longitudinal normals are the result of intersection of different sheets of the cone $[\mathbf{n}, \Lambda\mathbf{n}]\,\Lambda^2\mathbf{n} = 0$.

From (39.24)-(39.32) we can readily consider the number and orientation of the acoustic axes in a trigonal crystal by analogy with the treatment (section 38) for tetragonal crystals. It is found [18] that there may be four or ten acoustic axes, the values of the elastic parameters determining which number applies.

Chapter 8

Reflection and Refraction of Elastic Waves

41. Boundary Conditions for Plane Elastic Waves

So far we have considered plane elastic waves in a homogeneous unbounded crystal, which is an idealized case, because we virtually always have waves in a crystal of restricted dimensions. This means that we have to consider the behavior of elastic waves at the surfaces bounding a crystal, which is much more complicated than the problem of waves in an unbounded homogeneous crystalline medium. For this reason we shall consider only the behavior of plane waves at an infinite plane boundary, on both sides of which lie unbounded homogeneous media differing in density and elastic properties.

This is a boundary problem in the theory of elastic waves, and the boundary conditions play the main part in solving it. These conditions are formulated in different ways in accordance with the properties of the media and of the contact between them. For instance, the two may be rigidly coupled at the interface, so the displacement vector must then be continuous from one medium to the other. Let **q** be unit vector normal to the plane of the interface and directed from the first medium into the second. Then, if the origin of the coordinate system is located on that plane, the equation of the latter in vector form is

$$\boldsymbol{rq} = 0, \tag{41.1}$$

in which **r** is the radius vector. The condition of continuity of the displacement vector may be put as

$$\boldsymbol{u}^{\mathrm{I}} - \boldsymbol{u}^{\mathrm{II}} = 0 \qquad (\boldsymbol{rq} = 0). \tag{41.2}$$

in which $\mathbf{u}^{\mathrm{I}}$ and $\mathbf{u}^{\mathrm{II}}$ are the displacement vectors of points in the two media at the interface.

The stresses at the interface must also be continuous. This condition may be derived from the $\sigma \mathbf{n} = \mathbf{P}$ of (3.11), as suitably modified. In place of σ we put σ^{I} (the stress tensor for the first medium), while $\mathbf{n}$ is replaced by $\mathbf{q}$, the normal to the interface. The density P of the external surface forces is here replaced by the stresses exerted on the first medium by the second, i.e., $\sigma^{\mathrm{II}}\mathbf{q}$. The second boundary condition is therefore

$$(\sigma^{\mathrm{I}} - \sigma^{\mathrm{II}})\,\boldsymbol{q} = 0 \qquad (\boldsymbol{rq} = 0). \tag{41.3}$$

Thus we have the boundary conditions of (41.2) and (41.3) when the two solids are in rigid contact, and these split up into six equations in terms of the components.

The boundary conditions take a different form for the free surface of a solid. By free is meant absence of contact with any medium whose elasticity and inertia could in any way influence the motion of the surface, so such a surface is, strictly speaking, one with a vacuum, or at least a gas at low pressure. Further, the term free surface implies freedom from external forces. The condition of (41.2) drops out for a free surface, because there is no second medium, and hence no conditions related to the displacement of the surface of the first medium. The absence of external surface forces also means that $\mathbf{P} = \sigma^{\mathrm{II}}\mathbf{q} = 0$, so there is left only the vector boundary condition

$$\sigma \boldsymbol{q} = 0 \qquad (\boldsymbol{rq} = 0), \tag{41.4}$$

which is equivalent to three equations in terms of components.

A feature common to all the boundary conditions of (41.2)-(41.4) is linearity and homogeneity in all the quantities representing deformations ($\mathbf{u}^{\mathrm{I}}$ and $\mathbf{u}^{\mathrm{II}}$) and stresses (σ^{I} and σ^{II}). As regards plane waves, these general features are sufficient to give us the basic laws of reflection and refraction. From (15.2), (15.3), (15.5), (15.6), (15.8), (15.12), and (21.32) we have

$$\left.\begin{aligned} \boldsymbol{u} &= \boldsymbol{u}^0 e^{i\varphi} = \boldsymbol{u}^0 e^{i\omega(\boldsymbol{mr} - t)}, \\ \sigma_{ij} &= c_{ijkl}\frac{\partial u_l}{\partial x_k} = i\omega c_{ijkl} m_k u_l^0 e^{i\omega(\boldsymbol{mr} - t)}, \end{aligned}\right\} \tag{41.5}$$

or

$$\sigma = \sigma^0 e^{i\varphi}, \quad \sigma^0_{ij} = i\omega c_{ijkl} m_k u^0_l. \tag{41.6}$$

We will apply (41.2)-(41.4) to the case where the deformations and stresses in both media consist of plane waves of the form of (41.5) and (41.6). Those in the first medium are denoted by

$$\left.\begin{aligned} u'_{(k)} &= u^{0'}_{(k)} e^{i\varphi'_k}, \\ \sigma'_{(k)} &= \sigma^{0'}_{(k)} e^{i\varphi'_k}, \\ \varphi'_k &= \omega'_k (m'_{(k)} r - t), \quad k = 1, 2, \ldots, k' \end{aligned}\right\} \tag{41.7}$$

while those in the second medium are

$$\left.\begin{aligned} u''_{(k)} &= u^{0''}_{(k)} e^{i\varphi''_{(k)}}, \\ \sigma''_{(k)} &= \sigma^{0''}_{(k)} e^{i\varphi''_{(k)}}, \\ \varphi''_{(k)} &= \omega''_k (m''_{(k)} r - t), \quad k = 1, 2, \ldots, k''. \end{aligned}\right\} \tag{41.8}$$

Conditions (41.2) and (41.3) then become

$$\sum_{k=1}^{k'} u^{0'}_{(k)} e^{i\varphi'_k} - \sum_{k=1}^{k''} u^{0''}_{(k)} e^{i\varphi''_k} = 0, \tag{41.9}$$

$$\left(\sum_{k=1}^{k'} \sigma^{0'}_{(k)} e^{i\varphi'_{(k)}} - \sum_{k=1}^{k''} \sigma^{0''}_{(k)} e^{i\varphi''_{(k)}}\right) q = 0. \tag{41.10}$$

The expressions attached to the phase factors $e^{i\varphi'_k}$, $e^{i\varphi''_k}$ are independent of coordinates and time, so (41.9) and (41.10) are linear relations between exponential functions, which should be identically satisfied for any instant and for all points on the interface. Exponential functions with different exponents are linearly independent, so (41.9) and (41.10) can apply only when all the phases φ'_k, φ''_k coincide at the interface:

$$\varphi'_1 = \varphi'_2 = \ldots = \varphi'_{k'} = \varphi''_1 = \varphi''_2 = \ldots = \varphi''_{k''} \quad (qr = 0) \tag{41.11}$$

or

$$\omega'_{k_1} (m'_{(k_1)} r - t) = \omega''_{k_2} (m''_{(k_2)} r - t). \tag{41.12}$$

no matter what k_1 and k_2 may be. Equation (41.12) applies for any t, so

$$\omega_1' = \omega_2' = \ldots = \omega_1'' = \omega_2'' = \ldots = \omega. \quad (41.13)$$

Thus all the waves in (41.9) and (41.10) must have the same frequency. Conditions (41.12) reduce to the equations

$$m'_{(1)} r = \ldots = m'_{(k')} r = m''_{(1)} r = \ldots = m''_{(k'')} r. \quad (41.14)$$

These may be put as

$$(m'_{(k_1)} - m'_{(k_2)}) r = (m'_{(k_1)} - m''_{(k_2)}) r = (m''_{(k_1)} - m''_{(k_2)}) r = 0$$

for all k_1 and k_2, with $\mathbf{qr} = 0$. Hence any radius vector perpendicular to **q** must be perpendicular to the difference of any two refraction vectors of (41.9) and (41.10), which is possible only if all these differences are parallel to **q**:

$$m'_{(k_1)} - m'_{(k_2)} \parallel m'_{(k_1)} - m''_{(k_2)} \parallel m''_{(k_1)} - m''_{(k_2)} \parallel q, \quad (41.15)$$

or, which is the same,

$$[m'_{(k_1)} - m'_{(k_2)}, q] = [m'_{(k_1)} - m''_{(k_2)}, q] = [m''_{(k_1)} - m''_{(k_2)}, q] = 0, \quad (41.16)$$

or, finally,

$$[m'_{(1)} q] = \ldots = [m'_{(k')} q] = [m''_{(1)} q] = \ldots = [m''_{(k'')} q]. \quad (41.17)$$

Equations (41.17) express the general laws of reflection and refraction for plane elastic waves at the interface between any two media in the most general covariant vector form. They imply that the vector product of the refraction vector of any wave by the normal **q** to the interface must equal any other such product. Let **a** be a vector equal to any of the vector products of (41.17):

$$a = [m'_{(k_1)} q] = [m''_{(k_2)} q]. \quad (41.18)$$

Then all the vectors **q**, $\mathbf{m}'_{(1)}, \ldots, \mathbf{m}'_{(k')}, \mathbf{m}''_{(1)}, \ldots, \mathbf{m}''_{(k'')}$ will be perpendicular to **a**, so they will all lie in a single plane (the plane of incidence), whose equation is

$$ar = 0. \tag{41.19}$$

This is one of the basic laws of reflection and refraction for elastic (and electromagnetic) waves at the interface between any two media.

Comparison of the absolute values of the vector products of (41.17) shows that, with $|m| = 1/v$,

$$\frac{\sin\psi_1'}{v_1'} = \frac{\sin\psi_2'}{v_2'} = \ldots = \frac{\sin\psi_{k'}'}{v_{k'}'} = \frac{\sin\psi_1''}{v_1''} = \ldots = \frac{\sin\psi_{k''}''}{v_{k''}''}. \tag{41.20}$$

Here $\psi_1', \ldots, \psi_{k''}''$ are the angles formed by the refraction vectors of the corresponding waves with the normals to the surface ($\cos\psi = \mathbf{nq}$); the sines of these angles are proportional to the corresponding phase velocities. This is the second basic law of reflection and refraction, which applies also to electromagnetic waves.

We form the vector product of $\mathbf{q}$ and equation (41.18):

$$[q[mq]] = m - mq \cdot q = [qa]; \tag{41.21}$$

in which $\mathbf{m}$ is any of the vectors $\mathbf{m}_{(k_1)}'$, $\mathbf{m}_{(k_2)}''$. We put

$$mq = \xi, \quad [qa] = [q[mq]] = b, \tag{41.22}$$

which shows that ξ is the projection of the refraction vector on the normal to the interface, while $\mathbf{b}$ is the projection of this vector on the interface plane. From (41.21),

$$m_{(k_1)}' = b + \xi_{k_1}' q, \quad m_{(k_2)}'' = b + \xi_{k_2}'' q. \tag{41.23}$$

Thus all waves occurring at the boundary between two media have their $\mathbf{b}$ vectorially equal. This is equivalent to both of the laws of reflection and refraction formulated above.

From (41.23) we have that all the refraction vectors differ only in their projections on the normal to the interface, so a straight line through the end of $\mathbf{b}$ parallel to $\mathbf{q}$ will be the locus of the ends of all the $\mathbf{m}$ vectors if these are laid off from a single point. This feature allows us to give a simple geometrical definition of the directions of the refraction vectors for all waves if the common projection **b** on the interface is known. Here we must remember that by definition (section 24) the refraction surface of (24.6) is the locus of the ends of the **m** vectors, so the end of each **m** vector must lie on a straight line parallel to **q** and also on one of the three sheets of the appropriate refraction surface. Hence to find $\mathbf{m}'_{(k)}$, $\mathbf{m}''_{(k)}$ we should construct curves for the intersection of the refraction surfaces in the two media in which lie all the **m** vectors (Fig. 14). The plane of incidence is defined by **q** and **b**, which must be known. We draw AB parallel to **q** through the end of **b** to find the ends of the **m** vectors as the points of intersection with all of the sheets of the refraction surfaces in the two media (L', T', L'', T''_1 T''_2, Fig. 14).

The problem is solved analytically as follows. Into the equation of (24.6) for the refraction surface,

$$|\Lambda^m - 1| = |\Lambda^m| - \overline{\Lambda}_t^m + \Lambda_t^m - 1 = 0 \qquad (41.24)$$

we substitute the general expression of (41.23) for the refraction vector:

$$\boldsymbol{m} = \boldsymbol{b} + \xi \boldsymbol{q}, \qquad (41.25)$$

in which parameter ξ remains undetermined, while **q** and **b** are given. Since (41.24) is of sixth degree in **m**, we get a complete equation of sixth degree for ξ. First we assume that all six roots of this equation are real,* which means that AB (Fig. 14) actually does intersect all sheets of the refraction surface above and below the interface. It is clear from geometry that the points of intersection lying above the interface (first medium) have $\xi < 0$, while those below (second medium) have $\xi > 0$. Then the three negative roots ξ of (41.24) and (41.25) should be taken for the first medium, the three positive roots being taken for the second. However, the problem is really more complicated, because the wave energy actually propagates along the rays, whose directions may differ substantially from those of the corresponding phase normals, which are parallel to the refraction vectors. In principle, this allows the refraction vectors to be directed into the second medium while the

---

* A definite physical meaning also attaches to complex ξ in (41.24) and (41.25), as we shall see (section 44).

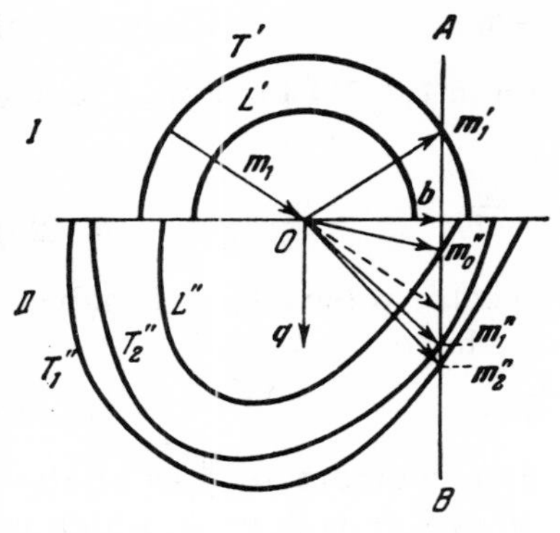

Fig. 14

wave propagates in the first, and vice versa. Such cases are rare, but they are possible for crystalline media with special orientations. We will assume that we have the simpler case where $\xi < 0$ corresponds to a wave in the first medium and $\xi > 0$ to one in the second.

Then for a given **b** the boundary conditions in the general case imply that six distinct plane elastic waves can occur near the interface: three in the first medium and three in the second. Stress is placed on the words for a given **b**, because this restriction is extremely important, since the boundary problem for plane waves can be posed in different ways. The usual approach is that a wave is given in the first medium as coming from infinity (the incident wave); only this wave would be present if there were no boundary, which gives rise to additional waves in both media, which in the first medium are termed reflected, while in the second they are termed refracted. The problem is to find the directions of propagation and the velocities (the refraction vectors) and also the vector amplitudes of the reflected and refracted waves, these quantities for the incident wave being given. The incident wave at the boundary resembles the others in that it must satisfy the general boundary conditions of (41.2)-(41.4), so its refraction vector, denoted by **m**, also must be expressed in the form of (41.25). Knowing **m**, we find from (41.18) and (41.22) that

$$\boldsymbol{a} = [\boldsymbol{mq}], \tag{41.26}$$

$$\boldsymbol{b} = [\boldsymbol{q}[\boldsymbol{mq}]]. \tag{41.27}$$

This defines the plane of incidence (i.e., **a**) and also **b**, so (41.24) and (41.25) give the displacement vectors of the three reflected and three refracted waves. The refraction vectors give us the velocities and directions of these waves, but the amplitudes remain to be determined. Here we use Christoffel's equation in the form of (24.15),

$$(\Lambda^m - 1)\,u = 0. \tag{41.28}$$

Now **m** satisfies (41.24), so the solution to (41.28) for **u** may [compare (17.27) and (24.21)] be put as

$$u \cdot u = A(\overline{\Lambda^m - 1}). \tag{41.29}$$

Multiplication of this by an arbitrary vector **p** (**up** $\neq 0$) and including **up** in the undetermined factor A, we have

$$u = A(\overline{\Lambda^m - 1})\,p. \tag{41.30}$$

Thus the refraction vectors give the displacement vectors of all six waves apart from the undetermined scalar factors A, which may be found by substituting (41.30) into (41.9) and (41.10), which now may be put as

$$u_0 + u_1 + u_2 + u_0' + u_1' + u_2' - (u_0'' + u_1'' + u_2'') = 0, \tag{41.31}$$

$$(\sigma_0 + \sigma_1 + \sigma_2 + \sigma_0' + \sigma_1' + \sigma_2' - (\sigma_0'' + \sigma_1'' + \sigma_2''))\,q = 0, \tag{41.32}$$

in which ($\mathbf{u}_0$, $\mathbf{u}_1$, $\mathbf{u}_2$), (σ_0, σ_1, σ_2) are, respectively, the vector amplitudes of the displacement and the tensor amplitudes of the stress of (41.6) for the incident waves (quasilongitudinal and quasitransverse). A single (double) prime denotes a reflected (refracted) wave. This gives six linear equations, which contain the six A of (41.30) as well as the given displacement vector of the incident wave. These are solved for A_0', A_1', A_2', A_0'', A_1'', A_2'' to get all the parameters of the reflected and refracted waves.

This result relates to the general case of rigid contact between two solids; the case of (41.4) (free boundary) differs only in that the second medium has no waves, so the two systems of (41.24) and (41.25) are replaced by one for the first medium only. There are only three reflected waves for a given incident wave, whose amplitude factors are given by the three equations

$$(\sigma_0 + \sigma_1 + \sigma_2 + \sigma_0' + \sigma_1' + \sigma_2')\,q = 0. \tag{41.33}$$

The problem is thereby much simplified, although the general mode of solution is as before.

The above relates to the usual approach (incident wave given, reflected and refracted waves to be found), but a different approach is possible, in which we do not specify any wave in advance but seek to determine the possible sets of plane waves on the two sides of an interface that are allowed by (41.28) together with (41.2) and (41.3) or (41.4). Here **b** is unknown as well as the ξ of (41.25). This is a more general formulation, but also a more complicated one, and the mode of solution is also different. The approach allows us to establish the occurrence of new types of wave essentially related to the boundary, which cannot propagate in an unbounded medium (Rayleigh surface waves). In section 46 we consider in more detail the boundary problem in this approach.

42. Reflection of Elastic Waves at the Free Boundary of an Isotropic Medium

Here we consider the simplest possible boundary problem for elastic waves. The refraction vector of the incident wave is as in (41.25); it must satisfy (41.24), which for an isotropic medium takes the form of (24.7'):

$$((a+b)\,\boldsymbol{m}^2-1)(a\boldsymbol{m}^2-1)^2=0, \tag{42.1}$$

whence we have for longitudinal waves

$$\boldsymbol{m}_0^2=\frac{1}{a+b} \tag{42.2}$$

and for transverse waves

$$\boldsymbol{m}_1^2=\frac{1}{a}. \tag{42.3}$$

The corresponding displacement vectors must satisfy (24.7):

$$(\Lambda^m-1)\,\boldsymbol{u}=(a\boldsymbol{m}^2+b\boldsymbol{m}\cdot\boldsymbol{m}-1)\,\boldsymbol{u}=0. \tag{42.4}$$

For a longitudinal wave ($\mathbf{u}=\mathbf{u}_0$)

$$a\boldsymbol{m}_0^2-1=\frac{a}{a+b}-1=-b\boldsymbol{m}_0^2, \tag{42.5}$$

so from (42.4)

$$b(\boldsymbol{m}_0 \cdot \boldsymbol{m}_0 - \boldsymbol{m}_0^2)\,\boldsymbol{u}_0 = b[\boldsymbol{m}_0[\boldsymbol{m}_0\boldsymbol{u}_0]] = 0, \tag{42.6}$$

whence

$$\boldsymbol{u}_0 = C_0\boldsymbol{m}_0. \tag{42.7}$$

(42.3) and (42.4) give for a transverse wave that

$$\boldsymbol{u}_1\boldsymbol{m}_1 = 0. \tag{42.8}$$

From (41.23)

$$\boldsymbol{m}_0 = \boldsymbol{b} + \xi_0\boldsymbol{q}, \quad \boldsymbol{m}_1 = \boldsymbol{b} + \xi_1\boldsymbol{q} \tag{42.9}$$

and $\mathbf{u}_1$ may be put as a linear combination of two vectors orthogonal to $\mathbf{m}_1$: $\mathbf{a} = [\mathbf{bq}]$ and $(\xi_1\mathbf{b} - \mathbf{a}^2\mathbf{q})$, with arbitrary coefficients C_1 and C_2:

$$\boldsymbol{u}_1 = C_1\boldsymbol{a} + C_2(\xi_1\boldsymbol{b} - \boldsymbol{a}^2\boldsymbol{q}). \tag{42.10}$$

The first of the two components of $\mathbf{u}_1$, namely $\mathbf{a}$, is perpendicular to the plane of incidence, while the second is parallel to it. Instead of $\xi_1\mathbf{b} - \mathbf{a}^2\mathbf{q}$ we may take also $[\mathbf{m}_1\mathbf{a}]$. The boundary conditions of (41.33) are put in expanded form via (41.6); replacement of the c_{ijkl} by λ_{ijkl} and elimination of the common factor $i\omega$ gives

$$\lambda_{ijkl}(m_k u_l + m'_{0k}u'_{0l} + m'_{1k}u'_{1l} + m'_{2k}u'_{2l})\,q_j = 0. \tag{42.11}$$

This is the most general boundary condition for plane elastic waves at the free surface of any homogeneous medium. Here $\mathbf{m}$ and $\mathbf{u}$ are the vectors of the incident wave, $\mathbf{m}'_0$ and $\mathbf{u}'_0$ are those for the reflected quasilongitudinal wave, and $\mathbf{m}'_1$, $\mathbf{u}'_1$, $\mathbf{m}'_2$, $\mathbf{u}'_2$ are those for the reflected quasitransverse waves. This condition could be generalized formally even further if $m_k u_l$ were replaced by the sum of several such expressions for several incident waves; but the linearity of (42.11) implies that the problem would reduce to solving for each wave separately.

We apply (42.11) to the particular case of an isotropic medium. First we determine $\mathbf{m}'_0$, $\mathbf{m}'_1$, $\mathbf{m}'_2$; from section 41, the ends of these vectors must lie at the points where sheets L and T of (42.2) and (42.3) meet a straight line parallel to $\mathbf{q}$ drawn through

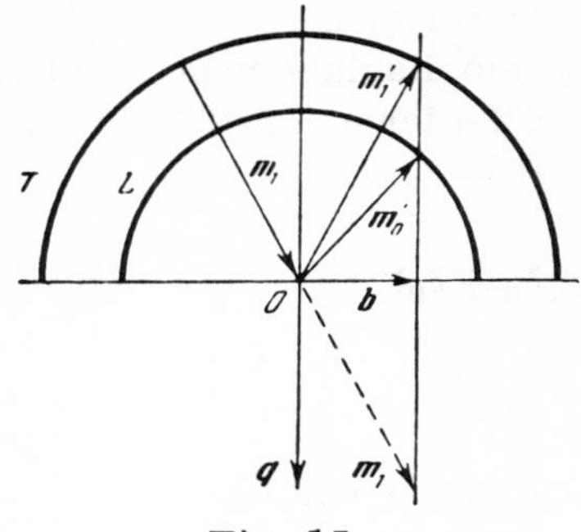

Fig. 15

the end of the vector $\mathbf{m}_1$ that is reflected at 0 (Fig. 15). The $\mathbf{m}_1$ of Fig. 15 is the refraction vector of a transverse wave, because it forms the radius of the larger sphere representing the transverse-wave sheet of (42.3) from the refraction surface. This gives us the vectors $\mathbf{m}_0'$ and $\mathbf{m}_1'$ for the reflected waves, to which correspond three independent displacement vectors: one for the longitudinal wave of (42.7) and two for the transverse wave of (42.10). An analytic solution for $\mathbf{m}_0$, and $\mathbf{m}_1$ requires joint solution of (42.2) and (42.3), respectively, with

$$m_0' = b + \xi_0' q, \quad m_1' = b + \xi_1' q. \tag{42.12}$$

Substitution of (42.12) into (42.2) and (42.3), with $\mathbf{b}^2 = \mathbf{a}^2 = [\mathbf{mq}]^2$, gives

$$\xi_0' = -\sqrt{\frac{1}{a+b} - a^2}. \tag{42.13}$$

$$\xi_1' = -\sqrt{\frac{1}{a} - a^2}. \tag{42.14}$$

The negative signs to the roots correspond to $\mathbf{m}_0'$ and $\mathbf{m}_1'$ directed upwards from the interface. Figure 15 shows that the normal of one of these waves (that of the same type as the incident wave) forms the same angle with the interface as does the normal of the incident wave. This corresponds to equality of the angles of incidence and reflection (one of the basic laws of reflection), but it is correct only when the first medium is isotropic, and then only for the reflected wave of the same type as the incident one. From Fig. 15 and (42.2), (42.3), (42.13), and (42.14) we have

$$\xi_0' = -\xi_0, \quad \xi_1' = -\xi_1, \tag{42.15}$$

in which ξ_0 and ξ_1 relate to incident longitudinal and transverse waves.

Formulas (42.12)-(42.14) give the refraction vectors of the reflected waves, so we have only to find the displacement vectors. Here we use (42.11), with substitution of the tensor of (26.31) for the λ_{ijkl} of an isotropic medium:

$$\lambda_{ijkl} = (b-a)\delta_{ij}\delta_{kl} + a(\delta_{ik}\delta_{jl} + \delta_{il}\delta_{jk}), \tag{42.16}$$

whence we have

$$\lambda_{ijkl} m_k u_l q_j = (b-a) m_k u_k q_l + a(u_k q_k m_l + m_k q_k u_l),$$

or

$$\lambda_{ijkl} m_k u_l q_j = [(b-a) mu \cdot q + a(u \cdot m + m \cdot u) q]_l. \tag{42.17}$$

The refraction vectors of the two reflected transverse waves coincide for an isotropic medium, so we may combine the last two terms in (42.11). Consider first the case of an incident longitudinal wave. For the displacements of the reflected waves, by analogy with (42.7) and (42.10), we put

$$u_0' = C_0' m_0', \quad u_1' = C_1' a - C_2'(a^2 q - \xi_1' b) \quad (u_1' m_1' = 0). \tag{42.18}$$

Substitution of (42.7), (42.17), and (42.18) into (42.11) gives

$$(b-a)(C_0 m_0^2 + C_0' m_0'^2) q +$$
$$+ a(2C_0 m_0 \cdot m_0 + 2C_0' m_0' \cdot m_0' + m_1' \cdot u_1' + u_1' \cdot m_1') q = 0, \tag{42.19}$$

or [see (42.2), (42.3), (42.12) and (42.15)]

$$\frac{b-a}{b+a}(C_0 + C_0') q + 2a[\xi_0 (C_0 - C_0') b + \xi_0^2 (C_0 + C_0') q] +$$
$$+ a[\xi_1' C_1' a - C_2'(a^2 - \xi_1'^2) b - 2\xi_1' a^2 C_2' q] = 0. \tag{42.20}$$

Vectors **a**, **q**, and **b** are mutually orthogonal in pairs and so linearly independent, so the coefficients to these must become zero individually. This gives us three equations:

$$\xi_1' C_1' = 0, \tag{42.21}$$

$$\left(\frac{b-a}{b+a} + 2a\xi_0^2\right)(C_0 + C_0') - 2a\xi_1' a^2 C_2' = 0, \tag{42.22}$$

$$2\xi_0(C_0 - C_0') - (a^2 - \xi_1'^2)C_2' = 0. \qquad (42.23)$$

Setting aside the case $\mathbf{a}=0$ or $\xi_0'=0$, we have from (42.21) that $C_1'=0$, so the reflected transverse wave has no displacement component perpendicular to the plane of incidence of (42.18). From (42.13)-(42.15) we have

$$\frac{b-a}{b+a} + 2a\xi_0^2 = a(\xi_1'^2 - a^2) = 1 - 2a\boldsymbol{a}^2. \qquad (42.23')$$

We solve (42.22) and (42.23) for C_0' and C_2' to get

$$C_0' = \frac{4a^2\xi_0\xi_1'\boldsymbol{a}^2 + (1 - 2a\boldsymbol{a}^2)}{4a^2\xi_0\xi_1'\boldsymbol{a}^2 - (1 - 2a\boldsymbol{a}^2)} C_0, \qquad (42.24)$$

$$C_2' = \frac{4a\xi_0(1 - 2a\boldsymbol{a}^2)}{4a^2\xi_0\xi_1'\boldsymbol{a}^2 - (1 - 2a\boldsymbol{a}^2)} C_0. \qquad (42.25)$$

Consider now the reflection of a transverse wave polarized perpendicular to the plane of incidence. Here

$$\boldsymbol{u}_1 = C_1\boldsymbol{a}, \quad \boldsymbol{m}_1 = \boldsymbol{b} + \xi_1\boldsymbol{q}, \quad \xi_1 = \sqrt{\frac{1}{a} - \boldsymbol{a}^2} = -\xi_1'. \qquad (42.26)$$

Condition (42.19) alters only on account of the terms related to the incident wave and becomes

$$\frac{b-a}{b+a}C_0'\boldsymbol{q} + aC_1\xi_1\boldsymbol{a} + a\left[2C_0'\boldsymbol{m}_0'\cdot\boldsymbol{m}_0' + \boldsymbol{m}_1'\cdot\boldsymbol{u}_1' + \boldsymbol{u}_1'\cdot\boldsymbol{m}_1'\right]\boldsymbol{q} = 0. \qquad (42.27)$$

As before, we get

$$\xi_1(C_1 - C_1') = 0, \qquad (42.28)$$

$$\left.\begin{aligned} (1 - 2a\boldsymbol{a}^2)C_0' + 2\xi_1\boldsymbol{a}^2C_2' &= 0, \\ 2a\xi_0'C_0' + (1 - 2a\boldsymbol{a}^2)C_2' &= 0. \end{aligned}\right\} \qquad (42.29)$$

The latter two equations may be solved for C_0' and C_2' different from zero only if the determinant is zero:

$$(1 - 2a\boldsymbol{a}^2)^2 - 4a\xi_0'\xi_1\boldsymbol{a}^2 = 0. \qquad (42.30)$$

But this equation cannot be satisfied for ξ_0 and ξ_1 real, because (42.13) and (42.15) give $\xi_0' < 0$, $\xi_1 > 0$, so both terms on the left are positive; hence $C_0' = C_2' = 0$, and only a transverse wave will be reflected, whose displacement is as for the incident wave [see (42.18) and (42.28)]:

$$u_1' = u_1. \tag{42.31}$$

A transverse incident wave with its displacement vector in the plane of incidence gives

$$u_2 = C_2(\xi_2 b - a^2 q), \quad \xi_2 = \xi_1 = -\xi_1'. \tag{42.32}$$

The boundary conditions of (42.11) now reduce to (42.21) and

$$\left.\begin{aligned} (1 - 2a a^2) C_0' + 2a\xi_2 a^2 C_2' &= 2a\xi_2 a^2 C_2, \\ 2a\xi_0' C_0' + (1 - 2a a^2) C_2' &= -(1 - 2a a^2) C_2, \end{aligned}\right\} \tag{42.33}$$

whence

$$\left.\begin{aligned} C_0' &= \frac{4a\xi_2 a^2 (1 - 2a a^2)}{(1 - 2a a^2)^2 - 4a^2 \xi_0' \xi_2 a^2} C_2, \\ C_2' &= \frac{4a^2 \xi_0' \xi_2 a^2 + (1 - 2a a^2)}{4a^2 \xi_0' \xi_2 a^2 - (1 - 2a a^2)} C_2. \end{aligned}\right\} \tag{42.34}$$

The case of normal incidence on the boundary demands special consideration; here $\mathbf{a} = 0$ for all waves, so the refraction vectors of all waves are parallel to $\mathbf{q}$. Secondly, the plane of incidence becomes a meaningless term, so the displacement of a transverse wave cannot be resolved into components parallel and perpendicular to that plane, so the displacement vectors should be substituted unchanged into the boundary conditions. Then for normal incidence

$$\left.\begin{aligned} & m_0 = -m_0' = \xi_0 q, \quad m_1 = -m_1' = \xi_1 q, \\ & \xi_0 = \frac{1}{v_0} = \sqrt{\frac{1}{a + b}}, \quad \xi_1 = \frac{1}{v_1} = \sqrt{\frac{1}{a}}. \end{aligned}\right\} \tag{42.35}$$

A longitudinal incident wave gives $\mathbf{u}_0 = C_0\mathbf{q}$, $\mathbf{u}_0' = C_0'\mathbf{q}$, $\mathbf{u}_1'\mathbf{q} = 0$, and the general boundary conditions

$$\sum [(b - a)\, mu \cdot q + a(m \cdot u + u \cdot m)\, q] = 0 \tag{42.36}$$

(with summation over all waves) become

$$(a+b)\xi_0(C_0-C_0')q - a\xi_1 u_1' = 0. \tag{42.37}$$

But $\mathbf{q}$ and $\mathbf{u}_1'$ are linearly independent (orthogonal), so both terms on the left are zero, i.e.,

$$u_0' = u_0, \quad u_1' = 0. \tag{42.38}$$

Thus only a longitudinal wave is reflected at a free boundary, the amplitude of the displacement vector remaining unchanged.

From (42.36) we have for normal incidence of a transverse wave that

$$-(a+b)\xi_0' C_0' q + a\xi_1(u_1 - u_1') = 0,$$

whence

$$u_0' = 0, \quad u_1' = u_1, \tag{42.39}$$

i.e., only a transverse wave with the same amplitude for the displacement vector is reflected. The reflected wave differs from the incident wave only in the direction of propagation.

The second special case occurs when $\xi_0' = 0$, which is possible only for a transverse wave with a large angle of incidence ψ_0, which must be such that the perpendicular to the plane from the end of $\mathbf{m}_1$ (Fig. 16) touches the longitudinal wave sheet at A, which lies on the interface. This case is impossible for an incidental longitudinal wave, because $|\mathbf{m}_0|<|\mathbf{m}_1|$. From (42.13) we have

$$\xi_0' = 0, \quad a^2 = \frac{1}{a+b}, \quad m_0' = b, \quad u_0' = C_0' b. \tag{42.40}$$

i.e., a longitudinal wave propagates along the boundary. Here

$$\left.\begin{aligned} \xi_1 &= -\xi_1' = \sqrt{\frac{1}{a} - \frac{1}{a+b}} = \sqrt{\frac{b}{a(a+b)}}, \\ u_1 &= C_1 a + C_2(\xi_1 b - a^2 q), \quad u_1' = C_1' a - C_2'(\xi_1 b + a^2 q). \end{aligned}\right\} \tag{42.41}$$

Substitution of (42.40) and (42.41) into (42.36) gives

$$(b-a)a^2 C_0' q + a\left[\xi_1(u_1 - u_1') - a^2(C_2 m_1 + C_2' m_1')\right] = 0. \tag{42.42}$$

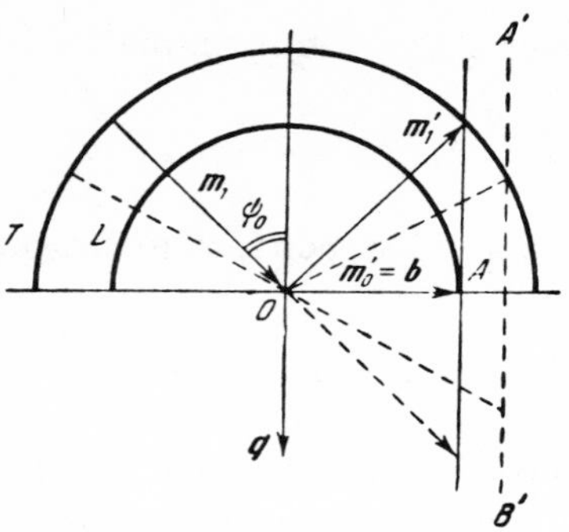

Fig. 16

or

$$a^2((b-a)C_0' - 2\xi_1 a(C_2 - C_2'))q + a\xi_1(C_1 - C_1')a +$$
$$+ a(\xi_1^2 - a^2)(C_2 + C_2')b = 0,$$

whence

$$C_0' = \frac{4a\xi_1}{b-a}C_2, \quad C_1' = C_1, \quad C_2' = -C_2. \tag{42.43}$$

The same result is obtained from (42.28) and (42.34) with $\xi_0 = 0$.

43. Reflection at the Free Boundary of a Crystal

Here we consider in more detail the general case of reflection of plane elastic waves at the free boundary of an anisotropic solid. We saw in section 41 that this problem is treated by first finding the refraction vectors for the reflected waves. We take as known the refraction vector **m** of the incident wave, as well as the vector amplitude **u** of this. Subscript k as usual takes the values 0 (quasilongitudinal), 1, or 2 (quasitransverse in the latter two). We also assume that the $\mathbf{m}_k'$ for the reflected waves have already been found by joint solution of (41.24) and (41.25). Then from Fig. 17

$$m = b + \xi q, \quad b = [q[mq]], \quad \xi = mq, \tag{43.1}$$

$$m_0' = b + \xi_0' q, \quad m_1' = b + \xi_1' q, \quad m_2' = b + \xi_2' q. \tag{43.2}$$

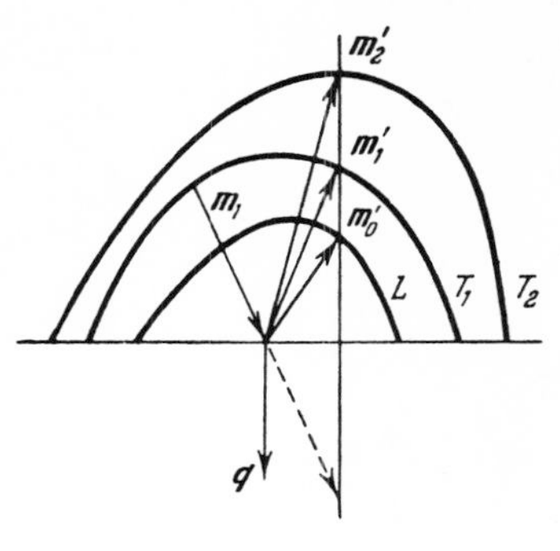

Fig. 17

We now have to find $\mathbf{u}_0'$, $\mathbf{u}_1'$, $\mathbf{u}_2'$ for the reflected waves. From the $\mathbf{m}_k'$ we can determine from (41.30) the $\mathbf{u}_k'$, apart from the amplitude factors C_k':

$$u_k' = C_k'\left(\overline{\Lambda^{m_k'} - 1}\right)p = C_k' u_k^{(1)}, \quad (43.3)$$

in which $\mathbf{p}$ is arbitrary, provided that $\overline{(\Lambda^{m_k'} - 1)}\mathbf{p} \neq 0$. In particular, provided that the last condition is obeyed, we may take $\mathbf{p} = \mathbf{q}$, $\mathbf{p} = \mathbf{a}$, or $\mathbf{p} = \mathbf{b}$.

In this case (41.32) becomes

$$(\sigma + \sigma_0' + \sigma_1' + \sigma_2')\, q = 0, \quad (43.4)$$

in which σ is the stress tensor for the incident wave, σ_k' being the same for the reflected waves. Omitting the common factor $i\omega e^{i\varphi_k}$, we put from (41.6) for the reflected waves that

$$(\sigma_k')_{lj} = c_{ljln}(m_k')_l (u_k')_n \quad (43.5)$$

and for the incident wave

$$\sigma = c_{ljln}(m)_l (u)_n. \quad (43.6)$$

Substitution of (43.2), (43.3), (43.5), and (43.6) into (43.4) gives [see also (42.11)]

$$C_0' U_0' + C_1' U_1' + C_2' U_2' + U = 0, \quad (43.7)$$

in which $\mathbf{U}$ and $\mathbf{U}_k'$ are defined as follows:

$$U = (M + \xi N)\, u, \quad U_k' = (M + \xi_k' N)\, u_k^{(1)}. \quad (43.8)$$

while tensors M and N are

$$M = (M_{ln}) = (c_{ljln} q_j b_l), \quad N = (N_{ln}) = (c_{ljln} q_j q_l) = \rho \Lambda^q. \quad (43.9)$$

Equation (43.7) is easily solved; it is multiplied by $[\mathbf{U}_1' \mathbf{U}_2']$ to give

$$C'_0 = -\frac{U[U'_1 U'_2]}{U'_0[U'_1 U'_2]}. \tag{43.10}$$

Similarly we have

$$C'_1 = -\frac{U[U'_2 U'_0]}{U'_0[U'_1 U'_2]}, \quad C'_2 = -\frac{U[U'_0 U'_1]}{U'_0[U'_1 U'_2]}. \tag{43.11}$$

Use of (43.8) with (12.25), (12.50), and (12.77) gives

$$U'_0[U'_1 U'_2] = V|\tilde{M} + SN|, \quad V = u_0^{(1)}[u_1^{(1)} u_2^{(1)}], \tag{43.11'}$$

in which tensor S is as follows:

$$S = \frac{1}{V}\left(\xi'_0[u_1^{(1)} u_2^{(1)}] \cdot u_0^{(1)} + \xi'_1[u_2^{(1)} u_0^{(1)}] \cdot u_1^{(1)} + \xi'_2[u_0^{(1)} u_1^{(1)}] \cdot u_2^{(1)}\right), \tag{43.11"}$$

this having the eigenvectors $[\mathbf{u}_1^{(1)}\mathbf{u}_2^{(1)}]$, $[\mathbf{u}_2^{(1)}\mathbf{u}_0^{(1)}]$, $[\mathbf{u}_0^{(1)}\mathbf{u}_1^{(1)}]$ and the corresponding eigenvalues ξ'_0, ξ'_1, ξ'_2. Expressions analogous to (43.11') and (43.11"), but with $\mathbf{u}_k^{(1)}$ and ξ'_k replaced by $\mathbf{u}$ and ξ , apply also to the numerators of (43.10) and (43.11).

The general solution is given by (43.3) and (43.8)-(43.11) for reflection from a free boundary if we know the refraction vectors of the reflected waves. From (42.16) and (43.9) we have for an isotropic medium that

$$M = (b - a)\, q \cdot b + ab \cdot q, \quad N = a + bq \cdot q. \tag{43.12}$$

For example, suppose $\mathbf{u}_0 = C_0\mathbf{m}_0$, $\mathbf{u}_0^{(1)} = \mathbf{m}'_0$, $\mathbf{u}_1^{(1)} = \mathbf{a}$, $\mathbf{u}_2^{(1)} = \xi'_1\mathbf{b} - \mathbf{a}^2\mathbf{q}$; then we get (42.24) and (42.25) from (43.8), (43.10), and (43.11).

Thus the main difficulty lies in finding the refraction vectors of the reflected waves, i.e., the parameters ξ'_k, which are found by joint solution of (41.24) and (41.25). Let

$$K_{in} = \frac{1}{2}\lambda_{ijln}(b_j q_l + q_j b_l), \tag{43.13}$$

then, with $\mathbf{m} = \mathbf{b} + \xi\mathbf{q}$, we have

$$\Lambda^m = \Lambda^b + 2\xi K + \xi^2\Lambda^q. \tag{43.14}$$

It is easily shown from (12.25) and (12.64) that

$$|\alpha+\beta+\gamma|=|\alpha|+|\beta|+|\gamma|+$$
$$+[\bar{\alpha}(\beta+\gamma)+\bar{\beta}(\gamma+\alpha)+\bar{\gamma}(\alpha+\beta)]_t+(\alpha\beta\gamma+\gamma\beta\alpha)_t+$$
$$+\alpha_t\beta_t\gamma_t-(\alpha_t\beta\gamma+\beta_t\gamma\alpha+\gamma_t\alpha\beta)_t. \quad (43.15)$$

Use of this with (43.14) converts (41.24) to

$$|\Lambda^m-1|=A_0\xi^6+2A_1\xi^5+A_2\xi^4+2A_3\xi^3+A_4\xi^2+2A_5\xi+A_6=0, \quad (43.16)$$

in which

$$\left.\begin{aligned} &A_0=|\Lambda^q|,\ A_1=(\bar{\Lambda}^q-K)_t,\ A_2=[\overline{\Lambda^q}(\Lambda^b-1)+4\bar{K}\Lambda^q]_t,\\ &A_3=4|K|+[(\overline{\Lambda^q+\Lambda^b}-\overline{\Lambda^q}-\overline{\Lambda^b})K]_t-\Lambda^q_tK_t+(\Lambda^qK)_t,\\ &A_4=[(\overline{\Lambda^b-1})\Lambda^q+4K(\Lambda^b-1)]_t,\\ &A_5=[(\overline{\Lambda^b-1})K]_t,\quad A_6=|\Lambda^b-1|. \end{aligned}\right\} \quad (43.17)$$

This equation is almost always extremely cumbrous, especially for crystals of low symmetry; it simplifies when the plane of incidence is a symmetry, whereupon equation (41.24) for the refraction surface splits up into (25.5) and (25.19):

$$\boldsymbol{m}\tau\boldsymbol{m}=1, \quad (43.18)$$

$$|\tau^m-1|=|\tau^m|-\tau^m_t+1=0, \quad (43.19)$$

in which

$$\tau=(\tau_{ab})=(\lambda_{3ab3}), \quad (43.20)$$

$$\tau^m=(\tau^m_{ab})=(\lambda_{acdb}m_cm_d),\quad a,\ b,\ c,\ d=1,\ 2, \quad (43.21)$$

and the x_3 axis is assumed perpendicular to the symmetry plane. Equation (43.18) gives the section of sheet T_1. Substitution of $\mathbf{m}=\mathbf{b}+\xi\mathbf{q}$ in (43.18) gives

$$\boldsymbol{q}\tau\boldsymbol{q}\xi^2+2\boldsymbol{b}\tau\boldsymbol{q}\xi+\boldsymbol{b}\tau\boldsymbol{b}-1=0, \quad (43.22)$$

whence

$$\xi = \xi_1' = -\frac{1}{q\tau q}\left(b\tau q + \sqrt{q\tau q - a\tau a}\right). \tag{43.23}$$

The sign of the square root has been chosen on the basis that $\xi_1' < 0$ for a reflected wave (Fig. 17).

In (43.19) we have to substitute as follows, by analogy with (43.14):

$$\tau^m = \tau^b + 2\xi\varkappa + \xi^2\tau^q, \tag{43.24}$$

in which

$$\varkappa_{ab} = \frac{1}{2}\lambda_{acdb}(b_c q_d + q_c b_d). \tag{43.25}$$

All tensors are given in the symmetry plane in these relations, so we apply the relationships of section 14. A relationship for the two-dimensional matrices α, β, and γ is found from (14.24):

$$|\alpha+\beta+\gamma| = |\alpha| + |\beta| + |\gamma| + \\ + \alpha_t\beta_t + \beta_t\gamma_t + \gamma_t\alpha_t - (\alpha\beta + \beta\gamma + \gamma\alpha)_t. \tag{43.26}$$

We use (43.24) and (43.26) to transform (43.19) to

$$B_0\xi^4 + 2B_1\xi^3 + B_2\xi^2 + 2B_3\xi + B_4 = 0, \tag{43.27}$$

in which

$$\left.\begin{aligned} &B_0 = |\tau^q|, \quad B_1 = \tau_t^q\varkappa_t - (\tau^q\varkappa)_t, \quad B_4 = |\tau^b - 1|, \\ &B_2 = 4|\varkappa| + \tau_t^q\tau_t^b - (\tau^q\tau^b)_t - \tau_t^q, \quad B_3 = \tau_t^b\varkappa_t - (\tau^b\varkappa)_t - \varkappa_t. \end{aligned}\right\} \tag{43.28}$$

The complete equation of fourth degree of (43.27) has to be solved to determine the refraction vectors in the general case, even when the plane of incidence is a symmetry plane.

The displacements are easily found for all waves in this case. Relations (43.18) and (43.23) apply for purely transverse waves, whose displacement vectors are perpendicular to the symmetry plane (section 17), i.e., to the plane of incidence, so we may put $\mathbf{u}_1 = C_1\mathbf{a}$, $\mathbf{u}_1' = C_1'\mathbf{a}$ for the corresponding incident and reflected waves. The displacement vectors of the other two waves lie in the plane of incidence and satisfy

$$(\tau^m - 1)\, u = 0, \tag{43.29}$$

so from (14.39) and (43.19)

$$u \cdot u = 1 - \frac{\tau^m - 1}{(\tau^m - 1)_t} = \frac{\tau_t^m - 1 - \tau^m}{\tau_t^m - 2}, \tag{43.30}$$

whence

$$u = C\left(\tau_t^m - 1 - \tau^m\right) p. \tag{43.31}$$

Thus the displacement vectors of all waves lie either in the plane of incidence or perpendicular to it. The same is true of the **U** of (43.8), since the tensors of (43.9) do not change in type on multiplication by vectors, in the sense of being perpendicular or parallel to the plane of incidence. The latter is evident also from considerations of symmetry, so all the vectors in (43.7) also belong to one of these two types, and (43.7) splits up into two separate equations: one for the vectors perpendicular to the plane of incidence,

$$C_1' U_1' + U_1 = 0 \tag{43.32}$$

and one for the vector parallel to that plane,

$$C_0' U_0' + C_2' U_2' + U = 0. \tag{43.33}$$

In the first case tensors M and N of (43.9) reduce to scalars (section 25):

$$M = q\tau b, \quad N = q\tau q, \tag{43.34}$$

so [see (43.8)]

$$U_1 = (q\tau b + \xi_1 q\tau q)\, C_1 a, \quad U_1' = (q\tau b + \xi_1' q\tau q) \cdot a \tag{43.35}$$

and the solution to (43.32) is

$$C_1' = -\frac{q\tau b + \xi_1 q\tau q}{q\tau b + \xi_1' q\tau q}\, C_1. \tag{43.36}$$

In the second case we obtain a two-dimensional problem, tensor M being analogous to tensor κ of (43.25); more precisely,* tensor κ is equal to the symmetrized tensor M. The solution to (43.33) is found from (43.10) and (43.11) with $\mathbf{U}_1' = \mathbf{a}$:

$$C_0' = -\frac{a[U_2'U]}{a[U_2'U_0']}, \quad C_2' = -\frac{a[UU_0']}{a[U_2'U_0']}, \tag{43.37}$$

with $\mathbf{U}$, $\mathbf{U}_0'$, $\mathbf{U}_2'$ defined by (43.8) and (43.31).

As an illustration we consider the reflection of elastic waves from the free surface of a hexagonal crystal. The simplest case is when the plane of incidence is perpendicular to L^6, and this differs from that for an isotropic medium only in that the refraction surface splits up into three circles (instead of two for an isotropic medium). If the symmetry axis is perpendicular to the boundary ($\mathbf{q} = \mathbf{e}$), any plane of incidence is a symmetry plane, and (43.23), (43.27), (32.28), and (32.42) may be applied to give

$$\left.\begin{aligned} &\tau^q = \tau^e = g_1 + g_2 + (g_2 + g_3 + g_4)\, q \cdot q, \\ &\tau_t^q = 2g_1 + 3g_3 + g_3 + g_4, \\ &\tau^b = g_1 a^2 + g_2 a^2 q \cdot q + g_3 b \cdot b, \quad \tau_t^b = (2g_1 + g_2 + g_3)\, a^2, \\ &\varkappa = \frac{1}{2} g_3 (b \cdot q + q \cdot b), \quad \varkappa_t = 0, \quad |\varkappa| = -\frac{1}{4} g_3^2 a^2. \end{aligned}\right\} \tag{43.38}$$

Then (see section 14) we have from (43.27) and (43.28) that

$$\left.\begin{aligned} &B_0 = \lambda_{33}\lambda_{44}, \quad B_1 = B_3 = 0, \quad B_4 = (\lambda_{11}a^2 - 1)(\lambda_{44}a^2 - 1), \\ &B_2 = \lambda_{33}(\lambda_{11}a^2 - 1) + \lambda_{44}(\lambda_{44}a^2 - 1) - g_3^2 a^2, \\ &(\lambda_{44}\xi^2 + \lambda_{11}a^2 - 1)(\lambda_{33}\xi^2 + \lambda_{44}a^2 - 1) - g_3^2 a^2 \xi^2 = 0. \end{aligned}\right\} \tag{43.39}$$

Thus (43.27) is here biquadratic; it gives $\xi_2' < \xi_0' < 0$ and also the corresponding refraction vectors for the quasilongitudinal ($\mathbf{m}_0' = \mathbf{b} + \xi_0'\mathbf{q}$) and quasitransverse ($\mathbf{m}_2' = \mathbf{b} + \xi_2'\mathbf{q}$) waves ($|\mathbf{m}_0'| < |\mathbf{m}_2'|$). We have from (32.33) for the purely transverse wave that

$$\tau = g_1 + g_5 + (g_2 - g_5)\, e \cdot e, \tag{43.40}$$

so

$$b\tau q = 0, \quad q\tau q = g_1 + g_2, \quad b\tau b = (g_1 + g_5)\, a^2$$

* The tensor $\alpha_S = (\alpha + \widetilde{\alpha})/2$ (section 10) is said to be symmetrized with respect to tensor α.

and from (43.23)

$$\xi_1 = -\xi_1' = \sqrt{\frac{1-(g_1+g_5)\,a^2}{g_1+g_2}}. \tag{43.41}$$

From (43.36) we have $C_1' = C_1$; substitution from (43.24) and (43.38) into (43.31), with $\mathbf{p} = \mathbf{q}$, gives for the other two waves

$$\boldsymbol{u} = (\lambda_{44}\xi^2 + \lambda_{11}a^2 - 1)\,\boldsymbol{q} - g_3\xi\boldsymbol{b}. \tag{43.42}$$

From (32.42), (43.9), (43.25), and (43.38)

$$M = \lambda_{13}\boldsymbol{q}\cdot\boldsymbol{b} + \lambda_{44}\boldsymbol{b}\cdot\boldsymbol{q}, \quad N = \tau^q = \lambda_{44} + (\lambda_{33} - \lambda_{44})\,\boldsymbol{q}\cdot\boldsymbol{q}. \tag{43.43}$$

The result from (43.8) is

$$\boldsymbol{U}_s' = \xi_s'(P_1 + P_2\xi_s'^2)\,\boldsymbol{q} + (Q_1 + Q_2\xi_s'^2)\,\boldsymbol{b}, \quad s = 0,\ 2, \tag{43.44}$$

in which

$$\left.\begin{aligned} P_1 &= \lambda_{33}(\lambda_{11}a^2 - 1) - \lambda_{13}(\lambda_{13} + \lambda_{44})\,a^2, \quad P_2 = \lambda_{33}\lambda_{44}, \\ Q_1 &= \lambda_{44}(\lambda_{11}a^2 - 1), \quad Q_2 = -\lambda_{13}\lambda_{44}. \end{aligned}\right\} \tag{43.45}$$

Then we have

$$\begin{aligned} \boldsymbol{a}\,[\boldsymbol{U}_2'\boldsymbol{U}_0'] = \boldsymbol{a}^2(\xi_0' - \xi_2')\,[P_1Q_1 + (P_2Q_1 - P_1Q_2)\,\xi_0'\xi_2' + \\ + P_2Q_1(\xi_0'^2 + \xi_2'^2) + P_2Q_2\xi_0'^2\xi_2'^2]. \end{aligned} \tag{43.46}$$

The numerators in (43.37) are found from this by simple replacement of ξ_0' or ξ_2' by ξ_0 or ξ_2 for quasilongitudinal or quasitransverse incident waves. From (43.44), we have for an incident quasilongitudinal wave that

$$\boldsymbol{U} = C_0\,[\xi_0(P_1 + P_2\xi_0^2)\,\boldsymbol{q} + (Q_1 + Q_2\xi_0^2)\,\boldsymbol{b}] \tag{43.47}$$

and (43.37) gives, with $\xi_0' = -\xi_0$ for $\mathbf{q} = \mathbf{e}$, that

$$\left.\begin{aligned} C_0' &= \frac{1}{D}(\xi_0' - \xi_2')\,[P_1Q_1 + (P_2Q_1 - P_1Q_2)\,\xi_0\xi_2' + P_2Q_1(\xi_0^2 + \xi_2'^2) + \\ &\qquad + P_2Q_2\xi_0^2\xi_2'^2], \\ C_2' &= -\frac{2\xi_0}{D}\,[P_1Q_1 + (P_1Q_2 + P_2Q_1)\,\xi_0^2 + P_2Q_2\xi_0^4], \\ D &= (\xi_0' + \xi_2')\,[P_1Q_1 - (P_2Q_1 - P_1Q_2)\,\xi_0\xi_2' + P_2Q_1(\xi_0^2 + \xi_2'^2) + P_2Q_2\xi_0^2\xi_2'^2]. \end{aligned}\right\} \tag{43.48}$$

The solution for an incident quasitransverse wave is obtained similarly.

An isotropic medium may be considered as a particular case of a hexagonal crystal; here we put $g_2 = g_4 = g_5 = 0$, $g_1 = \lambda_{44} = a$, $g_3 = \lambda_{33} - \lambda_{44} = b$, whereupon (43.48) becomes (42.24) and (42.25).

44. The Complex Refraction Vector and Inhomogeneous Plane Waves

So far the refraction vector has been taken as real, but the equations of motion of (15.4) may be satisfied by the functions of (15.5) subject to more general assumptions. In fact, plane waves of the form

$$\mathbf{u} = \mathbf{u}^0 e^{i\omega(\mathbf{mr} - t)} \tag{44.1}$$

can satisfy (15.4) and (15.21) for an isotropic medium even if the refraction vector $\mathbf{m}$ is complex:

$$\mathbf{m} = \mathbf{m}' + i\mathbf{m}''. \tag{44.2}$$

The conditions that $\mathbf{m}'$ and $\mathbf{m}''$ must satisfy are found from (24.7') and (24.8) after substitution from (42.2). In particular,

$$\mathbf{m}^2 = \mathbf{m}'^2 - \mathbf{m}''^2 + 2i\mathbf{m}'\mathbf{m}'' = \frac{1}{a+b}. \tag{44.3}$$

The right side is real, so we must have

$$\mathbf{m}'\mathbf{m}'' = 0, \tag{44.4}$$

$$\mathbf{m}'^2 - \mathbf{m}''^2 = \frac{1}{a+b}. \tag{44.5}$$

Thus the real and imaginary parts of the complex refraction vector for an isotropic medium must be mutually orthogonal. Substitution of (44.2) into (44.1) gives

$$\mathbf{u} = \mathbf{u}^0 e^{-\omega \mathbf{m}''\mathbf{r}} e^{i\omega(\mathbf{m}'\mathbf{r} - t)}. \tag{44.6}$$

The physical significance of $\mathbf{m}'$ is precisely that of the ordinary refraction vector of (21.32), since this appears in the phase as before, so we may put

$$m' = \frac{n'}{v}, \tag{44.7}$$

with $\mathbf{n}'$ as unit vector of the phase (wave) normal perpendicular to the planes of equal phase, while v is the phase velocity. We see from (44.6) that $\mathbf{m}''$ defines the change in amplitude. Only a decreasing amplitude is physically possible for a wave propagating to unrestricted distances, so a wave of the form of (44.6) cannot exist in an unbounded homogeneous medium, since its amplitude, and hence its energy, would increase without limit for $\mathbf{m}''\mathbf{r} < 0$, $|\mathbf{r}| \to \infty$. We introduce the unit vector

$$n'' = \frac{m''}{|m''|}. \tag{44.8}$$

This vector defines the direction of most rapid variation in the amplitude with $\mathbf{r}$, i.e., during propagation of the wave. The planes defined by

$$n''r = \text{const}, \tag{44.9}$$

are planes of equal amplitude (amplitude planes), so the vector $\mathbf{n}''$ perpendicular to them is naturally called the amplitude normal, by analogy with the phase normal $\mathbf{n}'$.

We have $\mathbf{n}' \perp \mathbf{n}''$ from (44.4), so the amplitude planes are always perpendicular to the phase planes in an isotropic medium.*

At a distance $\mathbf{r}$ from the origin such that

$$rn'' = \lambda, \tag{44.10}$$

in which λ is the wavelength:

$$\lambda = \frac{v}{\nu} = \frac{2\pi}{\omega\,|m'|}, \tag{44.11}$$

the amplitude has decreased by a factor $e^{2\pi\,|\mathbf{m}''|/|\mathbf{m}'|}$, while the energy (square of the amplitude) has decreased by $e^{4\pi|\mathbf{m}''|/|\mathbf{m}'|}$.

* Here it is assumed that the medium does not absorb the elastic waves.

The ratio

$$\varkappa = \frac{|m''|}{|m'|}, \tag{44.12}$$

characterizes the rate of decrease of amplitude with distance and is termed the absorption coefficient. The vector $\mathbf{m}'' = |\mathbf{m}''|\mathbf{n}''$ is called the extinction vector.

Thus the complex refraction vector consists of the ordinary refraction vector and an extinction vector, whose directions define the directions of most rapid change of phase and amplitude, and whose lengths define the rates of spatial variation in the phases and amplitude, respectively. This may be taken as a definition of the complex refraction vector.

Plane waves having a complex refraction vector are termed inhomogeneous if their phase and amplitude normals are not parallel. This name is related not to the change in amplitude in space generally (such a change would occur also for $\mathbf{n}' \| \mathbf{n}''$) but to the change within a given phase plane. Similarly, homogeneous plane waves are ones of the form of (44.1) with real refraction vectors ($\mathbf{m}'' = 0$) and also ones with complex $\mathbf{m}$ such that

$$m'' = Cm'. \tag{44.13}$$

The occurrence or otherwise of these inhomogeneous waves is closely related to the boundary conditions, which have to be considered in solving (15.4), as for any other differential equations. The boundary conditions can be specified only at infinity if the medium is unbounded and entirely homogeneous (all points equivalent). Such boundary conditions for plane waves may be formulated explicitly as follows: the amplitude must be bounded throughout space, including at infinity. In particular, a solution to (15.4) in the form of (15.5) for an infinite homogeneous medium implies the assumption that the wave fills the whole of infinite space, i.e., that the displacement at infinity is represented by expressions such as (15.5) with the same $\mathbf{u}^0$ and $\mathbf{k}$. This is physically possible for homogeneous waves with a real refraction vector; but a complex refraction vector means that the amplitude (and hence the energy) must somewhere at infinity become infinitely great, which is physically meaningless. Hence waves of the type of (44.6) cannot satisfy the equations of motion together with the boundary conditions at infinity for an unbounded medium; hence the presence of a boundary to the medium is essential to the occurrence and properties of inhomogeneous waves. All the same, once such a wave has arisen in some way at a boundary, equations (15.4) describe its subsequent propagation within the medium, so it is reasonable to examine the possible properties of these waves on the basis of the equations of motion without consideration of the boundary conditions, since such a wave under all circumstances will have properties compatible with (15.4). We shall consider the case of generation of inhomogeneous plane elastic waves at the free boundary of an isotropic medium.

In Chapters 3-5 we consider wave propagation assuming the direction of **n** given, the displacement vectors and phase velocities of the three corresponding isonormal waves being the unknowns. Here of the two parts (**n** and v) of the refraction vector

$$\boldsymbol{m} = \frac{\boldsymbol{n}}{v} \tag{44.14}$$

the first is taken as known while the second (the phase velocity) is to be found. This means that combination of the two in **m** at first sight appears formal. In fact, in Chapters 3 and 5-7 we mostly dealt with Christoffel's equation in the usual form:

$$(\Lambda^n - v^2)\boldsymbol{u} = 0, \quad |\Lambda^n - v^2| = 0, \tag{44.15}$$

where **n** and v appear separately. However, the refraction vector gave simpler and more compact expressions in relation to the energy flux, the ray velocity, and the wave surfaces; moreover, it is essential to the discussion of the reciprocal-velocity surface. In previous parts of this chapter we have seen that the refraction vector plays a very basic part in relation to reflection and refraction, where it was found necessary to represent the vector in the form of (41.25) rather than (44.14):

$$\boldsymbol{m} = \boldsymbol{b} + \xi \boldsymbol{q}, \tag{44.16}$$

while the normal (44.15) is unsuitable for finding ξ, the more general (24.6) being needed:

$$|\Lambda^m - 1| = 0. \tag{44.17}$$

Finally, we encountered the vector in the complex form of (44.2) in relation to decaying plane waves. It is clear that the form of (44.14) cannot be used for inhomogeneous waves, although the real and imaginary parts have direct physical meanings. The form of representation of the vector thus varies from one problem to another, only the vector as a whole remaining significant.

The vector is a fundamental concept of the theory of any wave process in any medium; in particular, vector **m** plays an equally important part in the propagation of electromagnetic waves in transparent, isotropic, anisotropic absorbing, magnetically anisotropic, or optically active media [17]. The refraction vector was first introduced in [37]; its most general definition is (44.1) for a plane monochromatic wave. If this vector is given, we may take

$$\boldsymbol{k} = \omega \boldsymbol{m} \tag{44.18}$$

as a definition of the wave vector.

We can now give the following general formulation of the problem in the theory of elastic waves. Consider a set of several homogeneous media each having its own density ρ and elastic-modulus tensor c_{ijkl}. These media can carry plane monochromatic elastic waves of the form of (44.1) if the vectors **m** and $\mathbf{u}^0$ within each medium satisfy

$$(\Lambda^m - 1)\, \boldsymbol{u}^0 = 0 \tag{44.19}$$

while conditions (41.31) and (41.32) or (41.33) are satisfied at any interface between two media. The **m** of (44.14) can be used only in the highly idealized case where we assume that (a) the homogeneous medium is unbounded and (b) the elastic wave is homogeneous; then (44.19) reduces to (44.15). Violation of (a), i.e., presence of boundaries, leads to the problems considered in previous parts of this chapter. Now we consider inhomogeneous waves, which cannot exist except in the presence of a boundary. Thus (b) may be violated only if (a) is violated; but (44.19) must be obeyed for waves within a medium in all cases, so we may first elucidate what restrictions it imposes on **m** and $\mathbf{u}^0$ before considering the boundary problem. The theory of inhomogeneous elastic waves is complicated and poorly developed, so we consider only an isotropic medium that does not absorb elastic waves.

For an isotropic medium (44.17) takes the form of (24.7'):

$$(m^2(a+b)-1)(m^2a-1)^2=0, \tag{44.20}$$

so with $\mathbf{m}=\mathbf{m}'+i\mathbf{m}''$ we get

$$m_0^2=\frac{1}{a+b}, \quad m_0'^2-m_0''^2=\frac{1}{a+b}, \quad m_0'm_0''=0, \tag{44.21}$$

$$m_1^2=\frac{1}{a}, \quad m_1'^2-m_1''^2=\frac{1}{a}, \quad m_1'm_1''=0. \tag{44.22}$$

The complex vectors $\mathbf{m}_0$ and $\mathbf{m}_1$ have $2\times3=6$ components each; on these six (44.20) imposes only the conditions of (44.21) or (44.22), and the relationships needed to define $\mathbf{m}_0$ and $\mathbf{m}_1$ completely can come only from the boundary conditions. The displacement vectors are found from (44.19), which for isotropic media takes the form of (24.7):

$$(\Lambda^m-1)\,u=(am^2-1)\,u+bum\cdot m=0. \tag{44.23}$$

Let $\mathbf{m}=\mathbf{m}_0$ in (44.21) and $\mathbf{u}=\mathbf{u}_0$; then (44.23) gives

$$\left(\frac{a}{a+b}-1\right)u_0+bu_0m_0\cdot m_0=0,$$

whence

$$u_0=(a+b)\,u_0m_0\cdot m_0=Cm_0. \tag{44.24}$$

Thus the displacement vector is proportional to $\mathbf{m}_0$; the coefficient of proportionality C can be arbitrary, as always in the solution of linear homogeneous equations. We called the wave longitudinal for $\mathbf{m}$ real, because the displacement vector lies along the wave normal. Here the term can be used only in a nominal sense, because $\mathbf{m}_0=\mathbf{m}_0'+i\mathbf{m}_0''$, but the name will be retained, though in inverted commas.

For the case of (44.22) we have from (44.23) that

$$u_1m_1=0. \tag{44.25}$$

Here $\mathbf{u}_1$ is orthogonal to $\mathbf{m}_1$ in the sense that the ordinary scalar product is zero,* so the wave may be called "transverse." As in the usual real case, condition (44.25) is satisfied by two linearly independent (complex) vectors $\mathbf{u}_1$. Elucidation of the properties of $\mathbf{u}_0$ and $\mathbf{u}_1$ is equivalent to elucidation of the state of polarization for inhomogeneous waves. The subsequent sections deal with this.

45. Invariant Characteristics of the Polarization of Plane Waves

Section 23 dealt with the elliptical polarization of plane waves; here we consider polarization in more detail. The vector of (44.1) in the most general case may be put as

$$u = u^{0}e^{i\varphi} = (u^{0\prime} + iu^{0\prime\prime})\,e^{i\varphi} = u' + iu'', \tag{45.1}$$

while the real part, which has the direct physical meaning of a displacement, is

$$u' = R = u^{0\prime}\cos\varphi - u^{0\prime\prime}\sin\varphi, \quad \varphi = \omega\,(mr - t). \tag{45.2}$$

It follows directly from (45.2) that the end of $\mathbf{u}'$ describes a plane curve, since it is a linear combination of two constant vectors $\mathbf{u}^{0\prime}$ and $\mathbf{u}^{0\prime\prime}$ with variable scalar coefficients cos φ and sin φ.

Equation (45.2) is the polarization curve in vector parametric form. We take $\mathbf{u}'$ as the radius vector R, φ being eliminated by forming the vector products by $\mathbf{u}^{0\prime}$ and $\mathbf{u}^{0\prime\prime}$ (see section 22). The result is

$$[Ru^{0\prime}] = [u^{0\prime}u^{0\prime\prime}]\sin\varphi, \quad [Ru^{0\prime\prime}] = [u^{0\prime}u^{0\prime\prime}]\cos\varphi.$$

Taking squares and adding, we have

$$[Ru^{0\prime}]^2 + [Ru^{0\prime\prime}]^2 = [u^{0\prime}u^{0\prime\prime}]^2. \tag{45.3}$$

Here $\mathbf{R}$ is not the usual three-dimensional radius vector but lies in

---

* Orthogonality for complex vectors $\mathbf{a}$ and $\mathbf{b}$ is frequently understood as meaning $\mathbf{a}^*\mathbf{b} = 0$.

the plane ($\mathbf{u}^{0'}$, $\mathbf{u}^{0''}$). It is clear that the curve of (45.3) cannot be a parabola or hyperbola, because (45.2) sets a bound to the length of $\mathbf{R}$. Thus (45.3) in the general case represents an ellipse. The square of the radius vector must be constant for circular polarization, which gives us (23.5):

$$(\boldsymbol{u}^{0'})^2 - (\boldsymbol{u}^{0''})^2 = \boldsymbol{u}^{0'} \boldsymbol{u}^{0''} = 0. \tag{45.4}$$

Then (45.3) becomes the equation of a circle of radius $|\mathbf{u}^{0'}| = |\mathbf{u}^{0''}|$, so (45.4) gives an invariant criterion for circular polarization.

If $\mathbf{u}^{0'}$ and $\mathbf{u}^{0''}$ are mutually perpendicular but unequal,

$$\boldsymbol{u}^{0'} \boldsymbol{u}^{0''} = 0, \quad |\boldsymbol{u}^{0'}| \neq |\boldsymbol{u}^{0''}|, \tag{45.5}$$

which gives a canonical ellipse, in which $\mathbf{u}^{0'}$ and $\mathbf{u}^{0''}$ define the magnitude and direction of the semiaxes. For example, we may take the direction of $\mathbf{u}^{0'}$ as that of the x axis and that of $\mathbf{u}^{0''}$ as the direction of the y axis, which gives

$$[\boldsymbol{R}\boldsymbol{u}^{0'}]^2 = (\boldsymbol{u}^{0'})^2 y^2, \quad [\boldsymbol{R}\boldsymbol{u}^{0''}]^2 = (\boldsymbol{u}^{0''})^2 x^2$$

and, from (45.5),

$$[\boldsymbol{u}^{0'} \boldsymbol{u}^{0''}]^2 = (\boldsymbol{u}^{0'})^2 (\boldsymbol{u}^{0''})^2.$$

Here (45.3) becomes

$$\frac{x^2}{(u^{0'})^2} + \frac{y^2}{(u^{0''})^2} = 1. \tag{45.6}$$

Vector $\mathbf{R}$ has a fixed direction for linear polarization; any change in φ produces a change parallel to $\mathbf{R}$ itself. The condition for this may be put as

$$\left[\boldsymbol{R}, \frac{d\boldsymbol{R}}{d\varphi}\right] = 0. \tag{45.7}$$

From (45.2) we have $d\mathbf{R}/d\varphi = \mathbf{R}' = -\mathbf{u}^{0'} \sin\varphi - \mathbf{u}^{0''} \cos\varphi$, so

$$[\boldsymbol{R}\boldsymbol{R}'] = -[\boldsymbol{u}^{0'} \boldsymbol{u}^{0''}], \tag{45.8}$$

hence (45.7) gives

$$[\boldsymbol{u}^{0'}\boldsymbol{u}^{0''}]=0. \tag{45.9}$$

Here the sum of the positive terms on the left in (45.3) must be zero, so

$$[\boldsymbol{R}\boldsymbol{u}^{0'}]=[\boldsymbol{R}\boldsymbol{u}^{0''}]=0,$$

so R is parallel to $\mathbf{u}^{0'}$ and $\mathbf{u}^{0''}$. Thus (45.9) is an invariant criterion for linear polarization. Condition (45.9) may also be put as

$$\boldsymbol{u}^{0''}=\alpha\boldsymbol{u}^{0'}, \tag{45.10}$$

where the real factor α may take the values 0 and ∞, which correspond to the particular cases $\mathbf{u}^{0''}=0$ and $\mathbf{u}^{0'}=0$.

This approach also gives a simple solution for the sense of rotation of **u**, which is defined by the direction of [RR'], which forms a right-hand screw with the direction of rotation of **u**.

We have the general case of elliptical polarization if neither (45.4), (45.5), or (45.9) is obeyed; here the magnitude and direction of the principal semiaxes of the displacement ellipse are given by (22.22) and (22.23), with **p** and **q** replaced by $\mathbf{u}^{0'}$ and $\mathbf{u}^{0''}$:

$$\left.\begin{aligned}
\boldsymbol{a}^2&=\frac{1}{2}\left\{(\boldsymbol{u}^{0'})^2+(\boldsymbol{u}^{0''})^2+\sqrt{((\boldsymbol{u}^{0'})^2+(\boldsymbol{u}^{0''})^2)^2-4[\boldsymbol{u}^{0'}\boldsymbol{u}^{0''}]^2}\right\},\\
\boldsymbol{b}^2&=\frac{1}{2}\left\{(\boldsymbol{u}^{0'})^2+(\boldsymbol{u}^{0''})^2-\sqrt{((\boldsymbol{u}^{0'})^2+(\boldsymbol{u}^{0''})^2)^2-4[\boldsymbol{u}^{0'}\boldsymbol{u}^{0''}]^2}\right\},\\
&\boldsymbol{a}\parallel[\boldsymbol{u}^{0''}[\boldsymbol{u}^{0'}\boldsymbol{u}^{0''}]]-\boldsymbol{a}^2\boldsymbol{u}^{0'}\parallel[\boldsymbol{u}^{0'}[\boldsymbol{u}^{0''}\boldsymbol{u}^{0'}]]-\boldsymbol{a}^2\boldsymbol{u}^{0''},\\
&\boldsymbol{b}\parallel[\boldsymbol{u}^{0''}[\boldsymbol{u}^{0'}\boldsymbol{u}^{0''}]]-\boldsymbol{b}^2\boldsymbol{u}^{0'}\parallel[\boldsymbol{u}^{0'}[\boldsymbol{u}^{0''}\boldsymbol{u}^{0'}]]-\boldsymbol{b}^2\boldsymbol{u}^{0''}.
\end{aligned}\right\} \tag{45.11}$$

Then

$$\boldsymbol{a}^2+\boldsymbol{b}^2=(\boldsymbol{u}^{0'})^2, \qquad \boldsymbol{a}^2\boldsymbol{b}^2=[\boldsymbol{u}^{0'}\boldsymbol{u}^{0''}]^2. \tag{45.12}$$

The general formulas of (45.11) include the particular cases of circular polarization and linear polarization. In the first case we see from (45.4) that the expression under the root becomes zero, and we get the radius of the circle of vibration as

$$|\boldsymbol{R}|=|\boldsymbol{u}^{0'}|=|\boldsymbol{u}^{0''}|. \tag{45.13}$$

In the second case $[\mathbf{u}^{0\prime}\mathbf{u}^{0\prime\prime}]=0$ and $\mathbf{b}^2=0$, so for the maximal length of **R** we get:*

$$|\boldsymbol{a}|=R=\sqrt{(\boldsymbol{u}^{0\prime})^2+(\boldsymbol{u}^{0\prime\prime})^2}=|\boldsymbol{u}|. \tag{45.14}$$

We have seen in section 22 that $\mathbf{u}^{0\prime}$ bisects all chords of the ellipse parallel to $\mathbf{u}^{0\prime\prime}$ and vice versa, so $\mathbf{u}^{0\prime}$ and $\mathbf{u}^{0\prime\prime}$ are conjugate semi-diameters of the ellipse.

Even more convenient and compact relations are obtained if we use $\mathbf{u}^0$ directly without resort to $\mathbf{u}^{0\prime}$ and $\mathbf{u}^{0\prime\prime}$. We can put (45.3) in the form

$$|[\boldsymbol{R}\boldsymbol{u}^0]|^2=\frac{1}{4}|[\boldsymbol{u}^0\boldsymbol{u}^{0*}]|^2. \tag{45.15}$$

Thus **R** is real and lies in the plane of $\mathbf{u}^{0\prime}$ and $\mathbf{u}^{0\prime\prime}$, or (which is the same) in the plane of $\mathbf{u}^0$ and $\mathbf{u}^{0*}$, so it must have the form

$$\boldsymbol{R}=\frac{1}{2}(\alpha\boldsymbol{u}^0+\alpha^*\boldsymbol{u}^{0*}). \tag{45.16}$$

We substitute (45.16) into (45.15) to get for the complex scalar α that

$$|\alpha|^2=1. \tag{45.17}$$

The principal axes of the ellipse of (45.15) correspond to turning points in R^2 subject to (45.17). As usual, we form the function

$$f(\alpha,\ \alpha^*)=\frac{1}{4}(\alpha^2\boldsymbol{u}^{02}+\alpha^{*2}\boldsymbol{u}^{0*2}+2\alpha\alpha^*|\boldsymbol{u}^0|^2)-\lambda(\alpha\alpha^*-1),$$

in which λ is an undetermined multiplier; here α and α^* must be

* For a complex vector, $|\mathbf{u}|$ denotes the modulus, i.e., the positive square root of the scalar product of the vector by its complex conjugate:

$$|\boldsymbol{u}|=\sqrt{\boldsymbol{u}\boldsymbol{u}^*}=\sqrt{(\boldsymbol{u}^{0\prime}+i\boldsymbol{u}^{0\prime\prime})(\boldsymbol{u}^{0\prime}-i\boldsymbol{u}^{0\prime\prime})}=\sqrt{(\boldsymbol{u}^{0\prime})^2+(\boldsymbol{u}^{0\prime\prime})^2}=|\boldsymbol{u}^0|$$

[see (45.1)].

considered as independent variables, since each is expressed via two independent parameters (the real and imaginary parts of α). We equate to zero the derivatives of f with respect to α and α^* to get

$$\boldsymbol{u}^{0^2}\alpha + (|\boldsymbol{u}^0|^2 - 2\lambda)\,\alpha^* = 0, \quad (|\boldsymbol{u}^0|^2 - 2\lambda)\,\alpha + \boldsymbol{u}^{0*^2}\alpha^* = 0. \qquad (45.18)$$

We multiply these by α and α^*; in addition, with (45.17), then gives

$$\lambda = \frac{1}{4}\left(\alpha^2\boldsymbol{u}^{0^2} + \alpha^{*^2}\boldsymbol{u}^{0*^2} + 2|\alpha|^2|\boldsymbol{u}^0|^2\right) = R^2_{\text{ext}}. \qquad (45.19)$$

On the other hand, we have an equation for λ:

$$\begin{vmatrix} \boldsymbol{u}^{0^2} & |\boldsymbol{u}^0|^2 - 2\lambda \\ |\boldsymbol{u}^0|^2 - 2\lambda & \boldsymbol{u}^{0*^2} \end{vmatrix} = |\boldsymbol{u}^{0^2}|^2 - (|\boldsymbol{u}^0|^2 - 2\lambda)^2 = 0, \qquad (45.20)$$

whence

$$\lambda = \frac{1}{2}(|\boldsymbol{u}^0|^2 \pm |\boldsymbol{u}^{0^2}|). \qquad (45.21)$$

The upper sign corresponds to the major semiaxis and the lower to the minor one, so

$$a^2 = \frac{1}{2}(|\boldsymbol{u}^0|^2 + |\boldsymbol{u}^{0^2}|), \quad b^2 = \frac{1}{2}(|\boldsymbol{u}^0|^2 - |\boldsymbol{u}^{0^2}|). \qquad (45.22)$$

Then we have

$$|\boldsymbol{u}^{0^2}| = a^2 - b^2, \quad |\boldsymbol{u}^0|^2 = a^2 + b^2, \qquad (45.23)$$

$$a^2b^2 = \frac{1}{4}(|\boldsymbol{u}^0|^4 - |\boldsymbol{u}^{0^2}|^2) = -\frac{1}{4}[\boldsymbol{u}^0\boldsymbol{u}^{0*}]^2 = \frac{1}{4}|[\boldsymbol{u}^0\boldsymbol{u}^{0*}]|^2. \qquad (45.24)$$

Here we use the following relationships for the complex vectors:

$$|\boldsymbol{u}|^2 = \boldsymbol{u}\boldsymbol{u}^*, \quad [\boldsymbol{u}\boldsymbol{u}^*]^2 = \boldsymbol{u}^2\boldsymbol{u}^{*^2} - (\boldsymbol{u}\boldsymbol{u}^*)^2 = |\boldsymbol{u}^2|^2 - |\boldsymbol{u}|^4. \qquad (45.25)$$

But

$$|\boldsymbol{u}^0|^2 = |\boldsymbol{u}|^2 = (\boldsymbol{u}^{0'})^2 + (\boldsymbol{u}^{0''})^2. \qquad (45.26)$$

$$|u^{0^2}| = |u^2| = |(u^{0'})^2 - (u^{0''})^2 + 2i u^{0'} u^{0''}| =$$

$$= \sqrt{[(u^{0'})^2 - (u^{0''})^2]^2 + 4(u^{0'} u^{0''})^2}, \tag{45.27}$$

so (45.22) coincides with (45.11); (45.18) gives α as

$$\frac{\alpha}{\alpha^*} = -\frac{|u|^2 - 2\lambda}{u^{0^2}} = -\frac{u^{0*^2}}{|u|^2 - 2\lambda}. \tag{45.28}$$

Multiplication of numerator and denominator by α and use of (45.17) and (45.21) gives

$$\alpha^2 = \pm \frac{|u^2|}{u^{0^2}}. \tag{45.29}$$

Then α for the major semiaxis is

$$\alpha = \pm \sqrt{\frac{|u^2|}{u^{0^2}}}, \quad \alpha^* = \pm \sqrt{\frac{|u^2|}{u^{0*^2}}} = \pm \sqrt{\frac{u^{0^2}}{|u^2|}}, \tag{45.30}$$

and for the minor semiaxis is

$$\alpha = \pm i \sqrt{\frac{|u^2|}{u^{0^2}}}, \quad \alpha^* = \mp i \sqrt{\frac{u^{0^2}}{|u^2|}}. \tag{45.31}$$

Here the upper or lower sign may be taken. For example, taking the upper sign in (45.30) and the lower in (45.31), we have by substitution into (45.16) that

$$R_{\max} = a = \frac{1}{2}\left(\sqrt{\frac{|u^{0^2}|}{u^{0^2}}}\, u^0 + \sqrt{\frac{|u^{0^2}|}{u^{0*2}}}\, u^{0*}\right), \tag{45.32}$$

$$R_{\min} = b = -\frac{i}{2}\left(\sqrt{\frac{|u^{0^2}|}{u^{0^2}}}\, u^0 - \sqrt{\frac{|u^{0^2}|}{u^{0*2}}}\, u^{0*}\right). \tag{45.33}$$

It is readily seen that $\mathbf{ab} = 0$.

Eqs. (45.32) and (45.33) mean that **a** and **b** (radius vectors of the major and minor semiaxes of the ellipse described by **u**) are given in magnitude and direction by the real and imaginary parts, respectively, of the following complex vector:

$$u_r^0 = \sqrt{\frac{|u^{0^2}|}{u^{0^2}}}\, u^0 = a + ib. \tag{45.34}$$

This vector differs from $\mathbf{u}^0$ only in multiplication by the scalar complex unimodular factor $\sqrt{|\mathbf{u}^{02}|/\mathbf{u}^{02}}$, which transforms $\mathbf{u}^0$ and the corresponding ellipse to the principal axes. Further, we have from (45.34) that

$$u_r^{0^2} = |u^0|^2. \tag{45.35}$$

As criterion for the direction of rotation we have, since

$$[uu^*] = -2i\,[u^{0'}u^{0''}], \tag{45.36}$$

that the real vector $i[\mathbf{u}^*\mathbf{u}]$ must form a right-handed screw with the direction of rotation of $\mathbf{u}$.

Formulas (45.15), (45.20)-(45.27), and (45.32)-(45.35) still apply if the amplitude $\mathbf{u}^0$ is replaced by the complete vector $\mathbf{u} = \mathbf{u}^0 e^{i\varphi}$, since the phase factor $e^{i\varphi}$ cancels out. The $\mathbf{u}_r$ derived from $\mathbf{u}$ via (45.34) is called reduced, and $\mathbf{u}_r$ is multiplied by the modulus of the factor when $\mathbf{u}$ is multiplied by an arbitrary complex scalar factor. Replacing $\mathbf{u}$ by $\mathbf{u}_1 = \beta\mathbf{u}$ in (45.34), we have

$$u_{1r} = (\beta u)_r = \sqrt{\frac{|(\beta u)^2|}{(\beta u)^2}}\, \beta u = |\beta|\, u_r. \tag{45.37}$$

This multiplication leaves the shape and orientation of the vibration ellipse unchanged, the size being increased by a factor $|\beta|$. Hence the reduced vector for the total field vector $\mathbf{u} = \mathbf{u}^0 e^{i\varphi}$ and for the vector amplitude $\mathbf{u}^0$ are the same.

The direction of rotation also does not alter when $\mathbf{u}$ is multiplied by an arbitrary complex scalar factor β, since $[\mathbf{u}\mathbf{u}^*]$ is then multiplied by $|\beta|^2$, so the sign of $i[\mathbf{u}\mathbf{u}^*]$ remains unchanged. Thus multiplication of $\mathbf{u}$ by an arbitrary fixed complex number alters nothing except the size of the vibration ellipse, so we can easily establish the relation between the polarizations of two vibrations $\mathbf{u}$ and $\mathbf{u}_1$ occurring in a single plane and satisfying $\mathbf{u}\mathbf{u}_1 = 0$. Let $\mathbf{n}$ be the normal to the $(\mathbf{u}, \mathbf{u}_1)$ plane; then $\mathbf{n}\mathbf{u} = \mathbf{n}\mathbf{u}_1 = 0$, and so $\mathbf{u}_1 = \beta[\mathbf{n}\mathbf{u}] = [\mathbf{n}\mathbf{u}_2]$, where $\mathbf{u}_2 = \beta\mathbf{u}$. But $\mathbf{n}$ is a unit real vector, so $\mathbf{R}_1 = [\mathbf{n}\mathbf{R}_2]$ from

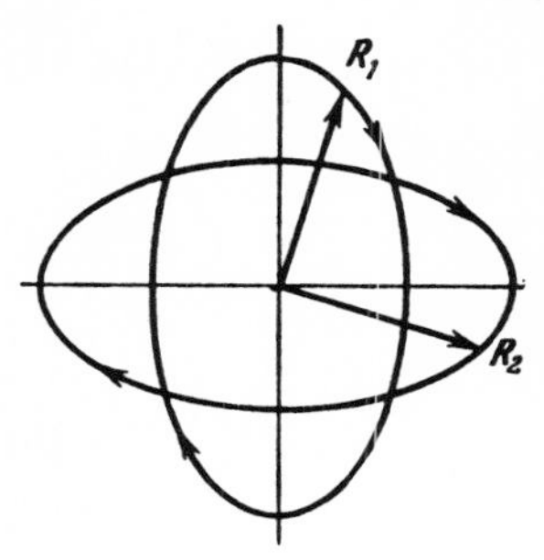

Fig. 18

(45.2) and $\mathbf{R}_1^2 = \mathbf{R}_2^2$, and so the real vectors $\mathbf{R}_1$ and $\mathbf{R}_2$ always remain equal and mutually perpendicular during their periodic variation. The curves they describe therefore have the same shape, size, and sense of traversal, but the two curves are mutually perpendicular (Fig. 18).

From (45.32), (45.33), and (45.36) it follows that replacement of **u** by its complex conjugate **u*** leaves the vibration ellipse unchanged in shape, size, and orientation, but that the sense of traversal is reversed.

The condition for linear polarization is given by (45.9) and (45.36) as

$$[uu^*] = 0. \tag{45.38}$$

For circular polarization we have

$$u^2 = 0, \tag{45.39}$$

so **u** must be zero or an isotropic vector.

The following general relationships apply to any complex vector:

$$|u|^4 + [uu^*]^2 = |u^2|^2, \tag{45.40}$$

$$|[uu_1]|^2 = |u|^2 \, |u_1|^2 - |uu_1^*|^2, \tag{45.41}$$

so for linear polarization we must have

$$|u|^2 = |u^2|, \tag{45.42}$$

and for circular polarization

$$|u|^4 + [uu^*]^2 = 0, \tag{45.43}$$

or

$$|u|^2 = |[uu^*]|. \tag{45.44}$$

The radius of the circle is given by (45.13) as

$$|R| = \frac{1}{2}|u + u^*| = \frac{1}{2}|u - u^*|,$$

or

$$|R| = \frac{1}{\sqrt{2}}|u|. \tag{45.45}$$

Normal **n** to the plane of **u** also enables us to obtain relationships that completely define circular polarization. Condition (45.39) may be considered as the condition for **u** to be orthogonal to itself. On the other hand, **un** = 0, so **u**, being orthogonal to **u** and **n**, may be put as

$$u = \beta [nu]. \tag{45.46}$$

Scalar multiplication by **u** gives $\mathbf{u}^2 = 0$, which is (45.39). Vector multiplication by **n** gives

$$[nu] = \beta [n [nu]] = -\beta u. \tag{45.47}$$

Comparison of (45.46) with (45.47) gives $\beta^2 = -1$, $\beta = \pm$ i, so

$$u = \pm i [nu]. \tag{45.48}$$

Scalar multiplication by **u*** gives

$$in [uu^*] = \pm |u|^2. \tag{45.49}$$

From (45.36), here the upper and lower signs correspond to opposite senses of description of the circle.

In (45.38)-(45.49) we can understand by **u** the total variable vector of the wave or the complex amplitude $\mathbf{u}^0$; the result is precisely the same.

The shape of the polarization curve may also be judged from

$$\gamma = \frac{|u^2|}{|u|^2}. \tag{45.50}$$

From (45.22), the ratio of the squares of the semiaxes of the polarization ellipse is

$$\frac{b^2}{a^2} = \frac{1-\gamma}{1+\gamma}, \tag{45.51}$$

i.e., is completely determined by γ. We have from (45.38) that for linear polarization

$$u^* = \beta u, \tag{45.52}$$

in which β is a complex scalar. Taking the modulus of both parts, and remembering that $|\mathbf{u}^*| = |\mathbf{u}|$, we have

$$|\beta| = 1. \tag{45.53}$$

We multiply (45.52) scalarly by **u** and take the modulus of both parts to get $|\mathbf{u}|^2 = |\mathbf{u}^2|$, which coincides with (45.42). Conversely, it is readily shown that (45.38) follows from (45.42), so γ takes the value one for linear polarization, while $\gamma = 0$ from (45.39) for circular polarization. Finally, for the general case of elliptical polarization we have $0 < \mathbf{b}^2 < \mathbf{a}^2$, so from (45.51) we have $0 < \gamma < 1$. A further criterion for the form of polarization is then

$$\left.\begin{array}{r l} \gamma = 0 & \text{— circular polarization,} \\ 0 < \gamma < 1 & \text{— linear polarization,} \\ \gamma = 1 & \text{— elliptical polarization.} \end{array}\right\} \tag{45.54}$$

Obviously, the necessary and sufficient condition for the polarization curves for two different vectors to coincide is that their γ coincide.

The general relations (45.32)-(45.54) for the polarization have the advantages of being invariant (independent of the coordinate system) and of being formulated for the complex displacement vector of the wave as a whole, so there is no need to separate **u** into amplitude and phase factors in order to use them.

These relationships for the polarization of the displacement, as represented by the complex vector **u**, can be transferred to any complex vector, no matter what its physical or mathematical significance.

The following are some general definitions [17] on this basis. We are given an arbitrary three-dimensional complex vector

$$A = A' + iA''; \tag{45.55}$$

which will be called

$$\text{linear if } [AA^*] = 0, \tag{45.56}$$

$$\text{nonlinear if } [AA^*] \neq 0. \tag{45.57}$$

In turn, a nonlinear complex vector **A** is called

$$\text{linear if } A^2 = 0, \tag{45.58}$$

$$\text{elliptical if } A^2 \neq 0. \tag{45.59}$$

Finally, an elliptical vector **A** is termed

$$\text{canonical if } A^{*^2} = A^2. \tag{45.60}$$

The meaning of these definitions will be quite clear from the arguments of this section; in particular, we can give the following brief but exact and complete definition of an inhomogeneous wave: a plane wave with a nonlinear refraction vector.

46. Inhomogeneous Waves at a Free Boundary

Section 44 allows us to characterize the polarization of inhomogeneous waves propagating in an isotropic solid. The "longitudinal" wave of (44.24) is elliptically polarized in the plane of the refraction vector $\mathbf{m}_0$, which from (45.60) is canonical, because (44.21) gives its square as real.

It is readily shown that the "transverse" wave of (44.25) can be linearly polarized, for the vector

$$u_1 = [m_1 m_1^*] = -2i[m_1' m_1''] \qquad ([u_1 u_1^*] = 0) \tag{46.1}$$

satisfies (44.25). The direction of the linear polarization is perpendicular to the plane of the complex nonlinear refraction vector $\mathbf{m}_1$; here section 45 (Fig. 18) shows that the polarization ellipse of vector $\mathbf{u}_1$ is orthogonal to the ellipse of $\mathbf{m}_1$ ($\mathbf{u}_1\mathbf{m}_1 = 0$). The general expression for $\mathbf{u}_1$ is a linear combination of these two vectors with arbitrary complex coefficients, so in the general case the polarization of $\mathbf{u}_1$ may range from linear to elliptical, the direction of the plane of $\mathbf{u}_1$ varying within wide limits. We introduce a unit vector $\mathbf{p}$ perpendicular to the plane of $\mathbf{m}_1$, which may be defined via

$$p = -\frac{i[m_1 m_1^*]}{|[m_1 m_1^*]|} \qquad (p^2 = 1, \quad pm_1 = 0, \quad p = p^*). \tag{46.2}$$

Then the general expression for the displacement of the "transverse" wave is

$$u_1 = C_1 p + C_2 [pm_1], \tag{46.3}$$

in which C_1 and C_2 are arbitrary complex coefficients. The square of $\mathbf{u}_1$ is [see (44.22)]

$$u_1^2 = C_1^2 + C_2^2 m_1^2 = C_1^2 + \frac{1}{a} C_2^2. \tag{46.4}$$

Taking $C_2 = \pm i\sqrt{a}C_1$, we get $\mathbf{u}_1^2 = 0$, so a "transverse" inhomogeneous wave in an isotropic medium can be circularly polarized. The plane of $\mathbf{u}_1$ (the plane of polarization) must have a definite orientation relative to $\mathbf{m}_1$. Here vector $\mathbf{u}_1$ takes the form

$$u_1 = C(p \pm i\sqrt{a}[pm_1]) = C((p \mp \sqrt{a}[pm_1'']) \pm i\sqrt{a}[pm_1']). \tag{46.5}$$

Figure 19 shows the corresponding vectors; it is clear from this that the plane of $\mathbf{u}_1$ passes through the vector $OB = \sqrt{a}[\mathbf{pm}_1']$ and the vector OA_1 or OA_2, which is equal to $\mathbf{p} \pm \sqrt{a}[\mathbf{pm}_1'']$. The inclination ϑ of the plane of polarization to the normal $\mathbf{p}$ is given by

$$\tan\vartheta = \sqrt{a}\,|[pm_1'']| = \sqrt{a}\,|m_1''|, \tag{46.6}$$

i.e., is proportional to the length of the extinction vector (section 44). The end of $\mathbf{u}_1$ describes a circle in this plane once per cycle. This shows that the plane of polarization of $\mathbf{u}_1$ in the "transverse" wave may lie at an angle to the plane of $\mathbf{m}_1$; this angle, and the angle of the plane of polarization to $\mathbf{p}$, may vary, C_1 and C_2 being the decisive quantities.

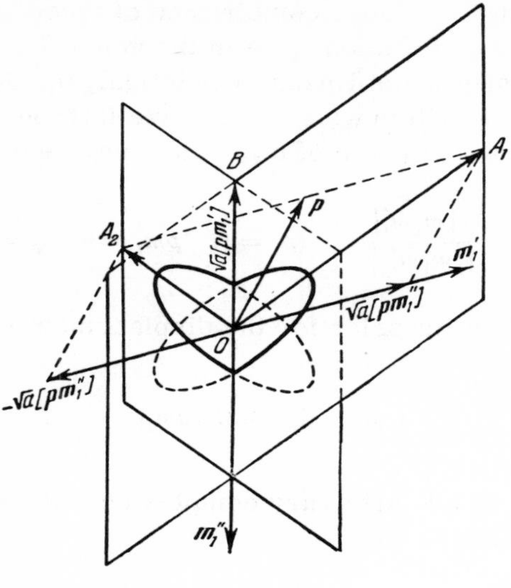

Fig. 19

Now we consider two particular boundary problems that arise for inhomogeneous elastic waves. The first is a logical extension of the case discussed at the end of section 42 (Fig. 16), where we saw that a transverse wave incident at a limiting angle ψ_0 defined by (42.40) via

$$\xi_0' = 0, \quad a^2 = \frac{1}{a+b}, \quad m_0' = b, \quad u_0' = C_0' b, \tag{46.7}$$

will give rise to a reflected longitudinal wave propagating parallel to the boundary of the isotropic medium. An angle of incidence greater than ψ_0 (Fig. 16) does not allow line A'B' to intersect or touch the longitudinal-wave sheet, so the point of intersection will be imaginary. Then for $\psi > \psi_0$ we have

$$a^2 = [m_1 q]^2 = m_1^2 \sin^2 \psi > m_1^2 \sin^2 \psi_0 = \frac{1}{a+b}, \tag{46.8}$$

so ξ_0' becomes purely imaginary:

$$\xi_0' = \sqrt{\frac{1}{a+b} - a^2} = \pm i \sqrt{a^2 - \frac{1}{a+b}} = \pm i\eta. \tag{46.9}$$

and the refraction vector of the reflected longitudinal wave becomes a complex nonlinear vector:

$$\boldsymbol{m}_0 = \boldsymbol{b} \pm i\eta\boldsymbol{q}, \quad \boldsymbol{m}_0^2 = \frac{1}{a+b}. \tag{46.10}$$

This is the case of an inhomogeneous wave of formulas (44.21) and (44.24); from (44.2) and (44.6),

$$\boldsymbol{m}_0'' = \pm i\eta\boldsymbol{q}, \quad \boldsymbol{u} = \boldsymbol{u}^0 e^{\mp\omega\eta\boldsymbol{q}\boldsymbol{r}} e^{i\omega(\boldsymbol{b}\boldsymbol{r}-t)}. \tag{46.11}$$

The amplitude of this wave will increase without limit away from the boundary ($\mathbf{qr} < 0$) if we take the upper sign, so the lower sign should be taken in (46.10), which causes $\mathbf{m}_0$ to be uniquely determined by the boundary conditions. The displacement vector of this wave is $\mathbf{u}_0' = \mathbf{C}_0'\mathbf{m}_0$, in which $\mathbf{C}_0'$ is given by (42.34); also, $\xi_0' = -i\eta$, so this coefficient is complex. Further, the above argument shows that this reflected inhomogeneous "longitudinal" wave is elliptically polarized, the polarization ellipse lying in the plane of incidence with its major axis b parallel to the boundary plane. The phase planes of this wave are perpendicular to $\mathbf{b} = \mathbf{m}_0'$, so they propagate along the boundary.

We assume the incident transverse wave to be polarized in the plane of incidence, since components normal to the plane of incidence are reflected independently (sections 42 and 43). The reflected transverse wave is also linearly polarized in the plane of incidence, because its displacement $\mathbf{u}_2' = C_2'(\xi_2'\mathbf{b} - \mathbf{a}^2\mathbf{q})$ is a linear vector. From (42.34) with $\xi_0' = -i\eta$, we have that the amplitude coefficient satisfies

$$|C_2'| = |C_2|. \tag{46.12}$$

This relation coincides with the corresponding result for the reflection of electromagnetic waves under analogous conditions ($\psi > \psi_0$, i.e., for internal reflection), but the analogy is restricted, because elastic waves are always totally reflected at a free boundary, since no wave is produced outside the medium.

The second case of inhomogeneous waves at a free boundary of an isotropic body is related to the different formulation of the boundary problem given at the end of section 41. So far we have taken the incident wave as given (via the refraction and displacement vectors), but a more general approach is to assume that all possible waves are present at the boundary and then to determine what combinations of these waves can satisfy the boundary conditions and the equations of motion, and what values of the parameters are required to do this. This problem will be considered first for the free boundary of an isotropic medium. From (41.17) and (41.23), all the refraction vectors must have the same projection **b** on the boundary plane, while their projections ξ on the normal **q** are given by $\mathbf{m}^2 = 1/(a+b)$ and $\mathbf{m}^2 = a$, so there are four possible values for ξ: two for longitudinal waves ($\pm\xi_0$):

$$\xi_0 = +\sqrt{\frac{1}{a+b} - a^2}, \quad \boldsymbol{m}_0^{\pm} = \boldsymbol{b} \pm \xi_0 \boldsymbol{q}, \tag{46.13}$$

which correspond to two longitudinal waves

$$\boldsymbol{u}_0^{\pm} = C_0^{\pm} \boldsymbol{m}_0^{\pm} = C_0^{\pm} (\boldsymbol{b} \pm \xi_0 \boldsymbol{q}). \tag{46.14}$$

and two ($\pm\xi_1$)

$$\xi_1 = +\sqrt{\frac{1}{a} - a^2}, \quad \boldsymbol{m}_1^{\pm} = \boldsymbol{b} \pm \xi_1 \boldsymbol{q}, \tag{46.15}$$

which correspond, from (42.18), to four independent transverse waves:

$$\boldsymbol{u}_1^{\pm} = C_1^{\pm} \boldsymbol{a}, \quad \boldsymbol{u}_2^{\pm} = C_2^{\pm} (a^2 \boldsymbol{q} \mp \xi_1 \boldsymbol{b}). \tag{46.16}$$

The boundary conditions of (41.17) for the refraction vectors are satisfied no matter what **a** (or **b**) may be. It remains to satisfy the boundary conditions of (41.33), which for isotropic media take the form of (42.36) and lead to the equation

$$\begin{aligned} &\frac{b-a}{b+a}(C_0^+ + C_0^-)\boldsymbol{q} + 2a\xi_0(C_0^+ \boldsymbol{m}_0^+ - C_0^- \boldsymbol{m}_0^-) + \\ &\quad + a\left[\xi_1 (C_1^+ - C_1^-)\boldsymbol{a} + \xi_1 (C_2^+ (a^2 \boldsymbol{q} - \xi_1 \boldsymbol{b}) - C_2^- (a^2 \boldsymbol{q} + \xi_1 \boldsymbol{b})) + \right. \\ &\quad\quad \left. + a^2 (C_2^+ \boldsymbol{m}_1^+ + C_2^- \boldsymbol{m}_1^-)\right] = 0. \end{aligned} \tag{46.17}$$

But **a**, **q**, and **b** are linearly independent, so this splits up into three equations:

$$\xi_1 (C_1^+ - C_1^-) = 0, \tag{46.18}$$

$$(1 - 2a\mathbf{a}^2)(C_0^+ + C_0^-) + 2a\xi_1 \mathbf{a}^2 (C_2^+ - C_2^-) = 0, \tag{46.19}$$

$$2a\xi_0 (C_0^+ - C_0^-) - (1 - 2a\mathbf{a}^2)(C_2^+ + C_2^-) = 0. \tag{46.20}$$

Here we do not consider as given *a priori* either $\mathbf{a}$ or any of the $C_s^{\pm}$ (s = 0, 1, 2), which is why the former symbols C_s and C_s' are not used, for these distinguish the given incident wave from the unknown reflected one. To these we join the equation

$$\xi_0^2 + \mathbf{a}^2 = \frac{1}{a+b}, \quad \xi_1^2 + \mathbf{a}^2 = \frac{1}{a}, \tag{46.21}$$

whereupon (46.18)-(46.21) together cover all possible cases of plane monochromatic waves at the free boundary of an isotropic medium. In particular, these contain the solutions to all the problems considered in section 42. In fact, if we specify $\mathbf{a}^2$ and an incident longitudinal wave ($C_0^+ = C_0$), and also assume that there are no other incident waves ($C_1^+ = C_2^+ = 0$), we get system (42.21)-(42.23). Putting $C_1^+ = C_1$, $C_0^+ = C_2^+ = 0$, we get the problem of (42.32)-(42.34). However, the system of (46.18)-(46.21) contains far more than this, for it allows us to examine all possible combinations of plane elastic waves at a free boundary. Equation (46.18) is unrelated to (46.19) and (46.20), so we may consider only the latter two. For example, we may ask whether there can be a wave combination such that $C_2^+ = C_2^- = 0$. From (46.19) and (46.20) it follows at once that this is possible either for $C_0^+ - C_0^- = 1 - 2a\mathbf{a}^2 = 0$ or for $C_0^+ + C_0^- = \xi_0 = 0$; the same result is obtained from (42.24) and (42.25).

Consider now the possible existence of the waves C_0^+, C_2^+ ($C_0^- = C_2^- = 0$). System (46.19) and (46.20) becomes

$$\left.\begin{aligned} (1 - 2a\mathbf{a}^2)\, C_0^+ + 2a\xi_1 \mathbf{a}^2 C_2^+ &= 0, \\ 2a\xi_0 C_0^+ - (1 - 2a\mathbf{a}^2)\, C_2^+ &= 0. \end{aligned}\right\} \tag{46.22}$$

For C_0^+, C_2^+ to be different from zero we must have

$$(1 - 2a\mathbf{a}^2)^2 + 4a^2 \xi_0 \xi_1 \mathbf{a}^2 = 0. \tag{46.23}$$

This is possible only if either $\mathbf{a}^2 < 0$ or $\xi_0 \xi_1 < 0$. The first can be disregarded, because $\mathbf{b} = [\mathbf{qq}]$ is a purely imaginary vector if $\mathbf{a}$ is purely imaginary; but then the extinction vector (section 44) will be parallel to the boundary plane, and the latter is taken as infinite, so the wave amplitude will increase indefinitely in some direction. As regards ξ_0 and ξ_1, they are, from the definitions of (46.13) and (46.15), positive if they are real; thus we only have to consider the case of ξ_0 and ξ_1 purely imaginary, which occurs if

$$\mathbf{a}^2 > \frac{1}{a}. \tag{46.24}$$

Thus we put

$$\xi_0 = \pm i \sqrt{\mathbf{a}^2 - \frac{1}{a+b}}, \quad \xi_1 = \pm i \sqrt{\mathbf{a}^2 - \frac{1}{a}}, \tag{46.25}$$

the signs of ξ_0 and ξ_1 being identical. Equation (46.23) gives after squaring that

$$(1 - 2a\mathbf{a}^2)^4 = 16a^4\mathbf{a}^4 \left(\mathbf{a}^2 - \frac{1}{a}\right)\left(\mathbf{a}^2 - \frac{1}{a+b}\right). \tag{46.26}$$

We expand the brackets and introduce the symbols

$$x = 2a\mathbf{a}^2, \quad g = \frac{a}{a+b} = \frac{v_1^2}{v_0^2}, \tag{46.27}$$

to get

$$2(1-g)x^3 - 2(3-2g)x^2 + 4x - 1 = 0. \tag{46.28}$$

Only real $x > 2$ are physically significant [see (46.24)]; it is found [38] that this equation has only one such root for any actual isotropic medium, the value being

$$2.19 \leqslant x \leqslant 2.62 \quad \text{for} \quad 0 \leqslant g \leqslant \frac{1}{2}. \tag{46.29}$$

The lower sign must be taken in (46.25) if the amplitude is not to increase above the boundary. Thus

$$\xi_0 = -i \sqrt{\frac{x}{2a} - \frac{1}{a+b}}, \quad \xi_1 = -i \sqrt{\frac{x-2}{2a}}. \tag{46.30}$$

From (46.22) we get $C_0^+/x\xi_1 = C_2^+/(x-1) = C$, so from (44.6), (46.14), and (46.16) we have

$$u = C\left[x\xi_1(b+\xi_0 q)e^{\omega|\xi_0|rq} + (x-1)\left(\frac{x}{2a}q - \xi_1 b\right)e^{\omega|\xi_1|rq}\right]e^{i\omega(br-t)}. \tag{46.31}$$

This represents a plane wave propagating parallel to the boundary surface along the direction of **b** with the phase velocity

$$v = \frac{1}{|b|} = \frac{1}{|a|} = \sqrt{\frac{2a}{x}}. \tag{46.32}$$

From (46.31), the amplitude $\mathbf{u}^0$ is of rather complicated structure:

$$u^0 = x\left(\frac{x-1}{2a}e^{\alpha_1} - |\xi_0\xi_1|e^{\alpha_0}\right)q + i|\xi_1|((x-1)e^{\alpha_1} - xe^{\alpha_0})b, \tag{46.33}$$

in which $\alpha_0 = \omega|\xi_0|\mathbf{rq}$, $\alpha_1 = \omega|\xi_1|\mathbf{rq}$. The wave is elliptically polarized in the (**q**, **b**) plane in the general case. The shape of the ellipse is dependent on the distance **rq** of the point in the medium from the boundary; it is readily deduced via the methods of section 45. The principal axes of the ellipse are parallel to **b** and **q**. Linear polarization is possible only at those points where $[\mathbf{u}^0\mathbf{u}^{0*}] = 0$, i.e., when

$$e^{\alpha_0-\alpha_1} = \frac{x-1}{2a|\xi_0\xi_1|} \tag{46.34}$$

or

$$e^{\alpha_0-\alpha_1} = \frac{x-1}{x}. \tag{46.35}$$

Now $|\xi_0| > |\xi_1|$ and $\mathbf{rq} \le 0$, so $\alpha_0 - \alpha_1 \le 0$, and hence $e^{\alpha_0-\alpha_1} \le 1$, and thus (46.35) must be obeyed at some distance from the boundary; the vibrations will be linearly polarized along **q** (normal to the boundary) at points on some plane parallel to the boundary. Condition (46.34) cannot be obeyed, because

$$\left(\frac{x-1}{2a|\xi_0\xi_1|}\right)^2 = \frac{(x-1)^2}{(x-2)(x-2g)} > 1,$$

so linear vibrations parallel to the boundary are impossible.

The amplitude decreases exponentially away from the boundary, so the energy of this wave is largely localized near the boundary; the waves of (46.31) are therefore called surface waves or Rayleigh waves, since they were first studied by Lord Rayleigh in 1887 [39].

It is simple to generalize the results of section 43 to the free boundary of any crystal. In place of (43.7) we may put (41.33) in the following form, by analogy with (46.17):

$$C_0^+U_0^+ + C_1^+U_1^+ + C_2^+U_2^+ + C_0^-U_0^- + C_1^-U_1^- + C_2^-U_2^- = 0, \quad (46.36)$$

in which the $U_s^\pm$ are defined by (43.8), while the corresponding $\xi_s^\pm$ are found as the possible solutions to (43.16) for some **b**; they are therefore functions of **b**. Thus (43.8) in principle allows us to find all the $U_s^\pm$ as functions of the c_{ijln}, **q**, and **b**. The most general boundary problem then becomes that of finding all possible $C_s^\pm$ and **b** that satisfy (46.36). If we specify **b** one of the C_s^+ (incident wave), while seeking to determine C_s^-, this will be the usual approach to reflection from the free boundary of a crystal, which was considered in section 43. If, on the other hand, we take **b** as undetermined, as also some of the C_s (putting the others equal to zero), we in this way can find other types of wave analogous to the Rayleigh waves of isotropic media.

Consider the following two cases. First consider the possibility of satisfying the boundary conditions of (46.36) in the presence of two waves $\mathbf{U}_0$ and $\mathbf{U}_2$. In that case (46.36) reduces to the equation

$$C_0U_0 + C_2U_2 = 0. \quad (46.37)$$

The following is the condition for the existence of finite C_0 and C_2 that satisfy this equation:

$$[U_0U_2] = 0. \quad (46.38)$$

This condition defines **b**, if the condition can be met; it also finally defines the expressions for $\mathbf{U}_0$ and $\mathbf{U}_2$, where **b** appears as a parameter, as well as C_0/C_2. This process has been carried through for an isotropic medium in (46.22)-(46.31). A scalar equation (see section 43) may be written in place of (46.38) for the case where the plane of incidence is a symmetry plane of the crystal:

$$\boldsymbol{a}\,[\boldsymbol{U}_0\boldsymbol{U}_2] = 0, \tag{46.39}$$

since $\mathbf{U}_0$ and $\mathbf{U}_2$ lie in the plane of incidence and $\mathbf{a}\,\|\,[\mathbf{U}_0\mathbf{U}_2]$. In the particular case of the orientation considered in section 43 for a hexagonal crystal, we get from (43.46) and (46.39) that

$$P_1Q_1 + (P_2Q_1 - P_1Q_2)\,\xi_0\xi_2 + P_2Q_1\,(\xi_0^2 + \xi_2^2) + P_2Q_2\xi_0^2\xi_2^2 = 0. \tag{46.40}$$

This equation here replaces (46.23). Taking an isotropic medium as a particular case of a hexagonal crystal (section 43), we can show simply that (46.40) becomes

$$(4ab\boldsymbol{a}^2 - (a+b))\,\xi_2 - (a+b)\xi_0 = 0, \tag{46.41}$$

which is a different form of the condition of (46.23) and leads to the same equation (46.28).*

Now consider the possibility of relation (46.36) for the case of three waves:

$$C_0\boldsymbol{U}_0 + C_1\boldsymbol{U}_1 + C_2\boldsymbol{U}_2 = 0, \tag{46.42}$$

which implies linear dependence of vectors $\mathbf{U}_0$, $\mathbf{U}_1$, $\mathbf{U}_2$, then the condition for finite C_0, C_1, C_2 becomes

$$\boldsymbol{U}_1\,[\boldsymbol{U}_0\boldsymbol{U}_2] = 0. \tag{46.43}$$

It is readily seen that (46.43) reduces to (46.39) for isotropic media, since $\mathbf{U}_1\,\|\,\mathbf{a}$ (section 43), as when the plane of incidence coincides with a symmetry plane of the crystal; (46.42) in this case splits up into $C_1 = 0$ and (46.37). However, we cannot assert *a priori* in the general case that (46.42) and (46.43) are equivalent to (46.37) and (46.39) and give nothing new. This aspect requires examination.

---

* This may be shown by multiplying (46.41) by $\xi_0 - \xi_2$ [see (43.46)] and expressing ξ_0^2, ξ_2^2 in terms of $\mathbf{a}^2$.

Chapter 9

Elastic Waves and the Thermal Capacity of a Crystal

47. Statistical Theory of the Thermal Capacity of a Solid

Dulong and Petit in 1819 discovered the rule that the atomic thermal capacity of any solid is 6 cal/deg. This rule agrees well with experiment for suitable high temperatures, and classical statistical physics readily explains it by considering each atom as a three-dimensional harmonic oscillator with quasielastic binding to some equilibrium position, the distribution of the energy over the degrees of freedom being invoked. This distribution indicates that each degree of freedom at equilibrium has the same energy kT/2, in which T is absolute temperature and k is Boltzmann's constant, which is 1.38044×10^{-16} erg/deg. The three degrees of freedom in the motion of a harmonic oscillator therefore have 3kT/2, but the total energy present is twice this, because the potential energy of an oscillator is equal to the kinetic energy per oscillator is therefore 3kT, and the energy per mole is

$$U = 3LkT = 3RT, \tag{47.1}$$

in which $L = 6.02486 \times 10^{23}$ mole^{-1} is Avogadro's number (the number of molecules in a gram-molecule) and R is about 2 cal/deg (the universal gas constant). Then the specific thermal capacity at constant volume is

$$C = C_V = \frac{dU}{dT} = 3R \approx 6 \text{ cal/deg}, \tag{47.2}$$

which is Dulong and Petit's rule.

However, there are marked deviations from this rule, especially at low temperatures; the thermal capacity also tends to zero as $T \to 0$. Einstein explained this by replacing the classical expression by Planck's quantum value for the mean energy of a one-dimensional oscillator of frequency ω:

$$\bar{\varepsilon} = \frac{h\omega}{e^{h\omega/kT} - 1}, \tag{47.3}$$

in which $h = 1.05439 \times 10^{-27}$ erg-sec is Planck's constant. Then the mean energy per mole is

$$U = \frac{3Lh\omega}{e^{h\omega/kT} - 1} = 3RT \frac{h\omega/kT}{e^{h\omega/kT} - 1}. \tag{47.4}$$

For $h\omega/kT \ll 1$, (high temperatures) we have $e^{h\omega/kT} \sim 1 + h\omega/kT + \ldots$, and (47.4) gives the result of (47.1); but (47.4) gives $C = dU/dT$ as approaching zero for $T \to 0$. Thus (47.4) is in general agreement with experiment for the limiting cases of high and low temperatures. Also, (47.4) shows that a rise in ω is equivalent to a fall in temperature; ω should be higher for solids in which the atoms are more firmly bound to their equilibrium positions, i.e., the harder solids, so these should tend, from (47.4), to have a lower thermal capacity at any given temperature. This is actually so (e.g., for diamond), but (47.4) still deviates from experiment.

Debye extended the theory of the specific heats of solids. Formula (47.4) represents the energy of a system of 3L independent harmonic oscillators all having the same frequency; but the atoms of a crystal lattice are very firmly bound, so their vibrations cannot be considered as independent. In fact, the crystal has to be considered as a system of coupled particles vibrating as a whole. In any case, the number of possible modes of vibration in such a system equals the number of degrees of freedom of the N particles, i.e., 3N; quantum theory gives the mean energy of any one mode as dependent only on frequency in accordance with (47.3), so the energy of a volume V must be deduced from the number of possible modes. The number between ω and $\omega + d\omega$ is dN_ω; the total energy is then the sum over all possible frequencies:

$$U = \int \bar{\varepsilon}_\omega \, dN_\omega = \int \frac{h\omega \, dN_\omega}{e^{h\omega/kT} - 1}. \tag{47.5}$$

In deducing U we may follow Debye and use a phenomenological approach in place of a microscopic examination of the modes of the coupled atoms; the crystal is then treated as a continuous elastic medium, and the spectrum of the proper modes of the individual atoms is replaced by the spectrum of elastic oscillations for the crystal as a whole, in which we consider only the low-frequency modes, whose wavelengths are large relative to the distance between atoms (acoustic waves). The discussion is usually simplified by considering a body in the form of a cube of side A. The proper modes consist of all possible standing waves that can persist in this cube. A standing wave will be set up by a wave traveling parallel to an edge A if this length contains an exact number of half-wavelengths; a wave normal **n** of any other direction requires that the projection of an edge on **n** be an integral number of half-wavelengths. This leads to the following conditions of periodicity at the faces of the cube:

$$n_1 A = l_1 \frac{\lambda}{2}, \quad n_2 A = l_2 \frac{\lambda}{2}, \quad n_3 A = l_3 \frac{\lambda}{2}, \tag{47.6}$$

in which $(n_1, n_2, n_3) = \mathbf{n}$, and l_1, l_2, and l_3 are positive integers. We introduce the wave vector

$$\boldsymbol{k} = \frac{2\pi}{\lambda} \boldsymbol{n} = \frac{\omega}{v} \boldsymbol{n}, \tag{47.7}$$

to put (47.6) as

$$l_i = \frac{A}{\pi} k_i. \tag{47.8}$$

Consider an elementary volume $(d\mathbf{k}) = dk_1 dk_2 dk_3$ in wave-vector space; from (47.8) we have that this corresponds to the following number of proper modes:

$$dN_{\boldsymbol{k}} = dl_1\, dl_2\, dl_3 = \frac{A^3}{\pi^3} dk_1\, dk_2\, dk_3 = \frac{V}{\pi^3} (d\boldsymbol{k}), \tag{47.9}$$

in which $V = A^3$ is the volume of the body. By analogy with (47.5) we get the energy of volume V as

$$U = \int \bar{\varepsilon}\, dN_{\boldsymbol{k}} = \frac{V}{\pi^3} \int \frac{h\omega\, (d\boldsymbol{k})}{e^{h\omega/kT} - 1}. \tag{47.10}$$

Converting to spherical coordinates in space **k**, we have

$$(d\boldsymbol{k}) = k^2\,dk\,d\Omega = \omega^2\,d\omega\,\frac{d\Omega}{v^3}. \tag{47.11}$$

Here we have used the fact that the phase velocity $v = v(\mathbf{n})$ is a function of the direction of the wave normal according to (15.21) and (15.22) but is independent of ω (i.e., we neglect dispersion, which causes λ_{iklm} to vary with frequency). Then $dk = d\omega/v$ for a given **n**. Further, for any given **n** there are three isonormal waves differing in velocity (v_1, v_2, and v_3) and in polarization, so the total number of proper modes must be found by summation over these three. Then the energy of (47.10) is replaced by

$$U = \frac{V}{8\pi^3}\int \bar{\varepsilon}\omega^2\,d\omega \int \left(\frac{1}{v_0^3} + \frac{1}{v_1^3} + \frac{1}{v_2^3}\right) d\Omega. \tag{47.12}$$

The factor 1/8 appears because the integration is carried over one octant for positive integers l_1, l_2, l_3. We put

$$I = \frac{1}{4\pi}\int \left(\frac{1}{v_0^3} + \frac{1}{v_1^3} + \frac{1}{v_2^3}\right) d\Omega = \left\langle \sum_{s=0}^{2} \frac{1}{v_s^3} \right\rangle, \tag{47.13}$$

to give (47.12) the form

$$U = \frac{1}{2\pi^2} VI \int \varepsilon\omega^2\,d\omega. \tag{47.14}$$

The integration with respect to frequency here runs from zero up to some maximum frequency, which is found from the condition that the number of proper modes equals the number of degrees of freedom. Let N_V be the number of molecules in volume V; then there are $3N_V$ degress of freedom, provided that each lattice node is filled by a complete molecule of the substance (molecular crystal). The parts of the molecule lie at different nodes in an ionic crystal, so the number of degrees of freedom is increased by a factor equal to the number of nodes per molecule. This is obvious, because the parts of the molecule at the various nodes (atomic groups, ions) may be considered as separate particles. Let s be the number of distinct structural nodes per molecule; then a volume V has $3sN_V$ degrees of freedom, and from (47.9) with (47.12)-

(47.14) we have the condition for the maximum frequency as

$$\int dN_k = \frac{1}{2\pi^2} VI \int\limits_0^{\omega_{max}} \omega^2\, d\omega = 3sN_V, \tag{47.15}$$

or

$$\omega_{max} = \left(\frac{18\pi^2 sN_V}{IV}\right)^{1/3}. \tag{47.16}$$

Then (47.14) becomes

$$U = \frac{1}{2\pi^2} VI \int\limits_0^{\omega_{max}} \frac{h\omega^3\, d\omega}{e^{h\omega/kT} - 1}. \tag{47.17}$$

We introduce the new variable of integration $x = h\omega/kT$ and put

$$\Theta = \frac{h\omega_{max}}{k}, \tag{47.18}$$

to get

$$U = \frac{1}{2\pi^2} VI \frac{(kT)^4}{h^3} \int\limits_0^{\Theta/T} \frac{x^3\, dx}{e^x - 1}. \tag{47.19}$$

This Θ is known as the characteristic temperature (Debye temperature); from (47.16) and (47.18)

$$I = \frac{18\pi^2 h^3 sN_V}{(k\Theta)^3 V}, \tag{47.20}$$

so (47.19) becomes

$$U = 9sN_V kT \left(\frac{T}{\Theta}\right)^3 \int\limits_0^{\Theta/T} \frac{x^3\, dx}{e^x - 1}. \tag{47.21}$$

The upper limit in the integral increases without limit as $T \to 0$; since

$$\int\limits_0^{\infty} \frac{x^3\, dx}{e^x - 1} = \frac{\pi^4}{15},$$

so we have that (47.21) for low temperatures gives us the energy of one mole of the substance ($N_V = L$) that

$$U = \frac{3\pi^4 sR}{5\Theta^3} T^4. \tag{47.22}$$

The molar thermal capacity is

$$C_V = \frac{dU}{dT} = \frac{12\pi^4 sR}{5\Theta^3} T^3. \tag{47.23}$$

This C_V tends to zero as $T \to 0$, in accordance with experiment.

Also, $\Theta/T \ll 1$ for high temperatures, and the integral is approximately $(\Theta/T)^3/3$, which gives Dulong and Petit's rule. Thus Debye's theory agrees well with experiment for the limiting cases of low and high temperatures; but Debye's formula (47.21) is how exact, because the ω_{max} of (47.15) has been defined on the basis that the set of modes corresponding to the degrees of freedom is represented solely by low frequencies starting at zero. This assumption becomes more accurate as the temperature falls, since the proportion of high-energy (high-frequency) excited quantized states increases with temperature. This means that measurements of specific heat and elastic constants near absolute zero play a special part in testing Debye's theory.

The quantities in (47.21) dependent on the properties of the substance are s, N_V, and Θ. Let $N_1 = N_V/V$ be the number of molecules (atoms) in unit volume. Then we may put

$$sN_1 = \frac{1}{V_a}, \tag{47.24}$$

in which V_a is the average volume per node (the atomic volume). Then from (47.20) and (47.24) we have the characteristic temperature as

$$\Theta = \frac{h}{k}\left(\frac{18\pi^2 sN_1}{I}\right)^{1/3} = \frac{h}{k}\left(\frac{18\pi^2}{IV_a}\right)^{1/3}. \tag{47.25}$$

The phase velocities are independent of direction in an isotropic solid, so the average of (47.13) is not needed, and we get, since $v_1 = v_2$, that

$$I = \frac{1}{v_0^3} + \frac{2}{v_1^3}. \tag{47.26}$$

However, it is a complicated problem to determine I for crystals, and this is considered in the sections that follow.

48. Computation of the Debye Temperature

This theory shows that any given solid has its own specific Debye temperature, which is given by (47.25) as

$$\Theta = C\,(IV_a)^{-1/3}, \tag{48.1}$$

in which C is a function of universal constants only:

$$C = \frac{h}{k}(18\pi^2)^{1/3}, \tag{48.2}$$

while the I of (47.13) equals the sum of the inverse cubes of the phase velocities averaged over all directions for the wave normal:

$$I = \left\langle \frac{1}{v_0^3} + \frac{1}{v_1^3} + \frac{1}{v_2^3} \right\rangle. \tag{48.3}$$

Section 15 gives the squares of the velocities as expressed via the reduced elastic moduli $\lambda_{\alpha\beta} = (1/\rho)c_{\alpha\beta}$. The $c_{\alpha\beta}$ listed in the tables are in units of 10^{11} dyne/cm^2, while ρ is in g/cm^3, so the velocities (and also $I^{-1/3}$) are in units of $10^{11/2}$ cm/sec. If V_a is put in cm^3, we get C as follows:

$$C = \frac{1.05439 \cdot 10^{-27} \cdot (18\pi^2)^{1/3} \cdot 10^{11/2}}{1.38044 \cdot 10^{-16}} = 1.3578 \cdot 10^{-5}.$$

Alternatively, for V_a in 10^{-21} cm^3, we get from (48.1) that

$$\Theta = 135.78\,(IV_a)^{-1/3}. \tag{48.4}$$

However, the tables do not always give V_a, but these can be deduced from ρ, Avogadro's number L, and the molecular or atomic weight A via (47.24):

$$V_a = \frac{1}{sN_1} = \frac{A}{\rho s L}. \tag{48.5}$$

Then instead of (48.4) we have

$$\Theta = 1146.8\left(\frac{\rho s}{AI}\right)^{1/3}. \tag{48.6}$$

Hence a theoretical derivation of the Debye temperature reduces to derivation of I from the known elastic constants. It has always been the case that the derivation of I for a crystal has involved cumbrous calculations, which recently have been deputed to computers [40]. Several methods have been proposed [43-50] for I, in view of its considerable interest, but detailed calculations have been performed only for a few cubic and medium-symmetry crystals, for which the symmetry simplifies the calculations. The entire calculations must be done afresh for each crystal because numerical methods are used with particular values for the elastic constants.

Sections 29 and 30 of Chapter 5 present general approximate methods for the phase velocities of elastic waves in crystals of all types, which not only very greatly reduce the volume of computation for I but also give general expressions for I in terms of the elastic constants for each system.

The simplest but roughest method for I is to replace the crystal by the isotropic medium on average most similar to it by the method of section 26; (26.12) and (26.13) give the elastic constants of this fictitious isotropic medium as

$$c_m = a_m + b_m = \langle \boldsymbol{n}\Lambda\boldsymbol{n}\rangle, \quad a_m = \frac{1}{2}\left(\langle\Lambda_t\rangle - \langle\boldsymbol{n}\Lambda\boldsymbol{n}\rangle\right), \tag{48.7}$$

or, from (26.18) and (26.19),

$$c_m = \frac{1}{15}\left(2\lambda_{iklk} + \lambda_{iikk}\right), \quad a_m = \frac{1}{30}\left(3\lambda_{iklk} - \lambda_{iikk}\right). \tag{48.8}$$

We introduce the components of the six-row matrix $\lambda_{\alpha\beta}$ (section 6) in place of the λ_{iklm} in the latter equations to get

$$c_m = \frac{1}{15}\left[3\left(\lambda_{11} + \lambda_{22} + \lambda_{33}\right) + 4\left(\lambda_{44} + \lambda_{55} + \lambda_{66}\right) + 2\left(\lambda_{12} + \lambda_{23} + \lambda_{31}\right)\right], \tag{48.9}$$

$$a_m = \frac{1}{15}\left[\left(\lambda_{11} + \lambda_{22} + \lambda_{33}\right) + 3\left(\lambda_{44} + \lambda_{55} + \lambda_{66}\right) - \left(\lambda_{12} + \lambda_{23} + \lambda_{31}\right)\right]. \tag{48.10}$$

For an isotropic medium with the parameters of (48.9) and (48.10),

$$v_0^2 = a_m + b_m = c_m, \quad v_1^2 = v_2^2 = a_m, \tag{48.11}$$

so from (47.26)

$$I = I_0 = c_m^{-3/2} + 2a_m^{-3/2}. \tag{48.12}$$

There is no difficulty in computing I from this [41], but the result is of low accuracy.

Values of I as accurate as may be desired may be obtained by using the approximate methods of sections 29 and 30 [53], but a modified form of the methods of those sections is more convenient. In sections 23 and 30 we used an isotropic medium most similar to the crystal for a given **n**, which gave a better approximation, but with parameters a and b dependent on **n**, We have to average over all **n** to find the I of (48.3), so the dependence of a and b on **n** complicates the problem. In other words, v_0, v_1, and v_2 are obtained more accurately if the parameters of the isotropic medium are functions of **n**, but then the averaging must be done approximately, which means that some accuracy is thereby lost, while the expressions become much more complicated.*

Therefore we consider an approximate theory entirely analogous to that of section 29 but based on the isotropic medium on average most similar to the crystal [54]. We represent tensor Λ for the crystal via (26.2) and (26.3) in the form

$$\Lambda = \Lambda_m + \Lambda'_m = a + b(\mathbf{n}\cdot\mathbf{n} + \alpha), \tag{48.13}$$

in which a and b are defined† by (26.13) or (48.7), where by analogy with (27.24)

$$\alpha = (1/b)\Lambda'_m = (1/b)(\Lambda - \Lambda_m) = (1/b)(\Lambda - a - b\mathbf{n}\cdot\mathbf{n}). \tag{48.14}$$

Then the following weaker conditions [see (26.11)] are obeyed instead of (27.25):

* But they are then much simpler [53] than when all other methods [43-50] are used.

† Subscript m to a and b has been omitted for simplicity.

$$\langle \alpha_t \rangle = \langle \mathbf{n}\alpha\mathbf{n} \rangle = 0. \tag{48.15}$$

This feature alters the relationships somewhat from those of sections 27 and 29. By analogy with (27.45) we put

$$v^2 = a + b\xi, \tag{48.16}$$

but the derivation that gave the characteristic equation of (27.46) now gives

$$\xi^3 - (1 + \alpha_t)\xi + (\alpha_t - \mathbf{n}\alpha\mathbf{n} + \bar{\alpha}_t)\xi - (\mathbf{n}\bar{\alpha}\mathbf{n} + |\alpha|) = 0. \tag{48.17}$$

From (48.13) and (48.16) we reduce $\Lambda\mathbf{u} = v^2\mathbf{u}$ to the form

$$(\mathbf{n}\cdot\mathbf{n} + \alpha)\mathbf{u} = \xi_i\mathbf{u}. \tag{48.18}$$

As in section 27, we substitute

$$\mathbf{u} = \mathbf{n} + \mathbf{u}', \quad \mathbf{u}'\mathbf{n} = 0 \tag{48.19}$$

and multiply scalarly by **n**, getting instead of (27.49) that

$$\xi = 1 + \mathbf{n}\alpha\mathbf{n} + \mathbf{n}\alpha\mathbf{u}'. \tag{48.20}$$

We use (48.19) and (48.20) to give (48.18) the following form, which is analogous to that of (29.1):

$$(1 + \mathbf{n}\alpha\mathbf{n} + \mathbf{n}^{\times 2}\alpha)\mathbf{u}' = \mathbf{k} - \mathbf{k}\mathbf{u}'\cdot\mathbf{u}', \tag{48.21}$$

in which the vector

$$\mathbf{k} = [\mathbf{n}[\alpha\mathbf{n}, \mathbf{n}] = (1/b)[\mathbf{n}[\Lambda\mathbf{n}, \mathbf{n}]] \tag{48.22}$$

is analogous to the **h** of (27.24) and (27.27) but differs from the latter merely in that the denominator b of (48.22) is constant and given by (48.7), not (27.7). By analogy with (29.2) and (27.57) we introduce the tensor

$$\beta = (1 + \mathbf{n}\alpha\mathbf{n} + \mathbf{n}^{\times 2}\alpha)^{-1} = \frac{1 + \mathbf{n}\alpha\mathbf{n} + \tau_t - \tau}{(1 + \mathbf{n}\alpha\mathbf{n})(1 + \mathbf{n}\alpha\mathbf{n} + \tau_t) + \bar{\tau}_t}, \tag{48.23}$$

in which

$$\tau = \mathbf{n}^{\times 2}\alpha = (\mathbf{n}\cdot\mathbf{n} - 1)\alpha, \quad \tau_t = \mathbf{n}\alpha\mathbf{n} - \alpha_t. \tag{48.24}$$

Then (48.21) may be given a form corresponding to (29.3):

$$\mathbf{u}' = \beta\mathbf{k} - \mathbf{k}\mathbf{u}'\cdot\beta\mathbf{u}'. \tag{48.25}$$

Obviously, we may now repeat all the arguments of section 29 that give approximate expressions for the displacement and velocity of the quasilongitudinal waves. The first iteration gives

$$\mathbf{u}'_{(1)} = \beta\mathbf{k}; \tag{48.26}$$

while the second approximation gives

$$\mathbf{u}'_{(2)} = \beta\mathbf{k} - \mathbf{k}\beta\mathbf{k}\cdot\beta^2\mathbf{k} \tag{48.27}$$

and so on. From (48.20) we have for ξ, which defines the velocity of the quasilongitudinal wave, that

$$\xi_0 = 1 + \epsilon, \qquad \epsilon = \mathbf{n}\alpha\mathbf{n} + \mathbf{n}\alpha\mathbf{u}', \tag{48.28}$$

while from (48.26) and (48.27) we have

$$\epsilon_{(1)} = \mathbf{n}\alpha\mathbf{n} + \mathbf{n}\alpha\beta\mathbf{k}, \tag{48.29}$$

$$\epsilon_{(2)} = \mathbf{n}\alpha\mathbf{n} + \mathbf{n}\alpha\beta\mathbf{k} - \mathbf{k}\beta\mathbf{k}\cdot\mathbf{n}\alpha\beta^2\mathbf{k}. \tag{48.30}$$

The expression for $\varepsilon_{(1)}$ is correct up to terms of the third order in α, inclusive; that for $\varepsilon_{(2)}$ is correct up to the fifth order; and that for $\varepsilon_{(k)}$ is correct up to order $2k+1$ in α (see section 29). From (48.23), (48.29), and (48.30) we have, retaining the terms of the appropriate order, that

$$\epsilon_{(1)} = \sigma + \mathbf{k}^2 + \mathbf{k}\alpha\mathbf{k} - \sigma\mathbf{k}^2 \quad (\sigma = \mathbf{n}\alpha\mathbf{n}) \tag{48.31}$$

$$\epsilon_{(2)} = \epsilon_{(1)} + \mathbf{k}^2(\sigma^2 - \overline{\tau}_t - \mathbf{k}^2) - (2\sigma + \tau_t)\,\mathbf{k}\alpha\mathbf{k} + \mathbf{k}\alpha\mathbf{k}(3\sigma^2 + 3\sigma\tau_t + (\tau_t)^2 - \overline{\tau}_t) +$$
$$+ \mathbf{k}^2(7\sigma^3 + 12\sigma^2\tau_t + 6\sigma(\tau_t)^2 + 3\sigma\overline{\tau}_t + \tau_t\overline{\tau}_t + (\tau_t)^3 - 3\mathbf{k}\alpha\mathbf{k} + 3\sigma\mathbf{k}^2). \tag{48.32}$$

From (48.16) we have for the I of (48.3) that

$$I = \langle (a + b\xi_0)^{-3/2} + (a + b\xi_1)^{-3/2} + (a + b\xi_2)^{-3/2} \rangle =$$
$$= a^{-3/2}\langle (1 + r_1\xi_1)^{-3/2} + (1 + r_1\xi_2)^{-3/2} \rangle + c^{-3/2}\langle (1 + r_2\epsilon)^{-3/2} \rangle, \tag{48.33}$$

in which

$$c = a + b, \quad r_1 = b/a, \quad r_2 = b/c. \tag{48.34}$$

The binomial theorem gives us for the first approximation that

$$I_1 = a^{-3/2} \left\langle 2 - \frac{3}{2} r_1(\xi_1 + \xi_2) + \frac{15}{8} r_1^2(\xi_1^2 + \xi_2^2) - \frac{35}{16} r_1(\xi_1^2 + \xi_2^3) \right\rangle +$$

$$+ c^{-3/2} \left\langle 1 - \frac{3}{2} r_2 \epsilon + \frac{15}{8} r_2^2 \epsilon^2 - \frac{35}{16} r_2^3 \epsilon^3 \right\rangle \tag{48.35}$$

and for the second approximation

$$I_2 = I_1 + \frac{63}{256} a^{-3/2} \left\langle 10 r_1^4(\xi_1^4 + \xi_2^4) - 11 r_1^5(\xi_1^5 + \xi_2^5) \right\rangle +$$

$$+ \frac{63}{256} c^{-3/2} \left\langle 10 r_2^4 \epsilon^4 - 11 r_2^5 \epsilon^5 \right\rangle. \tag{48.36}$$

The relation between the roots of an algebraic equation and the coefficients gives us from (48.17) that

$$\xi_0 + \xi_1 + \xi_2 = 1 + \alpha_t, \quad \xi_0(\xi_1 + \xi_2) + \xi_1 \xi_2 = \bar{\alpha}_t - \tau_t,$$

so from (48.28)

$$\rho_1 = \xi_1 + \xi_2 = \alpha_t - \epsilon, \quad \rho_2 = \xi_1 \xi_2 = \bar{\alpha}_t - \tau_t - (1 + \epsilon)(\alpha_t - \epsilon). \tag{48.37}$$

The symmetrical functions of the two variables ξ_1^k and ξ_2^k in (48.35) and (48.36) can be expressed in terms of ρ_1 and ρ_2, i.e., in terms of ε via (48.37). In fact,

$$\xi_1^2 + \xi_2^2 = \rho_1^2 - 2\rho_2,$$

$$\xi_1^3 + \xi_2^3 = \rho_1^3 - 3\rho_1\rho_2 = \rho_1(\xi_1^2 + \xi_2^2 - \rho_2),$$

$$\xi_1^4 + \xi_2^4 = \rho_1^4 - 4\rho_1^2\rho_2 + 2\rho_2^2 = (\xi_1^2 + \xi_2^2)^2 - 2\rho_2, \tag{48.38}$$

and, in general, we have the simple recurrence formula

$$\xi_1^{k+1} + \xi_2^{k+1} = \rho_1(\xi_1^k + \xi_2^k) - \rho_2(\xi_1^{k-1} + \xi_2^{k-1}).$$

From (13.37), (48.24), and (48.37) we have

$$\xi_1^2 + \xi_2^2 = (\alpha_t)^2 - \epsilon^2 + 2(\sigma - \epsilon - \bar{\alpha}_t), \tag{48.39}$$

$$\xi_1^3 + \xi_2^3 = (\alpha_t - \epsilon)[\alpha_t(\alpha_t + \epsilon) - 3(\epsilon - \sigma + \bar{\alpha}_t) - 2\epsilon^2], \tag{48.40}$$

and so on. We substitute an approximation for ε to get $\xi_1^k + \xi_2^k$ in the same approximation, where the k-th approximation is correct up to terms of degree 2k+1 in α, inclusive; unwanted terms are deleted from (48.39) and (48.40). For example, in the first approximation we get from (48.31) that $\varepsilon_{(1)}^2 = \sigma^2 + 2\sigma K^2$ and $\varepsilon_{(1)}^3 = \sigma^3$, so from (13.37) we get that (48.39) and (48.40) become

$$\xi_1^2 + \xi_2^2 = (\alpha^2)_t - (\mathbf{n}\alpha\mathbf{n})^2 - 2(\mathbf{k}^2 + \mathbf{k}\alpha\mathbf{k}), \qquad (48.41)$$

$$\xi_1^3 + \xi_2^3 = (\alpha_t - \mathbf{n}\alpha\mathbf{n})[\alpha_t(\alpha_t + \mathbf{n}\alpha\mathbf{n}) - 3(\mathbf{k}^2 + \bar{\alpha}_t) - 2(\mathbf{n}\alpha\mathbf{n})^2]. \qquad (48.42)$$

We see from (48.33)-(48.36) that, once ε and $\xi_1^k + \xi_2^k$ have been found, the problem reduces to averaging the expressions appearing in I over all directions of **n**. We see from (48.14) and (48.22) that the tensor

$$\alpha = \frac{1}{b}(\Lambda^n - a - b\mathbf{n}\cdot\mathbf{n})$$

contains the components of **n** as their squares, while $\sigma = \mathbf{n}\alpha\mathbf{n}$ contains these to the fourth degree, $\mathbf{k}^2$ contains them to the eighth degree, and $\mathbf{k}\alpha\mathbf{k}$ contains them up to degree 12. Thus the expressions in I in the first approximation contain homogeneous polynomials in the n_i of degree up to 12; the k-th approximation similarly contains them up to degree 4(2k+1), inclusive. This means that we must envisage calculations of means over products of various numbers of components of the unit vector in the final calculation of I. The next section deals with the covariant solution to this problem.

49. Averaging of the Products of Components of Unit Vector

The I of (48.18)-(48.31) has to be found via the means of products of large numbers of n_i. Here this problem is discussed in general form via a covariant approach.

Section 11 shows that the set of products $n_k n_l$ for all k and l forms a dyad (a tensor), and the averaging over all directions of **n** cannot alter the tensor character of this quantity, for an average is essentially a summation, and a sum of tensors is still a tensor. Hence we can say that $\langle n_k n_l \rangle = \alpha_{kl}$ or

$$\langle \mathbf{n}\cdot\mathbf{n} \rangle = \alpha, \qquad (49.1)$$

in which α is some tensor, which should not alter in response to any orthogonal transformation of the coordinates. Such a transformation S (with $\tilde{S} = S^{-1}$, see section 11) converts **n** to $\mathbf{n}' = S\mathbf{n}$, while α becomes $\alpha' = S\alpha S^{-1}$, and (49.1) becomes $\alpha' = \langle \mathbf{n}' \cdot \mathbf{n}' \rangle$. But $\langle \mathbf{n}' \cdot \mathbf{n}' \rangle = \langle \mathbf{n}' \cdot \mathbf{n} \rangle$, since in both parts we have an integral over all directions of **n** or **n'**. Therefore $\alpha' = S\alpha S^{-1} = \alpha$, so α remains unchanged under any transformation S. There is only one tensor (leaving aside scalar factors) that has this property, namely unit tensor, so α must be proportional to unit tensor: $\alpha_{kl} = c\delta_{kl}$. Thus we have

$$\langle n_k n_l \rangle = c\delta_{kl}. \tag{49.2}$$

The constant c is readily found by taking the trace of both parts. On the left we have $\langle n_k n_k \rangle = \langle n^2 \rangle = \langle 1 \rangle = 1$, for the result of averaging a constant quantity is to leave it unchanged. On the right we have 3c, so $c = 1/3$, and we finally have

$$\langle n_k n_l \rangle = \frac{1}{3}\delta_{kl}. \tag{49.3}$$

Similarly, the mean of the product of four components of **n** can be expressed only via combinations of unit tensors, there being three such combinations in this case:

$$\langle n_i n_k n_l n_m \rangle = a\delta_{ik}\delta_{lm} + b\delta_{il}\delta_{km} + c\delta_{im}\delta_{kl}. \tag{49.4}$$

The left side of this is unaltered by any permutation of the subscripts, so on the right we must have $a = b = c$. To find the common factor c we put $i = k$, $l = m$; on the left we get $\langle n_i^2 n_l^2 \rangle = 1$, and on the right $\delta_{ii}\delta_{ll} + 2\delta_{il}\delta_{il} = 15$, so $c = 1/15$ and*

$$\langle n_i n_k n_l n_m \rangle = \frac{1}{15}(\delta_{ik}\delta_{lm} + \delta_{il}\delta_{km} + \delta_{im}\delta_{kl}). \tag{49.5}$$

These arguments are clearly applicable to the mean of the product of any number of n_i; here no calculation is needed in order

---

* Direct calculations of the mean via $\langle n_k n_l \ldots \rangle = 1/4\pi \int n_k n_l \ldots d\Omega$, in which $n_1 = \sin\vartheta\cos\varphi$, $n_2 = \sin\vartheta\sin\varphi$, $n_3 = \cos\vartheta$, naturally gives the same results, but via far more cumbrous expressions.

to see that the mean of the product of an odd number of n_i is zero, because in that case the result of averaging cannot be represented as a combination of unit tensors. This can be seen in another way, since to every term $n_i n_k \ldots n_s$ there corresponds (if the number of n_i is odd) a similar term with the reverse sign, which corresponds to the direction of **n** reversed (i.e., to reversal of the signs of all the components of **n**). These terms therefore cancel out.

Subscripts i, k, l,... take only the values 1, 2, 3, so any product of the components of **n** may be put in the form $n_1^{k_1} n_2^{k_2} n_3^{k_3}$. Here the corresponding mean is zero if any of the integers k_1, k_2, and k_3 is odd, for then all the products of the components of the unit tensor into which the mean is resolved will contain the factor $\delta_{ss'}$ with $s \neq s'$, which is zero.

This shows that the mean of 2s components n_k may be put as

$$M_{2s} = \langle n_{k_1} n_{k_2} \ldots n_{k_{2s}} \rangle = C_{2s} \sum^{2s}. \tag{49.6}$$

Here C_{2s} is a numerical factor, while $\sum^{2s}$ is the sum of all possible distinct products of s components $\delta_{kk'}$ of unit tensor, in which k and k' are taken from the set $k_1, k_2, \ldots, k_{2s}$. Similarly,

$$M_{2(s+1)} = \langle n_{k_1} n_{k_2} \ldots n_{2s} n_{2s+1} n_{2s+2} \rangle = C_{2s+2} \sum^{2s+2}. \tag{49.7}$$

It is readily seen that for $\sum^{2s+2}$ we have

$$\sum^{2s+2} = \delta_{k_{2s+2} k_{2s+1}} \sum^{2s} + \delta_{k_{2s+2} k_1} \sum^{2s}{}'_{k_1} + \delta_{k_{2s+2} k_2} \sum^{2s}{}'_{k_2} + \ldots \delta_{k_{2s+2} k_{2s}} \sum^{2s}{}'_{k_{2s}}, \tag{49.8}$$

in which $\sum^{2s}{}'_{k}$ denotes the $\sum^{2s}$ of (49.6) with k replaced by k_{2s+1}. In (49.7) and (49.8) we put $k_{2s+2} = k_{2s+1}$ and sum (convolute) with respect to these indices; then $n_{k_{2s+2}} = n_{k_{2s+1}}$ and the left side of (49.7) becomes equal to M_{2s}. On the other hand, (49.8) shows that $\sum^{2s+2}$ becomes $(2s+3) \sum^{2s}$. Then comparison of (49.6) and (49.7) with $k_{2s+2} = k_{2s+1}$ gives

$$C_{2s} \sum^{2s} = C_{2s+2} (2s+3) \sum^{2s}. \tag{49.9}$$

whence we have

$$C_{2s+2}=\frac{C_{2s}}{2s+3}. \tag{49.10}$$

From (49.3) and (49.5) we have $C_2=1/3$, $C_4=1/15=C_2/(2\cdot 1+3)$; from (49.10) we get the following series of values for C_{2s}:

$$C_2=\frac{1}{3},\quad C_4=\frac{1}{3\cdot 5},\quad C_6=\frac{1}{3\cdot 5\cdot 7},\ \ldots,\ C_{2s}=\frac{1}{3\cdot 5\cdot 7\ldots(2s+1)}. \tag{49.11}$$

For the products of successive odd (or even) numbers we have

$$1\cdot 3\cdot 5\ \ldots\ (2s+1)=(2s+1)!!,\qquad 2\cdot 4\cdot 6\ \ldots\ 2s=(2s)!!, \tag{49.12}$$

so

$$C_{2s}=\frac{1}{(2s+1)!!} \tag{49.13}$$

and

$$M_{2s}=\langle n_{k_1}n_{k_2}\ \ldots\ n_{k_{2s}}\rangle=\frac{1}{(2s+1)!!}\sum^{2s}. \tag{49.14}$$

From (49.8) we readily obtain the number of terms of the form $\delta_{k_1k_2}\delta_{k_3k_4}\cdots\delta_{k_{2s-1}k_{2s}}$ in $\sum^{2s}$; let this number be p_{2s}. Then from (49.8)

$$p_{2s+2}=(2s+1)\,p_{2s}. \tag{49.15}$$

Now $p_2=1$ [see (49.3)], so $p_4=1\times 3$, $p_6=1\times 3\times 5$, and so on. In general,

$$p_{2s}=(2s-1)!! \tag{49.16}$$

Then we readily obtain explicit expressions for the means $\langle n_1^{2k_1}n_2^{2k_2}n_3^{2k_3}\rangle$ that differ from zero. From (49.6) and (49.13)

$$\langle n_1^{2k_1}n_2^{2k_2}n_3^{2k_3}\rangle=\frac{1}{[2(k_1+k_2+k_3)+1]!!}\sum^{2(k_1+k_2+k_3)}. \tag{49.17}$$

To find $\sum^{2(k_1+k_2+k_3)}$, we use (49.8) and consider $\sum^{2[(k_1+1)+k_2+k_3]}$; here $k_{2s+2}=k_{2s+1}=1$, and in the terms of (49.8) the Kronecker symbols to the $\sum'$, will differ from zero in $2k_1$ cases. Here we have $\sum'=\sum$, so we get

$$\sum^{2[(k_1+1)+k_2+k_3]} = (2k_1+1) \sum^{2(k_1+k_2+k_3)} . \tag{49.18}$$

The k_1, k_2, and k_3 are equivalent, so analogous formulas apply for k_2 or k_3 changed by unity. From (49.17) and (49.18) we have

$$\langle n_1^{2(k_1+1)} n_2^{2k_2} n_3^{2k_3}\rangle = \frac{1}{[2(k_1+1+k_2+k_3)+1]!!} \sum^{2[(k_1+1)+k_2+k_3]} =$$

$$= \frac{2k_1+1}{2(k_1+k_2+k_3)+3} \cdot \frac{\sum^{2(k_1+k_2+k_3)}}{[2(k_1+k_2+k_3)+1]!!} =$$

$$= \frac{2k_1+1}{2(k_1+k_2+k_3)+3} \langle n_1^{2k_1} n_2^{2k_2} n_3^{2k_3}\rangle. \tag{49.19}$$

Here we put $k_2 = k_3 = 0$; then $\langle n_1^{2k_1+2}\rangle = (2k_1+1)/(2k_1+3)\langle n_1^{2k_1}\rangle$, so with $k_1 = 0, 1, 2, \ldots$, we get

$$\langle n_1^{2k_1}\rangle = \frac{1}{2k_1+1} . \tag{49.20}$$

Of course, this is true also for n_2 and n_3, so we can put

$$\langle n_k^{2s}\rangle = \frac{1}{2s+1}, \tag{49.21}$$

in which there is no summation with respect to k. From (49.19) and (49.20), with the roles of k_2 and k_1 interchanged, we get for $k_3 = 0$ that

$$\langle n_1^{2k_1} n_2^2\rangle = \frac{1}{2k_1+3}\langle n_1^{2k_1}\rangle = \frac{(2k_1-1)!!}{(2k_1+1)!!}, \quad \langle n_1^{2k_1} n_2^4\rangle = \frac{(2k_1-1)!!\,3!!}{(2k_1+5)!!}$$

and so on. It is readily seen that the corresponding general formula is

$$\langle n_1^{2k_1} n_2^{2k_2}\rangle = \frac{(2k_1-1)!!\,(2k_2-1)!!}{[2(k_1+k_2)+1]!!} . \tag{49.22}$$

Here it is assumed that

$$(-1)!!=1. \tag{49.23}$$

The equivalence of k_1, k_2, and k_3 gives us from (49.21) and (49.22) that in the most general case

$$\langle n_1^{2k_1} n_2^{2k_2} n_3^{2k_3} \rangle = \frac{(2k_1-1)!!\,(2k_2-1)!!\,(2k_3-1)!!}{[2(k_1+k_2+k_3)+1]!!}. \tag{49.24}$$

To illustrate these relationships, and also with a view to application to cubic crystals, we calculate the means of various powers of the quantity of (36.4):

$$N = \boldsymbol{n} \nu \boldsymbol{n} = (\nu^2)_t = n_1^4 + n_2^4 + n_3^4. \tag{49.25}$$

From (49.21) we have

$$\langle N \rangle = \frac{3}{5} = 0.6. \tag{49.26}$$

Consider power l of (49.25):

$$N^l = (n_1^4 + n_2^4 + n_3^4)^l = \sum_{l_1+l_2+l_3=l} \frac{l!}{l_1!\,l_2!\,l_3!} n_1^{4l_1} n_2^{4l_2} n_3^{4l_3}. \tag{49.27}$$

The n_k with different subscripts are completely equivalent in the averaging, so the mean $\langle n_1^{4l_1} n_2^{4l_2} n_3^{4l_3} \rangle$ is dependent only on how l is split up in the sum $l_1+l_2+l_3$, not on which n_k the l_1, l_2, and l_3 refer to. Let $A_{l_1 l_2 l_3} = \langle n_1^{4l_1} n_2^{4l_2} n_3^{4l_3} \rangle$; the result is unaltered if on the right in this we perform any permutation of the subscripts to the components of **n** (while leaving unchanged the triplet l_1, l_2, l_3). Therefore $\langle N^l \rangle$ will, in the general case, contain terms of six types, in accordance with the distinct ways of splitting l into the sum $l_1+l_2+l_3$:

$$\left.\begin{array}{lll} \text{I.} & l_1 = l,\ l_2 = l_3 = 0 & 3A_l, \\ \text{II.} & l_1 = l_2,\ l_3 = 0 & 3\dfrac{l!}{(l_1!)^2} A_{l_1 l_1}, \\ \text{III.} & l_1 \neq l_2,\ l_3 = 0 & 6\dfrac{l!}{l_1!\,l_2!} A_{l_1 l_2}, \\ \text{IV.} & l_1 = l_2 = l_3 & \dfrac{l!}{(l_1!)^3} A_{l_1 l_1 l_1}, \end{array}\right\} \tag{49.28}$$

$$\begin{array}{lll} \text{V.} & l_1 = l_2 \neq l_3 & 3\dfrac{l!}{(l_1!)^2\, l_3!} A_{l_1 l_1 l_3}, \\ \text{VI.} & l_1 \neq l_2 \neq l_3 & 6\dfrac{l!}{l_1! l_2! l_3!} A_{l_1 l_2 l_3}. \end{array} \Bigg\} \quad (49.28)$$

For instance, for $l = 5$ we have the five possible ways: $l = 5+0+0 = 4+1+0 = 3+2+0 = 3+1+1 = 2+2+1$, of which the first is of type I, the second of type III, and the fourth and fifth of type V. The result is

$$\langle N^5 \rangle = 3A_5 + 6\frac{5!}{4!1!} A_{41} + 6\frac{5!}{3!2!} A_{32} +$$

$$+ 3\frac{5!}{3!1!1!} A_{311} + 3\frac{5!}{2!2!1!} A_{221}. \quad (49.29)$$

This may be verified by using the property that the sum of the coefficients to the A_l, $A_{l_1 l_2}$, $A_{l_1 l_2 l_3}$ must be 3^l. Once the separation as of (49.29) has been performed, the A_l, $A_{l_1 l_2}$, $A_{l_1 l_2 l_3}$ are found from (49.21), (49.22), and (49.24). The means $\langle |\nu| N^l \rangle = \langle n_1^2 n_2^2 n_3^2 N^l \rangle$ are found similarly, the separation of (49.28) persisting, but with each $A_{l_1 l_2 l_3}$ replaced by $\langle n_1^{4l_1+2} n_2^{4l_2+2} n_3^{4l_3+2} \rangle$.

50. Debye Temperatures of Cubic Crystals [53, 54]

The method and formulas of section 48 are entirely applicable to any crystal. In this section we consider cubic crystals, where the a, b, c of (48.35) and (48.36) take the following values;

$$\left. \begin{array}{l} a = c_1 + 0.2c_3, \qquad b = c_2 + 0.4c_3, \\ c = a + b = c_1 + c_2 + 0.6c_2. \end{array} \right\} \quad (50.1)$$

From (36.3), (36.20), and (48.14) we have

$$\alpha = r_b \left(\nu - \frac{1 + 2\,\mathbf{n} \cdot \mathbf{n}}{5} \right), \quad \nu = \begin{pmatrix} n_1^2 & 0 & 0 \\ 0 & n_2^2 & 0 \\ 0 & 0 & n_3^2 \end{pmatrix}, \quad (50.2)$$

in which

$$r_b = \frac{c_3}{b}. \quad (50.3)$$

From (48.22) and (48.24) we have

$$\left.\begin{aligned} &\alpha_t = 0, \quad \tau_t = \mathbf{n}\alpha\mathbf{n} = \sigma = r_b(N - 0.6),\\ &N = \mathbf{n}\nu\mathbf{n} = (\nu^2)_t = n_1^4 + n_2^4 + n_3^4,\\ &\mathbf{k} = r_b[\mathbf{n}[\nu\mathbf{n}, \mathbf{n}]] = r_b(\nu - N)\mathbf{n},\\ &\mathbf{k}^2 = r_b^2(\mathbf{n}\nu^2\mathbf{n} - (\mathbf{n}\nu\mathbf{n})^2) = r_b^2(n_1^6 + n_2^6 + n_3^6 - N^2) \end{aligned}\right\} \tag{50.4}$$

and from (36.21)

$$(\alpha^2)_t = -2\bar{\alpha}_t = (1/25)\, r_b^2(1 + 5N). \tag{50.5}$$

Further, we need the following quantities in (48.31) and (48.32):

$$\bar{\tau}_t = (\overline{\mathbf{n}^{\times 2}\alpha})_t = \mathbf{n}\bar{\alpha}\mathbf{n} = \mathbf{n}\alpha^2\mathbf{n} + \bar{\alpha}_t = r_b^2\left(\mathbf{n}\nu^2\mathbf{n} + \frac{17 - 65N}{50}\right), \tag{50.6}$$

$$\mathbf{k}\alpha\mathbf{k} = r_b^2(\mathbf{n}\nu^3\mathbf{n} - 2N\mathbf{n}\nu^2\mathbf{n} + N^2). \tag{50.7}$$

The Hamilton–Cayley theorem for $\nu^3 = \nu^2 - \bar{\nu}_t\nu + |\nu|$, so we can put the latter relation as

$$\mathbf{k}\alpha\mathbf{k} = r_b^3[|\nu| + (1 - 2N)\mathbf{n}\nu^2\mathbf{n} + (1/2)N(3N - 1)]. \tag{50.8}$$

To find I we take an incomplete second approximation, retaining in the $\varepsilon_{(2)}$ of (48.32) terms up to the fourth degree in α. Then

$$\epsilon_{(2)} = \epsilon_{(1)} + \mathbf{k}^2(\sigma^2 - \bar{\tau}_t - \mathbf{k}^2) - 3\sigma\mathbf{k}\alpha\mathbf{k}. \tag{50.9}$$

Correspondingly, we take in place of (48.36) that

$$I_2 = I_1 + (630/256)\ \langle a^{-3/2} r_1^4(\xi_1^4 + \xi_2^4) + c^{-3/2} r_2^4 \epsilon^4 \rangle. \tag{50.10}$$

Then the following means appear in the I_1 of (48.35) and the I_2 of (50.10), which may be found by the methods of the preceding section:

$$\begin{aligned} &\langle N\rangle = 3/5 = 0.6 && \langle N^2\rangle = 41/105 = 0.390476\\ &\langle N^3\rangle = 274.2/1001 = 0.273926 && \langle N^4\rangle = 498.2/2431 = 0.204936\\ &\langle \mathbf{n}\nu^2\mathbf{n}\rangle = 3/7 = 0.428571 && \langle(\mathbf{n}\nu^2\mathbf{n})^2\rangle = 241/1001 = 0.240759\\ &\langle N\mathbf{n}\nu^2\mathbf{n}\rangle = 23/77 = 0.298701 && \langle N^2\mathbf{n}\nu^2\mathbf{n}\rangle = 61/273 = 0.223443\\ &\langle|\nu|\rangle = 1/105 = 0.009524 && \langle N|\nu|\rangle = 1/231 = 0.004329. \end{aligned} \tag{50.11}$$

Substitution of the above expressions in (48.31), (48.35), (48.37), and (48.38) gives as first approximation for cubic crystals that

$$I_1 = a^{-3/2}\{2 + r_a r_b[0.1r_1(1-0.06r_a) + (57.2 - 8.4r_a + 0.48r_b/1001)]\}$$

$$+ c^{-3/2}\{1 - [r_b r_c/1001][57.2(1 - r_2) + 0.5r_b - r_c(7.2 - 6.7r_2)]\}, \quad (50.12)$$

in which

$$r_a = c_3/a, \quad r_c = c_3/c. \quad (50.13)$$

Then (50.10) gives in the incomplete second approximation that

$$I_2 = I_1 + [r_b^4/1001][a^{-3/2}(0.17r_1 + 0.26r_1^2 + 4.18r_1^3 + 5.72r_1^4) -$$

$$- c^{-3/2}(0.17r_2 + 0.7r_2^2 + 5.86r_2^3 - 5.28r_2^4)]. \quad (50.14)$$

These formulas will be tested by derivation of the Debye temperatures of some cubic crystals whose θ and elastic constants have been measured near absolute zero [40], i.e., for the conditions where Debye's theory applies most closely. Table VII gives the results; the first line gives the result from (50.12), while the second one gives the second approximation of (50.14). The third gives the observed Debye temperature, while the fourth gives the range of values computed by other workers. It is clear that (50.14) gives results as accurate as those listed in the last line of the table [40], which were obtained by extremely lengthy and laborious calculations, often performed by computer. In the case of lithium (one of the cubic crystals of highest anisotropy), the results from the present formulas are in considerably better agreement with experiment than those obtained by other workers.

Table VIII gives the Debye temperatures calculated from (50.12) and (50.14) for all cubic crystals whose elastic constants are known* ; the penultimate column gives Δ_m (relative mean-square elastic anisotropy). The order of approximation needed for the wave velocities increases with this quantity. The I have been calculated from (50.12) for crystals with $\Delta_m \leq 0.10$, but (50.14) was used for more anisotropic crystals.

---

*Apart from certain alloys, for which data on the densities are not available.

Table VII

	Al	Au	LiF	Ag	NaCl	Cu	KBr	Li
θ_1, °K	428.4	163.8	735.2	230.1	322.1	351.4	177.9	382.6
θ_2, °K	428.4	162.4	734.6	227.9	321.4	347.3	172.0	366.4
θ_{exp}	375-426	164.6-164.8	722-743	225.3-226.5	320	343.8-346.7	174	369
θ_a	427.4-428.7	161.6-162.2	734-734.6	226.4-227.1	321-321.9	344.4-345.4	171.7-172.8	334.6-338.4

The first approximation appears adequate for 60 crystals out of the 82 (73%), while the second approximation gives a correction of only 0.1-0.2% to the θ_1 from the first approximation for crystals of anisotropy from 0.10 to 0.14. This correction is meaningful when the elastic constants are known very precisely (LiF, Au), but in most cases it lies below the limits of error of the elastic constants (GaSb, GaAs, Ge, NH_4Cl, $Ba(NO_3)_2$, etc.). The correction in the second approximation becomes substantial if the elastic anisotropy is higher than this; e.g., the values of the anisotropy for KBr, RbI, Li, Th, and Na are, respectively, 0.29, 0.30, 0.33, and 0.34, and here the correction is 3-5% of θ_1; but even here the correction sometimes fails to exceed the limits of error of the elastic constants. For example, for KI ($\Delta_m = 0.26$) the correction should be about 3%, but this is not significant, because the elastic constants are known only with an accuracy sufficient to prove two significant figures in the result $\theta = 100$. Sometimes the literature gives conflicting values for elastic constants (C, V, W, NH_4Cl, NH_4Br, RbCl); several values for the Debye temperature are given in the table for these.

51. Debye Temperatures of Hexagonal Crystals

The general method of section 48 may be applied to the Debye temperature of a transversely isotropic medium; but this type of medium has the specific feature of a purely transverse wave for any **n**, whose phase velocity in terms of the elastic constants is given as a rational expression by equation (32.3). This means that much simpler and more precise relationships can be derived

Table VIII. Debye Temperatures for Cubic Crystals

No.	Crystal	Chemical formula	ρ, g/cm^3	V_a, 10^{-21} cm^3	$a_m \times 10^{11}$, cm^2/sec^2	$b_m \times 10^{11}$, cm^2/sec^2	$c_3 \times 10^{11}$, cm^2/sec^2	Δm	θ, °K
1	Aluminum	Al	2.734	0.01637	1.06	3.58	-0.351	0.029	428.4
2		$K-Al-SO_4$	1.753	0.1123	0.460	1.012	-0.165	0.041	147
3		$Rb-Al-SO_4$	1.884	0.1147	0.428	0.9568	-0.0987	0.026	141
4		$Cs-Al-SO_4$	1.999	0.1180	0.410	1.169	-0.051	0.012	137
5		$Tl-Al-SO_4$	2.322	0.1143	0.332	0.7997	-0.0939	0.031	124
6		$NH_4-Al-SO_4$	1.642	0.1146	0.471	1.111	-0.117	0.027	148
7		$CH_3NH_3-Al-SO_4$	1.589	0.1221	0.377	1.475	0.0447	0.0093	131
8		$K-Ga-SO_4$	1.898	0.1131	0.412	0.9002	-0.177	0.049	139
9		$Rb-Ga-SO_4$	2.025	0.1155	0.396	0.8633	-0.124	0.036	135
10		$Cs-Ga-SO_4$	2.127	0.1192	0.375	1.096	-0.0451	0.012	131
11	Alums	$NH_4-Ga-SO_4$	1.784	0.1154	0.424	0.9733	-0.137	0.036	140
12		$CH_3NH_3-Ga-SO_4$	1.717	0.1233	0.338	1.330	0.0513	0.012	124
13		$Rb-In-SO_4$	2.107	0.1199	0.369	0.7992	-0.114	0.036	129
14		$Cs-In-SO_4$	2.212	0.1231	0.362	0.9901	-0.0371	0.010	127
15		$Cs-Fe-SO_4$	2.065	0.1200	0.395	1.101	-0.0620	0.016	134
16		$K-Al-SeO_4$	1.986	0.1187	0.371	0.8404	-0.0957	0.028	130
17		$Rb-Al-SeO_4$	2.113	0.1208	0.356	0.8157	-0.0596	0.018	127
18		$Cs-Al-SeO_4$	2.224	0.1235	0.329	0.8536	-0.0243	0.0075	121
19		$NH_4-Al-SeO_4$	1.888	0.1203	0.381	0.9153	-0.0847	0.023	131
20		$CH_3NH_3-Al\text{-}SeO_4$	1.827	0.1275	0.368	0.9895	0.351	0.096	125
21		$Cs-Ga-SeO_4$	2.342	0.1249	0.313	0.7862	-0.0478	0.016	118
22	Ammonium bromide	NH_4Br	2.436	0.03338	0.372	0.84	0.45	0.14	193
23	Ammonium chloride	NH_4Cl	1.526	0.02910	0.70	1.48	0.760	0.13	278
24	Barium fluoride	BaF	4.83	0.0180	0.516	1.35	0	0	288
25	Barium nitrate	$Ba(NO_3)_2$	3.24	0.0446	0.484	1.165	0.537	0.12	201
26	Calcium fluoride	CaF_2	3.18	0.0136	1.37	3.23	1.41	0.11	506
27	Cesium bromide	CsBr	4.45	0.0397	0.20	0.426	0.17	0.10	136

TABLE VIII (continued)

No.	Crystal	Chemical formula	ρ, g/cm^3	V_a, 10^{-21} cm^3	$a_m \times 10^{11}$, cm^2/sec^2	$b_m \times 10^{11}$, cm^2/sec^2	$c_3 \times 10^{11}$, cm^2/sec^2	Δm	θ, °K
28	Cesium chloride	$CsCl$	3.99	0.00350	0.26	0.543	0.281	0.13	159
29	Cesium iodide	CsI	4.52	0.0477	0.16	0.338	0.11	0.081	114
30	Chromite	$FeCr_2O_4$	4.6	0.012	2.4	5.3	-1.2	0.058	695
31	Cobalt	Co	8.739	0.01120	0.855	2.620	0	0	437
32	Copper	Cu	9.018	0.01169	0.6577	1.794	-1.245	0.19	347.3
33	Diamond	C	3.51	0.00568	13.3	16.3	-6.44	0.073	2091
34	Galena	PbS	7.5	0.026	0.37	0.91	0.19	0.054	214
35	Gallium antimonide	$GaSb$	5.619	0.02829	0.633	1.22	-0.682	0.13	266
36	Gallium arsenide	$GaAs$	5.307	0.02263	0.917	1.73	-1.01	0.14	344
37	Germanium	Ge	5.35	0.0225	1.05	1.76	-1.00	0.13	371
38	Hexamethylene tetraamine	$C_6H_{12}N_6$	1.339	0.1738	0.412	0.762	0.13	0.041	120
39	Gold	Au	19.488	0.01680	0.1727	0.9829	-0.3023	0.10	162.4
40	Indium antimonide	$InSb$	5.789	0.03393	0.418	0.949	-0.517	0.14	203
41	Iron	Fe	7.874	0.01178	1.128	2.57	-1.73	0.16	465
42	Lead	Pb	11.3437	0.03033	0.089	0.406	-0.190	0.15	92
43	Lithium	Li	0.5471	0.02106	1.358	3.036	-3.793	0.30	366.4
44	Lithium bromide	$LiBr$	3.47	0.0208	0.449	0.89	-0.507	0.11	248
45	Lithium chloride	$LiCl$	2.068	0.01702	0.982	1.85	-1.11	0.14	391
46	Lithium fluoride	LiF	2.646	0.00814	2.09	3.34	-1.80	0.12	735
47	Lithium iodide	LiI	4.061	0.02736	0.271	0.554	-0.308	0.13	176
48	Magnesium oxide	MgO	3.58	0.00935	3.72	5.57	-3.04	0.11	934
49	Magnetite	Fe_3O_4	5.15	0.0106	1.78	3.73	-0.528	0.035	633
50	Molybdenum	Mo	10.2	0.0156	1.19	3.04	0.58	0.051	458
51	Nickel	Ni	8.9	0.011	1.0	2.4	-1.6	0.17	450
52	Palladium	Pd	11.9	0.0149	0.436	1.76	-0.821	0.14	264
53	Potassium	K	0.862	0.0753	0.203	0.483	-0.515	0.27	96.2

54	Potassium bromide	KBr	2.819	0.03504	0.37	0.749	0.9152	0.28	172
55	Potassium chloride	KCl	1.984	0.03120	0.530	1.076	1.07	0.23	224
56	Potassium cyanide	KCN	1.553	0.03481	1.0	0.975	0.030	0.048	315
57	Potassium fluoride	KF	2.5257	0.01910	0.707	1.50	1.00	0.16	314
58	Potassium iodide	KI	3.13	0.0440	0.214	0.46	0.489	0.26	124
59	Pyrite	FeS_2	5.00	0.0199	2.90	2.76	3.96	0.22	615
60	Rubidium bromide	RbBr	3.351	0.04097	0.230	0.49	0.577	0.29	129
61	Rubidium chloride	RbCl	2.799	0.03587	0.319	0.686	0.745	0.26	162
62	Rubidium fluoride	RbF	2.88	0.0301	0.50	1.12	0.91	0.20	224
63	Rubidium iodide	RbI	3.554	0.04961	0.1718	0.37	0.4637	0.30	103
64	Silicon	Si	2.328	0.02003	2.93	5.18	-2.47	0.11	646
65	Silver	Ag	10.635	0.01684	0.3525	1.140	0.6396	0.16	227.9
66	Silver bromide	AgBr	6.473	0.0241	0.140	0.674	0.137	0.065	136
67	Silver chloride	AgCl	5.56	0.0214	0.153	0.846	0.205	0.080	146
68	Sodium	Na	0.97	0.039	0.391	0.682	-1.10	0.34	164
69	Sodium bromate	$NaBrO_3$	3.339	0.03752	0.482	1.09	0.162	0.038	217
70	Sodium bromide	NaBr	3.203	0.02667	0.36	0.73	0.300	0.099	208
71	Sodium chlorate	$NaClO_3$	2.49	0.0355	0.570	1.24	0.498	0.10	237
72	Sodium chloride	NaCl	2.241	0.02165	0.7805	1.409	0.9415	0.15	321
73	Sodium fluoride	NaF	2.79	0.01250	1.11	2.14	0.552	0.061	472
74	Sodium iodide	NaI	3.667	0.03394	0.24	0.48	0.20	0.099	157
75	Spinel	$MgAl_2O_4$	3.6	0.0094	3.5	6.8	-4.7	0.16	884
76	Strontium nitrate	$Sr(NO_3)_2$	2.986	0.03923	0.464	1.17	-0.124	0.028	211
77	Thallium bromide	TlBr	7.557	0.03090	0.121	0.338	0.104	0.084	115
78	Thallium chloride	TlCl	7.00	0.0284	0.136	0.382	0.14	0.10	125
79	Thorium	Th	11.2	0.0344	0.271	0.552	-0.779	0.33	137
80	Tungsten	W	19.3	0.0158	0.789	1.87	-0.03	0.0043	371
81	Vanadium	V	6.022	0.01405	0.786	2.84	0.402	0.042	388
82	Zinc blende	ZnS	4.087	0.0198	0.78	2.318	-1.14	0.14	329

for θ than those given by direction application of the method of section 48.

From (32.3) we have $v_1^2 = \lambda_{66} + (\lambda_{44} - \lambda_{66}) n_3^2$, so

$$\langle 1/v_1^3 \rangle = \frac{1}{4\pi} \int \frac{d\Omega}{v_1^3} =$$

$$= \frac{1}{4\pi} \int\limits_{\varphi=0}^{2\pi} \int\limits_{\vartheta=0}^{\pi} \frac{\sin\vartheta\, d\vartheta\, d\varphi}{v_1^3} = \frac{1}{2} \int\limits_{-1}^{1} \frac{dx}{\sqrt{[\lambda_{66} + (\lambda_{44} - \lambda_{66}) x^2]^3}},$$

in which $x = \cos\vartheta = \mathbf{ne} = n_3$. This integral is elementary, and we get

$$\langle 1/v_1^3 \rangle = 1/(\lambda_{66}\sqrt{\lambda_{44}}). \tag{51.1}$$

This for a hexagonal crystal

$$I = \langle 1/v_0^3 + 1/v_1^3 + 1/v_2^3 \rangle = 1/(\lambda_{66}\sqrt{\lambda_{44}}) + I_{02}, \tag{51.2}$$

in which I_{02} denotes the average sum of the inverse cubes of the phase velocities for the quasilongitudinal and quasitransverse waves:

$$I_{02} = \langle 1/v_0^3 + 1/v_2^3 \rangle. \tag{51.3}$$

The squares of these velocities are eigenvalues of the tensor of (32.28):

$$\lambda^n = c_0 + c_1 \mathbf{n} \cdot \mathbf{n} + c_2 \mathbf{e} \cdot \mathbf{e}, \tag{51.4}$$

in which

$$c_0 = g_1 + g_2 n_3^2, \quad c_1 = g_3, \quad c_2 = g_2 + g_4 n_3^2. \tag{51.5}$$

Correspondingly, the displacements $\mathbf{u}_0$ and $\mathbf{u}_2$ are solutions to

$$\lambda^n \mathbf{u} = v^2 \mathbf{u}. \tag{51.6}$$

Therefore the problem here becomes two-dimensional.

We will use the method of section 48 (comparison with the most similar isotropic medium), but this will be applied to the two-dimensional tensor Λ. We thus seek the tensor

$$\lambda_0 = a_0 + b_0\mathbf{n}\cdot\mathbf{n}, \tag{51.7}$$

most similar on average to the tensor of (51.4), the averaging being extended to all directions of **n** lying in the meridional plane P.

By analogy with (26.11) we get the relations

$$\langle \lambda^n - \lambda_0 \rangle_t = 0, \qquad \langle \mathbf{n}(\lambda^n - \lambda_0)\mathbf{n} \rangle = \mathbf{v}, \tag{51.8}$$

which, however, here give

$$2a_0 + b_0 = \langle \lambda^n \rangle_t, \qquad a_0 + b_0 = \langle \mathbf{n}\lambda^n\mathbf{n} \rangle, \tag{51.9}$$

which differ from equations (26.12) for the three-dimensional case. We seek solutions in the form

$$a_0 = \langle \lambda^n_t - \mathbf{n}\lambda^n\mathbf{n} \rangle, \qquad b_0 = \langle 2\mathbf{n}\lambda^n\mathbf{n} - \lambda^n_t \rangle. \tag{51.10}$$

Use of (51.4) and (51.5) gives

$$a_0 = g_1 + g_2 + (2/15)\,g_4, \qquad b_0 = g_3 - (1/3)\,g_2 + (1/15)\,g_4. \tag{51.11}$$

These relationships differ substantially from those of (33.2).

Then the tensor of (51.4) may be put as

$$\lambda^n = a_0 + b_0(\mathbf{n}\cdot\mathbf{n} + \alpha), \tag{51.12}$$

in which

$$\alpha = (1/b_0)(\lambda^n - \lambda_0) = -\left(f_2(1 - n_3^2) + \frac{2}{15}f_4\right) + \frac{1}{3}\left(f_2 - \frac{1}{5}f_4\right)\mathbf{n}\cdot\mathbf{n} + (f_2 + f_4n_3^2)\,\mathbf{e}\cdot\mathbf{e}. \tag{51.13}$$

The notation here is

$$f_2 = g_2/b_0, \quad f_4 = g_4/b_0. \tag{51.14}$$

Tensor α characterizes the relative difference of $\lambda^{\mathbf{n}}$ from λ_0. As a measure of this difference we use, as before, the quantity

$$\langle \alpha^2 \rangle_t = (4/225)(35\, f_2^2 + 30\, f_2 f_4 + 8\, f_4^2). \tag{51.15}$$

From (51.8) we have

$$\langle \alpha_t \rangle = \langle \mathbf{n}\alpha\mathbf{n} \rangle = 0. \tag{51.16}$$

We put

$$v^2 = a_0 + b_0 \xi. \tag{51.17}$$

From (25.23) we have

$$v^4 - \lambda_t^n v^2 + |\lambda^n| = 0. \tag{51.18}$$

From (51.12) we get

$$\lambda_t^n = 2a_0 + b_0(1 + \alpha_t), \tag{51.19}$$

$$|\lambda^n| = (1/2)[(\lambda_t^n)^2 - (\lambda^{n^2})_t] = a_0^2 + a_0 b_0 (1 + \alpha_t) + b_0^2(\alpha_t - \mathbf{n}\alpha\mathbf{n} + |\alpha|). \tag{51.20}$$

We substitute (51.17), (51.19), and (51.20) into (51.18) to get

$$\xi^2 - (1 + \alpha_t)\xi + \alpha_t - \mathbf{n}\alpha\mathbf{n} + |\alpha| = 0. \tag{51.21}$$

As in section 48, we arrive at the equations

$$(\mathbf{n}\cdot\mathbf{n} + \alpha)\mathbf{u} = \xi\mathbf{u}, \quad \mathbf{u} = \mathbf{n} + \mathbf{u}', \quad \mathbf{u}'\mathbf{n} = 0, \tag{51.22}$$

$$\xi = 1 + \mathbf{n}\alpha\mathbf{n} + \mathbf{n}\alpha\mathbf{u}', \tag{51.23}$$

$$\gamma\mathbf{u}' = \mathbf{k} - \mathbf{k}\mathbf{u}'\cdot\mathbf{u}', \tag{51.24}$$

$$\gamma = 1 + \mathbf{n}\alpha\mathbf{n} + (\mathbf{n}\cdot\mathbf{n} - 1)\alpha, \quad \mathbf{k} = \alpha\mathbf{n} - \mathbf{n}\alpha\mathbf{n}\cdot\mathbf{n}, \tag{51.25}$$

but here all the vectors, including **u'** are two-dimensional and lie in the (**e**, **n**) plane, so we can use the formulas of section 14 for two-dimensional tensors, which give

$$\beta = \gamma^{-1} = \frac{\bar{\gamma}}{|\gamma|} = \frac{\gamma_t - \gamma}{|\gamma|} = \frac{1 + 2\mathbf{n}\alpha\mathbf{n} - \alpha_t + (1 - \mathbf{n}\cdot\mathbf{n})\alpha}{(1 + \mathbf{n}\alpha\mathbf{n})(1 + 2\mathbf{n}\alpha\mathbf{n} - \alpha_t)}. \tag{51.26}$$

Then (51.24) becomes

$$\mathbf{u}' = \beta\mathbf{k} - \mathbf{k}\mathbf{u}'\cdot\beta\mathbf{u}' \tag{51.27}$$

and the first and second approximations for the displacement of the quasilongitudinal wave are

$$\mathbf{u}'_{(1)} = \beta\mathbf{k}, \quad \mathbf{u}'_{(2)} = \beta\mathbf{k} - \mathbf{k}\beta\mathbf{k}\cdot\beta^2\mathbf{k}. \tag{51.28}$$

From (51.23) we have for the quasilongitudinal wave that

$$\xi_0 = 1 + \epsilon, \quad \epsilon = \mathbf{n}\alpha\mathbf{n} + \mathbf{n}\alpha\mathbf{u}', \tag{51.29}$$

in which, from (51.28),

$$\epsilon_{(1)} = \mathbf{n}\alpha\mathbf{n} + \mathbf{n}\alpha\beta\mathbf{k}, \quad \epsilon_{(2)} = \epsilon_{(1)} - \mathbf{k}\beta\mathbf{k}\cdot\mathbf{n}\alpha\beta^2\mathbf{k}. \tag{51.30}$$

From (51.21) we have that the sum of the roots is $\xi_0 + \xi_2 = 1 + \alpha_t$, i.e.,

$$\xi_2 = \alpha_t - \epsilon. \tag{51.31}$$

Thus we may find ξ_2 to the same accuracy when ξ_0 has been found to a given accuracy.

Now from (51.17) we obtain for the I_{02} of (51.3) that

$$I_{02} = \langle (a_0 + b_0\xi_0)^{-3/2} \rangle + \langle (a_0 + b_0\xi_2)^{-3/2} \rangle =$$

$$= a_0^{-3/2} \langle (1 + r_1\xi_2)^{-3/2} \rangle + c_0^{-3/2} \langle (1 + r_2\epsilon)^{-3/2} \rangle, \tag{51.32}$$

in which

$$c_0 = a_0 + b_0, \quad r_1 = b_0/a_0, \quad r_2 = b_0/c_0.$$

The binomial expansion gives us for the first approximation (i.e., up to the third degree in ε and ξ_2) that

$$I_{02}^{(1)} = a^{-3/2} \langle 1 - \frac{3}{2} r_1 \xi_2 + \frac{15}{8} r_1^2 \xi_2^2 - \frac{35}{16} r_1^3 \xi_2^3 \rangle + $$

$$+ c_0^{-3/2} \langle 1 - \frac{3}{2} r_2 \epsilon + \frac{15}{8} r_2^2 \epsilon^2 - \frac{35}{16} r_2^3 \epsilon^3 \rangle. \qquad (51.33)$$

The second approximation gives

$$I_{02}^{(2)} = I_{02}^{(1)} + \frac{63}{256} \left[a_0^{-3/2} \langle 10 r_1^4 \xi_2^4 - 11 r_1^5 \xi_2^5 \rangle + c_0^{-3/2} \langle 10 r_2^4 \epsilon^4 - 11 r_2^5 \epsilon^5 \rangle \right]. \qquad (51.34)$$

From (51.13) we have

$$\alpha_t = (2 f_2 + f_4)(n_3^2 - 1/3), \quad \mathbf{n}\alpha\mathbf{n} = 2 f_2 (n_3^2 - 1/3) + f_4 (n_3^4 - 1/5). \qquad (51.35)$$

Now from (49.20)

$$\langle n_3^{2k} \rangle = 1/2k + 1, \qquad (51.36)$$

so (51.16) is obeyed.

Taking only the first approximation, we have from (51.25), (51.26), and (51.30) that

$$\epsilon_{(1)} = \sigma + \mathbf{k}^2/(1 + 2\sigma - \alpha_t) = \sigma + \mathbf{k}^2(1 - 2\sigma + \alpha_t). \qquad (51.37)$$

Here the expressions have been shortened by putting $\sigma = \mathbf{n}\alpha\mathbf{n}$. To the same degree of accuracy we get from (51.31) that

$$\epsilon_{(1)}^2 = \sigma^2 + 2\sigma \mathbf{k}^2, \quad \epsilon_{(1)}^3 = \sigma^3, \quad \xi_2 = \alpha_t - \sigma - \mathbf{k}^2(1 - 2\sigma + \alpha_t),$$

$$\xi_2^2 = (\alpha_t - \sigma)^2 + 2\mathbf{k}^2(\alpha_t - \sigma), \quad \xi_2^3 = (\alpha_t - \sigma)^3. \qquad (51.38)$$

From (51.13) and (51.25) we have

$$\mathbf{k} = (f_2 + f_4 n_3^2) n_3 (\mathbf{e} - n_3 \mathbf{n}), \quad \mathbf{k}^2 = (f_2^2 + 2 f_2 f_4 n_3^2 + f_4^2 n_3^4) n_3^2 (1 - n_3^2), \qquad (51.39)$$

$$\langle \mathbf{k}^2 \rangle = f_2^2 p_1 + 2 f_2 f_4 p_2 + f_4^2 p_3 = (2/315)(21 f_2^2 + 18 f_2 f_4 + 5 f_4^2), \qquad (51.40)$$

in which

$$p_s = \langle n_3^{2s}(1 - n_3^2)\rangle = \frac{2}{(2s+1)(2s+3)}. \qquad (51.41)$$

From (51.2) and (51.33) we get for the first approximation that

$$I = I_0 + (3/2)(r_1 a_0^{-3/2} - r_2 c_0^{-3/2})\langle\epsilon\rangle +$$

$$+ (5/16) r_1^2 a_0^{-3/2}\langle 6\xi_2^2 - 7r_1\xi_2^3\rangle + (5/16) r_2^2 c_0^{-3/2}\langle 6\epsilon^2 - 7r_2\epsilon^3\rangle. \qquad (51.42)$$

Here I_0 denotes the zero approximation:

$$I_0 = 1/(\lambda_{66}\sqrt{\lambda_{44}}) + a_0^{-3/2} + c_0^{-3/2}. \qquad (51.43)$$

From (51.35)-(51.41) we get

$$I = I_0 + (r_1 a_0^{-3/2} - r_2 c_0^{-3/2})(A_0 + 2A_1 - 2A_2) +$$

$$+ r_1^2 a_0^{-3/2}(A_3 + 5A_1 + r_1 A_4) + r_2^2 c_0^{-3/2}(A_5 + 5A_2 - r_2 A_6), \qquad (51.44)$$

in which

$$A_0 = (3/2)\langle\mathbf{k}^2\rangle = (1/105)(21 f_2^2 + 18 f_2 f_4 + 5 f_4^2),$$

$$A_1 = (3/4)\langle(\alpha_t - \sigma)\mathbf{k}^2\rangle = (f_4/175)\left(f_2^2 + \frac{34}{33} f_2 f_4 + \frac{335}{1287} f_4^2\right),$$

$$A_2 = (3/4)\langle\sigma\mathbf{k}^2\rangle = (2/105)\left(f_2^3 + \frac{11}{5} f_2^2 f_4 + \frac{239}{165} f_2 f_4^2 + \frac{43}{143} f_4^3\right),$$

$$A_3 = (15/8)\langle(\alpha_t - \sigma)^2\rangle = f_4^2/70,$$

$$A_4 = -(35/16)\langle(\alpha_t - \sigma)^3\rangle = 19 f_4^3/96525,$$

$$A_5 = (15/8)\langle\sigma^2\rangle = 2\left(\frac{1}{3} f_2^2 + \frac{2}{7} f_2 f_4 + \frac{1}{15} f_4^2\right),$$

$$A_6 = (35/16)\langle\sigma^3\rangle = \left(\frac{2}{3} f_2\right)^3 + (8 f_4/15)\left(f_2^2 + \frac{19}{33} f_2 f_4 + \frac{7}{65} f_4^2\right). \qquad (51.45)$$

The second approximation is restricted to terms of fourth degree in α, as in the previous section, so from (51.30) and (51.26) we get for $\varepsilon_{(2)}$ that

$$\epsilon_{(2)} = \sigma + \mathbf{k}^2/(1 + 2\sigma - \alpha_t) - \mathbf{k}^2\cdot\mathbf{n}\alpha\beta^2\mathbf{k}/[(1+\sigma)(1 + 2\sigma - \alpha_t)]. \qquad (51.46)$$

From (14.8), we have for a two-dimensional tensor that $\beta^2 = \beta_t\beta - |\beta|$, so

$$\mathbf{n}\alpha\beta^2\mathbf{k} = \beta_t\mathbf{n}\alpha\beta\mathbf{k} - |\beta|\mathbf{n}\alpha\mathbf{k} = \beta_t\mathbf{k}^2/(1 + 2\sigma - \alpha_t) - \mathbf{k}^2/|\gamma|.$$

But

$$\beta_t = \gamma_t/|\gamma| = \frac{2 + 3\sigma - \alpha_t}{(1 + \sigma)(1 + 2\sigma - \alpha_t)},$$

so

$$\mathbf{n}\alpha\beta^2\mathbf{k} = \mathbf{k}^2/(1 + 2\sigma - \alpha_t)^2.$$

Retaining terms up to the fourth degree in (51.46), we have

$$\epsilon_{(2)} = \sigma + \mathbf{k}^2[1 - (2\sigma - \alpha_t) + (2\sigma - \alpha_t)^2] - \mathbf{k}^4 = \epsilon_{(1)} + \mathbf{k}^2[(2\sigma - \alpha_t)^2 - \mathbf{k}^2]. \tag{51.47}$$

This has terms of fourth degree additional to those of (51.37), and these must be incorporated in (51.33). Moreover, (51.34) gives the terms

$$(630/256)\,[a_0^{-3/2} r_1^4 \langle \xi_2^4 \rangle + c_0^{-3/2} r_2^4 \langle \epsilon^4 \rangle]. \tag{51.48}$$

Then, denoting the second approximation for ξ_2 by ξ_2', we have in this approximation that

$$\epsilon_{(2)}^2 = \epsilon_{(1)}^2 + \mathbf{k}^2[\mathbf{k}^2 - 2\sigma(2\sigma - \alpha_t)],$$

$$\epsilon_{(2)}^3 = \epsilon_{(1)}^3 + 3\sigma^2\mathbf{k}^2, \quad \epsilon_{(2)}^4 = \sigma^4,$$

$$\xi_2' = \xi_2 - \mathbf{k}^2[(2\sigma - \alpha_t)^2 - \mathbf{k}^2], \quad \xi_2'^2 = \xi_2^2 + \mathbf{k}^2[\mathbf{k}^2 + 2(\alpha_t - \sigma)(2\sigma - \alpha_t)],$$

$$\xi_2'^3 = \xi_2^3 - 3(\alpha_t - \sigma)^2\mathbf{k}^2, \quad \xi_2'^4 = (\alpha_t - \sigma)^4. \tag{51.49}$$

Then (51.42) and (51.48) give the addition to (51.44) arising from the second approximation as

$$I' = r_1 a_0^{-3/2}[B_0(5r_1 - 4) + B_1(8 - 20r_1 + 35r_1^2) + 8B_2 + r_1^3 B_3]$$
$$+ r_2 c_0^{-3/2}[B_0(5r_2 + 4) - B_2(8 + 20r_2 + 35r_2^2) - 8B_1 + r_2^3 B_4], \tag{51.50}$$

in which

$$\left.\begin{aligned} B_0 &= (3/8)\langle \mathbf{k}^2[\mathbf{k}^2 + 2\sigma(\alpha_t - \sigma)]\rangle, \\ B_1 &= (3/16)\langle(\alpha_t - \sigma)^2\mathbf{k}^2\rangle, \end{aligned}\right\} \tag{51.51}$$

$$\left.\begin{aligned} B_2 &= (3/16)\left\langle \sigma^2 \mathbf{k}^2 \right\rangle, \\ B_3 &= (630/256)\left\langle (\alpha_t - \sigma)^4 \right\rangle, \\ B_4 &= (630/256)\left\langle \sigma^4 \right\rangle. \end{aligned}\right\} \tag{51.51}$$

Table IX gives the calculated Debye temperatures for all hexagonal crystals whose elastic constants are presently known. Column 9 gives $\delta = \langle\alpha^2\rangle_t$, which characterizes the relative anisotropy of the λ^n of (51.4). Comparison of columns 9 and 10 shows that the zero approximation is adequate for barium titanate, beryllium, yttrium, and cadmium sulfate; the first approximation is almost always adequate for the others. In fact, even for a crystal as anisotropic as zinc ($\delta = 0.146$) there is only about 1% difference between the Debye temperature given by (51.44) and that obtained by laborious numerical methods [40].

The mean volume per lattice node must be known in order to calculate the Debye temperature from (48.1). This is given by (48.5):

$$V_a = \frac{A}{\rho L s}$$

in which A is the atomic (molecular) weight, ρ is the density, L is Avogadro's number, and s is the number of structural nodes per molecule; the last is easily deduced from the lattice structure for simple compounds, but ones of complicated structure present difficulties, and it is not always possible to determine s unambiguously. Examples are cancrinite and beryl, for which no θ are given for this reason. Of course, these simple methods for I provide a means of deducing s for complicated molecules; the numerous examples of Table VII (cubic crystals), and also the examples of magnesium and zinc (Table IX), show that this method gives very good agreement with experiment [40], the discrepancies being 1-3%, although any change in s produces a substantial change in θ. Hence the observed θ and calculated I enable us to evaluate s when other methods are difficult to apply. In fact, we have from (48.1) and (48.5) that

$$s = IA\theta^3/135.78^3\rho L. \tag{51.52}$$

Table IX. Debye Temperatures of Hexagonal Crystals

No.	Crystal [formula]	ρ, g/cm^3	Va, 10^{-21} cm^3	$a_0 \times 10^{11}$, cm^2/sec^2	$b_0 \times 10^{11}$, cm^2/sec^2	$\delta = \langle \alpha^2 \rangle_t$	I_0	I	θ, °K
1	Barium titanate [$BaTiO_3$]	5.72	0.0135	0.753	2.12	0.0002	3.21	3.21	386
2	Beryl [$Be_3Al_2(SiO_2)_6$]	2.66	–	3.40	6.80	0.022	0.353	0.358	–
3	Beryllium [Be]	1.82	0.00822	8.57	8.66	0.02	0.0997	0.100	1449
4	Cadmium [Cd]	8.648	0.02158	0.229	0.886	0.13	15.1	17.2	189
5	Cadmium sulfide [CdS]	4.82	0.0249	0.323	1.34	0.0008	11.4	11.4	207
6	Cancrinite [$(Na_2Ca)_4(AlSiO_4)_6CaO_3$]	2.650	–	1.06	1.41	0.11	2.32	2.37	–
7	Cobalt [Co]	8.84	0.0111	1.09	2.34	0.0196	2.38	2 43	454
8	Ice [H_2O]	0.90	0.0111	0.428	1.02	0.0066	8.26	8.36	299
9	Magnesium [Mg]	1.779	0.02269	1.132	2.396	0.0027	1.913	1.920	385.9
10	β-Quartz [SiO_3]	2.44	0.01255	1.45	2.79	0.0036	1.142	1.145	559
11	Yttrium [Y]	4.472	0.03301	0.590	1.10	0.0034	5.12	5.12	246
12	Zinc [Zn]	7.279	0.01491	0.5458	1.495	0.146	4.117	4.565	332.5

Then s is the integer closest to the value of the right side. Thus measurement of θ and calculation of I together provide a solution to this difficult problem in chemical crystallography.

The θ for cobalt is of particular interest, since this can exist in cubic and hexagonal modifications. The result (Table VIII) for cubic cobalt is $\theta = 437°K$, while for the hexagonal form (Table IX) it is $\theta = 454°K$. The change from cubic to hexagonal form thus causes a 4% change in θ.

The formulas for the Debye temperature of hexagonal crystals provide a means of substantially simplifying the calculations for crystals of lower symmetry, since we can use the method of section 34 (comparison with the most similar transversely isotropic medium).

Literature Cited

1. H. Jaffe and C. Smith, Phys. Rev., 121:1604 (1961); K. S. Aleksandrov and L. N. Ryabinkin, Dokl. Akad. Nauk SSSR, 142:1298 (1962).
2. H. Huntington, Solid State Physics, 7:213 (1958).
3. F. R. Gantmakher, Theory of Matrices. Gostekhizdat (1953).
4. N. E. Kochin, Vector Calculus and the Principles of Tensor Calculus. Izd. Akad. Nauk SSSR (1951).
5. W. Cady, Piezoelectricity, an Introduction to the Theory and Applications of Electromechanical Phenomena in Crystals. [Russian translation], IL (1949).
6. E. E. Flint, Principles of Crystallography. Moscow (1952).
7. V. V. Novozhilkov, Theory of Elasticity. Sudpromgiz (1958).
8. A. I. Mal'tsev, Principles of Linear Algebra. Gostekhizdat (1948).
9. F. I. Fedorov, Tr. Inst. Fiz. i Mat. Akad. Nauk BSSR, 2:230 (1957).
10. F. I. Fedorov, Vestn. Mosk. Gos. Univ., 6:36 (1964).
11. R. Courant and D. Hilbert, Methods of Mathematical Physics, Vol. 1. [Russian translation], GITTL (1933).
12. A. G. Khatkevich, Kristallografiya, 6:700 (1961).
13. M. Musgrave, Proc. Roy. Soc., A226:339, 356 (1954).
14. J. Neighbours, J. Acoust. Soc. America, 26:865 (1954).
15. F. I. Fedorov, Kristallografiya, 8:213 (1963).
16. A. K. Sushkevich, Principles of Higher Algebra. Gostekhizdat (1941).
17. F. I. Fedorov, Optics of Anisotropic Media. Izd. Akad. Nauk BSSR (1958).
18. A. G. Khatkevich, Kristallografiya, 7:742 (1962).
19. A. G. Khatkevich, Dissertation, Minsk (1963).

20. M. V. Vol'kenshtein, Molecular Optics. Gostekhizdat (1951).
21. Lord Rayleigh, Theory of Sound. [Russian translation], Gostekhizdat (1940); 1937 English edition, Dover, New York.
22. A. G. Khatkevich, Kristallografiya, 7:916 (1962).
23. M. Born, Optics. [Russian translation], ONTI DNTVU (1937).
24. M. Musgrave, in: Progress in Solid Mechanics, Vol. II. Ed. by Sneddon and Hill. Amsterdam (1961).
25. G. Miller and M. Musgrave, Proc. Roy. Soc., A236:352 (1956).
26. B. van der Waarden, Modern Algebra, Vol. 2. [Russian translation], Gostekhizdat (1947).
27. A. A. Savelov, Plane Curves. Fizmatgiz (1960).
28. S. P. Finikov, Textbook of Differential Geometry. Gostekhizdat (1952).
29. P. Waterman, Phys. Rev., 113:1240 (1950).
30. F. I. Fedorov, Kristallografiya, 8:398 (1963).
31. J. Synge, Proc. Roy. Irish Acad., A58:13 (1956).
32. F. I. Fedorov, Dokl. Akad. Nauk SSSR, 155:792 (1964).
33. V. A. Koptsik and Yu. A. Sirotin, Kristallografiya, 6:766 (1961).
34. K. S. Aleksandrov and T. V. Ryzhova, Kristallografiya, 6:289 (1961).
35. K. S. Aleksandrov, Dissertation, Moscow (1957).
36. F. I. Fedorov, Dokl. Akad. Nauk SSSR, 149:1060 (1963).
37. F. I. Fedorov, Dokl. Akad. Nauk SSSR, 84:1171 (1952).
38. H. Kolsky, Stress Waves in Solids. [Russian translation], IL (1955).
39. J. Rayleigh, Lond. Math. Soc. Proc., Vol. 17 (1887).
40. G. Alers and J. Neighbours, Rev. Mod. Phys., 31:675 (1959).
41. P. Jayarama Reddy, Physica, 29:63 (1963).
42. W. Voigt, Lehrbuch der Kristallphysik. Leipzig (1900).
43. J. de Launay, Solid State Physics, II:286 (1956).
44. S. Quimby and P. Sutton, Phys. Rev., 91:1122 (1953); P. Sutton, Phys. Rev., 99:1826 (1955).
45. D. Betts, A. Batia, and M. Wyman, Phys. Rev., 104:37 (1956).
46. E. Post, Canad. J. Phys., 31:112 (1953).
47. M. Blackman, Phil. Mag., 42:(7) 1441 (1951).
48. L. Hopf and G. Lechner, Vorh. Deutsch. Phys. Ges., 16:643 (1914).
49. W. Overton and J. Gaffney, Phys. Rev., 98:969 (1955).
50. W. Houston, Rev. Mod. Phys., 20:164 (1948).
51. J. Synge, J. Math. Phys., 35:323 (1957).

52. T. G. Bystrova and F. I. Fedorov, Izv. Akad. Nauk BSSR, Ser. Fiz.-Mat., 1:35 (1965).
53. F. I. Fedorov, Kristallografiya, 10:167 (1965).
54. F. I. Fedorov, Dokl. Akad. Nauk SSSR, 164:804 (1965).

Index